King Croesus' Gold

Excavations at Sardis and the History of Gold Refining

King Croesus' Gold

Excavations at Sardis and the History of Gold Refining

Andrew Ramage and Paul Craddock

with contributions by

M.R. Cowell, A.E. Geçkinli, D.R. Hook, M.S. Humphrey
K. Hyne, N.D. Meeks, A.P. Middleton and H. Özbal

Published for The Trustees of

The British Museum by

BRITISH MUSEUM 𝖨𝖨𝖨𝖨𝖨𝖨 PRESS

In association with
ARCHAEOLOGICAL EXPLORATION OF SARDIS
Harvard University Art Museums

To our wives, Nancy and Brenda

© 2000 The Trustees of the British Museum and
the President and Fellows of Harvard College

First published in 2000 by British Museum Press
A division of The British Museum Company Ltd
46 Bloomsbury Street, London WC1B 3QQ
in association with the Archaeological Exploration
of Sardis (as Monograph 11)

ISBN 0–7141–0888–X

Designed and set in Apollo by Andrew Shoolbred

Printed in Great Britain at Cambridge University Press

14159763

fco

Contents

Contributors

M.R. Cowell
Department of Scientific Research, British Museum,
Great Russell Street, London WC1B 3DG

P.T. Craddock
Department of Scientific Research, British Museum

A.E. Geçkinli
Faculty of Chemical and Metallurgical Engineering, Istanbul Technical
University, 80626 Maslak, Istanbul

D.R. Hook
Department of Scientific Research, British Museum

M.S. Humphrey
Department of Scientific Research, British Museum

K. Hyne
Formerly Department of Scientific Research, British Museum

N.D. Meeks
Department of Scientific Research, British Museum

A.P. Middleton
Department of Scientific Research, British Museum

H. Özbal
Bosphorus University, 80815 Bebek, Istanbul

A. Ramage
Department of the History of Art, Cornell University, Ithaca, New York 14853

Acknowledgements

We are indebted to many people for their encouragement and unstinting help, across a broad spectrum of academic disciplines, both in the humanities and the sciences.

Archaeological

The late Professor G.M.A. Hanfmann and Professor Crawford H. Greenewalt, Jr, his successor as Field Director, deserve particular thanks for recognising the importance of the discovery and supporting the field excavations and the subsequent analytical investigations in the laboratory. We had much encouragement also in the early interpretation of the finds from the late Professor Cyril S. Smith of the Massachusetts Institute of Technology. He believed in the main conclusions but played the helpful sceptic by pointing out anomalies and discrepancies.

We also stand in the debt of Sidney M. Goldstein, a conservator at Sardis in 1968 and a fellow graduate student. He helped in the beginning, discussing the interpretation of the little burnt rings that turned out to be hearths. He shared Andrew Ramage's excitement in the discovery and they both realised its importance even though they were not able to define the precise activity at first. In his capacity as a conservator, he organised the finds and took on the duty of compiling lists of the different categories and individual pieces of gold and ceramic materials collected in 1968 and 1969. These lists form the basis for the identification of items mentioned in the day-to-day field records.

Richard E. Stone, the Senior Conservator in 1968, must be recognised as the person most responsible for guiding Ramage and Goldstein towards the specific interpretation of gold refining for what was clearly some sort of metalworking process. He was at that time a graduate student in the Conservation Program at the Institute for Fine Arts, New York University.

We should also recognise the contribution of our Turkish colleagues in the field, in particular the team of workmen responsible for the actual digging. One of them in particular, the late Huseyin Bal, can be commended for a sharp eye and a delicate trowel. The Archaeological Museum in Manisa has been the recipient of their finds, and we are also grateful to the late Kemal Ziya Polatkan, director at the time of the discovery, and to successive directors (Kubilây Nayır and, since 1986, Hasan Dedeoğlu), who have been most helpful in permitting Goldstein and Ramage to study the pieces of gold and other refinery materials stored in their care, and for allowing the gold to travel to Istanbul for scientific study. Celalettin Şentürk and Teoman Yalçınkaya facilitated the physical transfer of gold specimens to the University of Istanbul in 1993.

Elizabeth Gombosi took many of the original field photographs and Michael Hamilton printed a large proportion. Elizabeth Wahle (in consultation with Sidney Goldstein) made many of the drawings reconstructing the process, and Catherine Alexander drew many of the tuyeres and refractory items. Sardis recorder for 1997, Ellen Roblee, and assistant recorder for 1998, John Vonder Bruegge, whether at Sardis or back at the office, helped in the retroactive inventory project of the technical material. Laura M. Gadbery, Associate Director and Head of the Sardis office in Cambridge, assisted in the preparation of the new object cards. To all of them, many thanks.

The original support and financial contributions that made the Archaeological Exploration of Sardis possible have by now been rehearsed several times, as has the generosity of the Republic of Turkey, through the Department of Antiquities and Museums, in affording us the privilege of working at Sardis. That is the background to all our endeavours. For the discovery and study of the gold refinery a few years may be picked out (1967–70, 1975, 1987 and 1993) when the support offered through the continuation of the excavations was crucial. These years cover the actual discovery and excavation of the refinery (1967–70), a short study season by Sidney M. Goldstein (1975) and two study seasons by Paul T. Craddock (1987 and 1993). Andrew Ramage was active at Sardis for all seasons except 1975 but did not engage in gold research other than that entailed in his original discovery until 1987. Major support in those years came from The Old Dominion Foundation, the Ford Foundation, the Loeb Classical Library Foundation of Harvard University and the National Endowment for the Humanities,[†] a federal agency whose

---

† NEH grants H67-0-56, H68-0-61, H69-0-23 supported general fieldwork in 1967, 1968 and 1969 respectively. RO-111-70-3966 supported fieldwork publications, as did RO-21414-87.

views do not necessarily concur with those set out here.

Individuals whose contributions have been generous and consistent include Mr and Mrs David Greenewalt, the late Dr Edwin H. Land and Mrs Land, Mr Thomas B. Lemann, Mrs Guy Smallwood and Mrs Gustavus F. Swift.

Scientific

Much of the scientific work was carried out at the Department of Scientific Research of the British Museum, and the various contributors to this volume wish to thank their colleagues within the Department for help and discussion over a wide range of topics. In particular, Dr I.C. Freestone is thanked for his help in interpreting the structure of the refractories. We also should thank our colleagues in other departments, notably the Department of Coins and Medals, for assistance in selecting the Lydian coins for scientific examination from the collections of the British Museum, and allowing them to be analysed. We are also grateful to Professor A.A. Gordus of the Department of Chemistry, University of Michigan, for allowing us to publish here the analyses of gold foils, etc. from the excavations and of Lydian coins from the Ashmolean Museum, Oxford, performed many years ago.

We wish to thank Dr Jack Ogden for help and advice over all matters auriferous, freely given from his encyclopaedic knowledge of the subject. We are grateful to Dr Justine Bayley, of the Ancient Monuments Laboratory of English Heritage, for imparting her knowledge of the excavated remains of parting, not least for teaching Paul Craddock to recognise the distinctive features of the debris of salt cementation. We must also thank Professor E.T. Hall, of the Research Laboratory for Archaeology and the History of Art, Oxford, both for his encouragement whilst a Trustee of the British Museum, and for his scientific advice, based on his own important experiments on gold

refining, which he has allowed us to quote. We also wish to thank Dr Alessandra Giumlia Mair for discussion and translation of the ancient texts in Chapters 2 and 3. We are grateful to the laboratories of Johnson Matthey Ltd for making available to us J.H.F. Notton's unpublished internal report on his gold-refining experiments.

The work on the gold foils from the Archaeological Museum in Manisa was conducted at the laboratories of the Department of Metallurgy and Materials Engineering of the Istanbul Technical University and partially supported by the Archaeometry Research Center of the Bosphorus University and the Archaeological Exploration of Sardis. Our thanks go to Dr V. Gürkan, who undertook the macrophotography of the gold foils, and to H. Sezer for his assistance with the SEM examination. The macrophotography of the material examined in the British Museum was undertaken by Anthony Milton of the British Museum Photographic Service.

Editorial

Our readers, Dr Oscar White Muscarella and Professor Mike Wayman, and our editors, John Day, Katherine Kiefer and Nina Shandloff, on both sides of the Atlantic, have persuasively but firmly helped us to smooth out the organisation of the book with its mixed historical and scientific components. Andrew Ramage's wife, Nancy Hirschland Ramage, was in on the discovery from the very beginning and made drawings of the cupels and of the furnaces. Many times recently she has read over his text and offered countless suggestions for improvement as well as encouraged him to persevere when things looked bleak.

Andrew Ramage
Paul Craddock
August 1999

Illustration acknowledgements

Acknowlegement is due to the following for permission to reproduce photographs and other illustrations:

Archaeological Exploration of Sardis for Figs 1.1 to 1.3, 1.5, 1.6, 1.9 to 1.11, 4.1 to 4.53, 4.55 to 4.57, A2.1 to A2.10

British Library Board for Fig. 2.6

Deutsches Archäologisches Institut Athens for Fig. 4.54

Department of Metallurgy and Materials Engineering, Istanbul Technical University for Figs 9.1 to 9.23

Hirmer Verlag München for Fig. 1.7

Trustees of the British Museum for Figs 1.4 and 1.8, 5.1 to 5.64, 6.1 to 6.10, 7.1, 9.24 to 9.34, 10.1 to 10.5, A1.1 to A1.10

Prologue

A. Ramage and P.T. Craddock

Excavation of the gold refinery at Sardis

The discovery of the gold refinery is one of the most important finds made in the course of the Harvard–Cornell excavations. The installations are at the moment unique in the classical world and of fundamental importance for the interpretation of several short or inconclusive passages from ancient authors around the world. The details of the sequential discoveries and day-to-day strategies are summarised below to produce a synthetic picture of the workshop and its procedures.

At the time of the excavation, the expedition conservator, who had extensive technical expertise, helped us to make the connection between multiple small hearths with associated lead oxide and the possibility of a gold-refining process. Subsequently, we saw the reddened masses of mud brick as furnaces for another, and even more vital stage in the parting process. This was on the basis of written accounts from the sixteenth century that apparently had retained a tremendous amount of traditional information and practice.

The place of Lydia in the centre of western Anatolia and the expansive policy of its kings meant that its capital, Sardis, became the focus of diplomacy and hostilities at the end of the seventh century and into the sixth and fifth centuries BC. Enormous wealth was generated during that period from conquest and from the exploitation of the alluvial gold in the River Pactolus and other rivers nearby. It is, therefore, of great significance to have found one of the installations where the natural gold–silver alloy was processed to provide pure gold and pure silver as is implied by the tradition in the ancient world that the Lydians were the first people to employ gold and silver for coins. The workshop was found in a domestic area, and the fittings and equipment give the impression of domestic items pressed into service for special requirements. Thus, the furnaces give the appearance of a battery of domestic ovens and the vessels for the parting process are derived from domestic jugs.

The dating of the refinery fits very closely with that of the reign of King Croesus (561–547 BC), who became a symbol for enormous wealth into our own time. The body of the text here contains an explanation of the chronology, as the circumstances of the discovery and the actual artefacts or workshop residue are described.

Detailed scientific examination of some of the finds, undertaken in the British Museum's Department of Scientific Research, elucidated the parameters of the refining process. When combined with the evidence found in historical texts describing various methods of refining gold, the archaeological evidence and the results of the scientific examination create a detailed picture of the gold-refining process used at Sardis.

The processes of gold refining

Gold occurs naturally as minute particles of metal in the primary quartz veins in which it formed. This is known as *primary* gold. Where these deposits have eroded, the gold can be released, reconcentrated by the action of water and deposited in beds of sands or gravels. This is *secondary* or *placer* gold. The gold, being of very different density to the other materials, is often concentrated within very specific areas of these sands or gravels. Sometimes these areas are quite small but still of very great value, containing large quantities of gold in near-surface deposits of easily worked sand or gravel in former or existing stream beds. It is believed that the Pactolus contained such deposits, and although nothing is known about the actual mining methods, their discovery and exploitation took place over a relatively short period, creating enormous wealth. However, the deposits were soon exhausted, and the riches of Croesus became but a memory to succeeding generations.

Almost all gold occurs naturally containing some silver. This can vary enormously but typically lies between about 5% and 40% by weight. The only other metal found in natural gold above trace levels is copper, which, however, is only rarely found in quantities greater than 1% or 2%.

With the advent of coinage, where both the weight and purity of the metal were guaranteed, it was clearly imperative to be able to refine the gold. The base metals, notably copper, could be removed from the gold by *cupellation*. The impure metal was melted together with lead and the resulting mixture subjected to a continuous blast of air at high temperature, typically in the region of 1100 °C. This completely oxidised the lead and any other base metals present, to form a molten mass of lead oxide – litharge – with the other metal oxides absorbed within it, but leaving the two noble metals, gold and silver, unaffected and separate.

Different methods were necessary to separate the gold and silver, an operation known in English as *parting*. Before the discovery of mineral acids, the only method of parting gold and silver was by cementation with acidic salts. In a *cementation* process, the principal reactants are in the solid state, and the usual salt used in antiquity seems to have been sodium chloride (common salt), but others, notably alum, ferric sulphate and potassium nitrate (nitre, also known as saltpetre), were also used alone or in combination. These were joined in the Middle Ages by *sulphur parting*, using either elemental sulphur or, more usually, an iron or antimony sulphide. The active agents were often supported on an inert medium or carrier, typically of clay or brick dust, which was especially useful if the reagents melted during the reaction. The mixture of active agents and the carrier was known as the *cement*. The finely divided, impure gold was placed, carefully surrounded by the cement, inside a ceramic container. This was known as the *parting vessel*, and was usually specified as being of earthenware, the relatively open structure of which would have been thermally more resilient and more porous than finewares. The parting vessel was then heated for prolonged periods at temperatures below the melting point of the impure gold, but the other reagents could be either solid or molten. Note that, although cementation processes are nominally solid state, at the operating temperature the active agents have a high vapour pressure, even though still solid, and in reality it was their vapours which attacked the solid metal. Thus, in the salt cementation process, which is believed to have been employed at Sardis, vapours of chloride ions and chlorine at elevated temperature attacked the surfaces of the impure gold, penetrating deeply into the interior of the metal along the grain boundaries. The process was performed for some hours, or even days according to some early descriptions, until the gold was purified. The silver was converted to silver chloride, which is also volatile, and was absorbed by the inert carrier, if present, and in the walls of the parting vessel and furnace. The silver could be recovered from these various materials by smelting them with lead to absorb the silver salts, followed by cupellation to release the now-pure silver.

Acid parting began in the Middle Ages, initially using nitric acid, which was replaced in the post-medieval period by sulphuric acid. Here, the solid impure gold was attacked by the hot aqueous solutions of the acid. Once again, the active reagents attacked preferentially down the grain boundaries. In order to facilitate the parting process, it was normal practice to add considerable quantities of silver to the impure gold, a procedure known as *graduation* or *quartation*. The reason for this seemingly quixotic practice of adding more of the metal they were trying to remove was to ensure that there was abundant silver within the body of the gold which the acid would eat away, creating considerable porosity throughout the gold, and thereby ensuring that all of the metal was exposed to the acid. In fact, this procedure originated earlier with the salt cementation process, where it was used as a special extra refinement, with the extra metal, usually copper, added *after* the first cementation. However, the greater porosity created in the gold was not as necessary for cementation parting, which operated at much higher temperatures than the acid parting process, with the active reagents being present as vapours rather than as liquids.

Summary of the historical evidence

A number of important points emerge from the broad survey contained in the historical Chapters 2 and 3 and in Appendices 3 to 5, together ranging over some very disparate subjects and sources.

Surface enrichment

The deliberate surface enrichment of gold artefacts seems to have been a widespread practice almost since the inception of goldworking. It also seems that corrosive salts

– such as common salt, alum, iron sulphates and nitre – were used, which were capable of removing some of the silver from the gold, as well as the base metals. These salts were subsequently to be the active agents in the cementation refining processes, but at high temperature rather than in an aqueous medium.

Everywhere, except possibly in India, the introduction of a gold coinage seems to have provided the stimulus for gold refining. Without the requirement of gold of guaranteed purity for the specific needs of coinage, there was no incentive to refine gold. The surface of finished gold artefacts could be enhanced quite satisfactorily without the loss of weight attendant on full refining. Surface enrichment could be achieved much more easily with mineral acids, but cementation methods continued well into the twentieth century, long after cementation had been abandoned for gold refining.

Cementation

The basics of cementation refining are very similar around the world. To a degree, this was inevitable, dictated by the physical chemistry of the process. A variety of active salts was used in antiquity, notably common salt and sulphates in the form of alum or iron sulphates. The chlorides and sulphates were used both alone and in combination; experiments performed in the eighteenth century suggest that a combination of salts could operate at lower temperatures. Nitrates in the form of impure saltpetre may have been used in ancient India, although apparently not in the West until the Middle Ages. Some recipes suggested the use of copper, lead or zinc sulphates, but these are unusual and uncertain.

There is little evidence for the use of either elemental sulphur or metal sulphides, such as marcasite or pyrites, for the refining of gold in classical antiquity, although they may have been components of recipes for surface treatments in late antiquity. The use of sulphur and sulphides, notably antimony sulphide (stibnite) only became prevalent in the medieval world. Similarly, strong mineral acids were only employed from the Middle Ages in Europe and the Middle East, and even then were rarely used alone for primary gold refining until well into the post-medieval period.

Some early recipes for gold-refining cements specify the addition of copper and even lead salts. These are repeated and expanded by many of the later accounts.

There is a suggestion in some of the ancient descriptions that two-stage processes were used. This is certainly true of the first reliable and detailed accounts, such as that of Theophilus in the twelfth century and those of authors, such as Ercker, during the Renaissance. In the second, more rigorous refining stage of the cementation, known as graduation, copper was often added. Almost all of the medieval and later recipes state that reagents have to be mixed with an inert support medium, powdered brick, sometimes with additional clay, being very popular, but no such materials are found in the earlier descriptions, notably that of Agatharchides, *c.* 116 BC.

The gold to be refined needed to have a large surface area exposed to the reagents. The granules of freshly won gold could be used as they were, or the gold could be hammered into thin foils, or be granulated by pouring the molten metal into water. The *Probierbüchlein*, compiled in the early sixteenth century, describes the addition of lead to the gold to facilitate the granulation.

The common descriptions through the ages of the parting vessel concur that it should be of earthenware rather than of fineware ceramic. Some descriptions actually specify a cooking pot, and Agricola recommends a vase-shaped pot, not dissimilar to the Lydian coarseware cooking pots at the Pactolus North refinery which seem to have been selected for use as parting vessels. Similarly, several accounts describe the use of earthenware potsherds for various tasks. Almost all the descriptions specify a lid should be fitted and sealed in place with clay, although in the Japanese and some of the later Indian processes, the cementation mixtures were not contained in a pot at all. The function of the lid was to prevent loss of the volatile silver salts, and both Indian and Japanese commentators note the loss of silver.

The various descriptions of the parting furnaces, which become ever more detailed from late antiquity through the Middle Ages, have strong similarities both amongst themselves and with the furnaces excavated at Sardis, suggesting a long continuity.

After parting, the gold was washed and melted, and the silver recovered from the spent cement. This was almost always done by cupellation – although note the reference in the *Probierbüchlein* to the amalgamation process for the recovery of silver from the parting cements. Cupellation was the standard process for the recovery and refining of silver throughout the Old World, and as such was not usually described in any detail in the early accounts of gold refining. The Indian saltpetre process is an exception, where it does seem that significant quantities of gold also had to be recovered from the cement along with the silver, and that a special process was necessary.

There are several excellent accounts of the salt cementation processes in the major nineteenth-century metallurgical texts. Percy's *Gold and Silver: Part One*, published in 1880, is pre-eminent, although by then the process had ceased to be used except, as Percy correctly stated, possibly in Japan. Percy's long description includes an important and detailed section on the process in classical antiquity (pp. 397–402), which was the first major discussion on the subject. He concluded that a method of separating gold from silver was known and practised in the time of Strabo:

> ... but centuries before that period the ancients must have been able to effect such a separation; for otherwise it is impossible to account for the remarkable purity of the gold of many of their coins.

Percy describes the chemistry of the processes in terms of fixed and immutable elements and of Dalton's atomic theory. His descriptions are thus very different from the theoretical explanations of the processes that had gone before. They are, in fact, the first modern descriptions, and as such form an appropriate starting point for the scientific investigation of the remains excavated at the Sardis refinery.

Amalgamation

The treatment of metals with mercury has a long history, stretching back into the first millennium BC, but there is no evidence for the extraction, in antiquity, of either gold or silver from their ores by amalgamation. There are descriptions from Roman times of the recovery of gold from gilded artefacts, but the earliest unequivocal descriptions of the treatment of gold ores with mercury to extract the gold are medieval. Silver in metallic form was already being recovered from its finely ground ores and slags in the later first millennium AD. The recovery of silver minerals from cupellation debris by amalgamation was a late medieval development, and the successful treatment of silver ores only began in the sixteenth century in the Americas.

It is possible that before mercury was used to extract gold, lead may have been used in the same manner. There is certain archaeological evidence from Roman times for the use of lead to extract the gold from the auriferous slags created by the smelting of pyritic gold. It is possible that the minute particles of gold from primary deposits could have been extracted from the finely ground ore with lead, although the literary evidence for this is tenuous and as yet there is no archaeological evidence.

Platinum group element inclusions

Platinum group element (PGE) inclusions, which are so apparent in some of the Lydian gold coins, were described by some ancient classical authors either as *adamas* or *adamans*. Pliny gives the most detailed descriptions, although confusingly describing gemstones at the same time. Plato mentions *adamans* and implies that it could be refined from the gold. The *Mappae Clavicula*, compiled in the post-Roman period, contains some elaborate recipes using lead, which describe the removal, recovery and even use of the inclusions.

Assaying

The ability to determine the composition of the gold was clearly essential to any meaningful refining process. Simple heating could reveal the presence of other metals in the gold if they were present in quantity, and the potential methods for quantitative determination were specific gravity, fire assay and the touchstone.

The principles of the specific gravity method seem to have been understood in antiquity, but were rarely, if ever, used. The method ascribed to Archimedes was correct in theory, but would not have been practicable. It was not until the development of proper precision balances during the Renaissance that the method could have been of practical application. Fire assay is a broad term covering both cupellation and parting methods. Cupellation alone could be used to estimate the quantities of base metals and as such seems to have been the standard method from remote antiquity in the Middle East, where the sources describe the weight loss of suspect gold on fire refining to constant weight. However, cupellation alone would be of no use to estimate the silver content, for which the fire assay would have had to include a cementation stage. This would involve considerable time, skilled effort and specialised equipment, and in practice does not seem to have been a common method.

From classical antiquity at least, until the recent past, the usual method of assay was by touchstone. Prior to the introduction of mineral acids in the Middle Ages, touchstones could only be really accurate with binary alloys of gold–silver or gold–copper and could not easily quantify alloys of gold with copper and silver in a combination that was usually intended to preserve the colour of the gold.

Thus, in practice, the ancient and medieval worlds had no routine effective method of determining the composition of gold adulterated with a combination of metals. This probably explains the proliferation of artificial and alchemic golds.

CHAPTER 1

Golden Sardis

A. Ramage

The wealth of the Lydian kings and their capital at Sardis was legendary from their own time until the present day. The gold itself is no legend, however, and small quantities can still be recovered from the sands and gravels of the River Pactolus and other nearby streams. Most of the stories about the wealth of Sardis centred around Croesus, who was the last king of an independent Lydia. He served as the model among the Greeks for an extraordinarily rich man who was not able to achieve happiness or success in spite of his wealth. He was, however, assumed to be responsible for changing the Lydians' use of coins made of an alloy of gold and silver to coins that were pure gold or pure silver.

The Harvard–Cornell excavations at Sardis have brought to light a series of installations that would have been capable of separating the major components of the mixed metal to enable the Lydians to produce the silver and gold coins for which they were so renowned.

Sardis in context

Gold and Sardis – the words are inextricably connected in the ancient record. Poets and historians frequently refer us to topographic details of Sardis and Lydia (Figs 1.1 and 1.2) and recount personal stories of its legendary kings. They cast the riches of Sardis, especially those of Croesus, who ruled the Lydian empire from 561 to 547 BC, in an allegorical and moralising mode. The name of Sardis carried with it the weight of many ancient comments and the lessons to be learned from the progression beyond wealth to luxury and pride. Very little is known about other Lydian cities from archaeological excavation, although the names of many places in Lydia are recorded by Herodotus and especially Xenophon and Strabo, and several towns that issued coins during the Roman era seem to have retained their

Lydian names. There are clumps of burial mounds within the region traditionally known as Lydia, but at present no definitive associations can be made with habitation sites of the Lydian period, which are very hard to discern in the modern topography.[1]

The lesson regarding wealth was exemplified by the Greek historian Herodotus' account of the meeting between Solon and Croesus, where the latter was disconcerted to find that riches were not to be equated with happiness. Croesus had asked Solon, a famous Athenian philosopher, who was the happiest mortal, and had expected him to say, 'Croesus'. But no such luck. Solon had answered with the name of Tellus, an obscure dead Athenian. Solon maintained that no one could be called happy before his death because there might be some untoward incident to turn things upside down. Such a reversal did in fact happen to Croesus, who lost his son in a hunting accident and forfeited his empire to King Cyrus the Great of Persia as a result of his own over-confidence. Even though ancient critics and modern historians have shown that this specific encounter with Solon would have been impossible, because Solon lived a generation before Croesus, it nevertheless gives an excellent flavour of the Greeks' attitude towards those who lived beyond their borders and did not speak Greek. The epithet *barbaroi*, literally denoting non-Greek-speaking groups, became more or less synonymous with Lydians and especially Persians. The Lydian kings and later the whole people, to the Greeks, were effete and given to luxury. This condition was, supposedly, brought about by self-indulgence. Scholars have speculated that this was a rhetorical manoeuvre to show the Greeks as morally superior; it runs all through Greek history writing. Other Greek writers had the same attitude towards Lydian wealth and specifically their gold, at least as far back as the time of King Gyges, who had usurped the kingly power in the early seventh century BC,

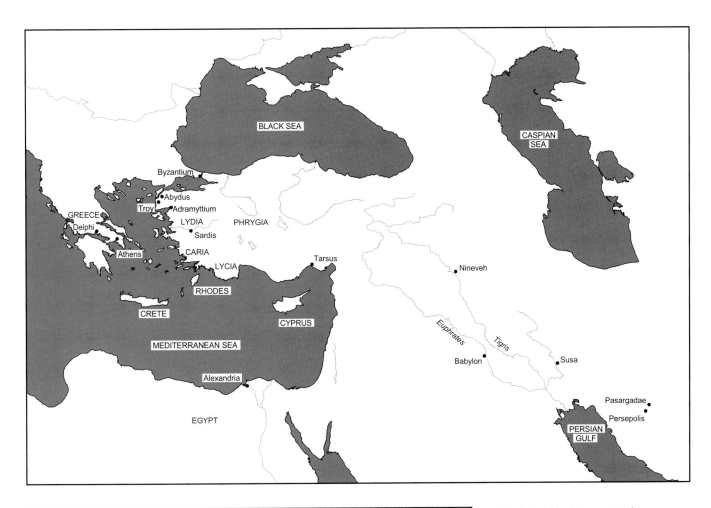

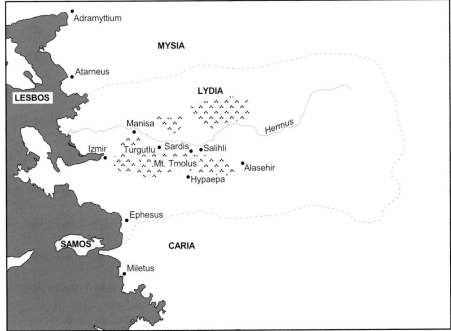

Fig. 1.1a *(above)* Eastern Mediterranean lands.

Fig. 1.1b *(left)* Sardis and central Lydia.

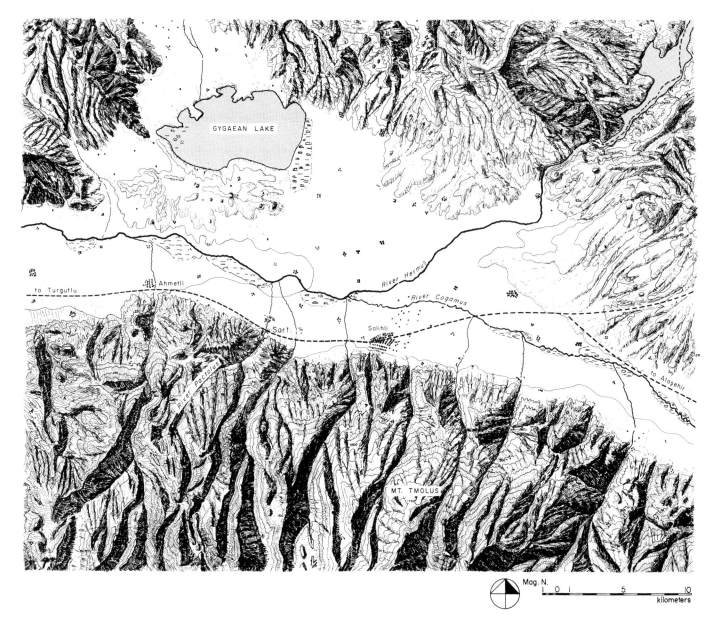

Fig. 1.2 Mount Tmolus and the region of Sardis.

for the contemporary Greek poet, Archilochus, pointedly spurns both wealth and power:[2]

> I do not care for the wealth of Gyges rich in gold. Envy has never taken hold of me. I am not vexed at the divine order nor do I long for a tyrant's power. These things are far from my eyes.

The poet Pindar brings out his own ambivalence towards gold in one of his victory odes (*Olympians* I.1), by declaring 'Water is best...' but going on to describe the shining attraction of gold. In context, this may have to be interpreted metaphorically, but the opposition of the two substances, water and gold, is in the same spirit as that of contrasting the simple Greeks and the luxurious Lydians.

The expansion of Lydian power at the expense of their neighbours to east and west seems to have been mainly the work of King Alyattes, the father of Croesus, in the late seventh and early sixth centuries. His predecessors, Gyges and Ardys, had been engaged in fighting off the nomadic Kimmerian tribes, who moved in from the Balkans and south Russia and ravaged western and central Anatolia for much of the seventh century.[3] The physical circumstances of Sardis are not referred to in Herodotus' romantic account of Gyges' coming to power and at present we cannot characterise a particular level in the excavations as belonging exactly to the period of Gyges.[4] It was left for Croesus to use and display his riches at Sardis and, by his over-confidence, to become the unwitting cause of the great conflict between the Persians and the Greeks, which was the central subject of Herodotus' enquiry in his *History*.

Of the actual luxury of the Lydian court at Sardis itself, we have only a few concrete examples in jewellery (Fig. 1.3), fine metalwork and ivories. The rest must be imagined: the fine fabrics, furniture, scents and culinary delights alluded to but hardly ever described by the ancient writers of any era. In fact, the gold and its effects seem to have monopolised the attention of ancient authors. Excavated items, with their usual bias against organic materials, have to form the largest source of information about ordinary Lydians because the lives and habits of the kings and aristocrats, if that is how they should be characterised, are the only topics worth mentioning.

Lydian metallurgy and the beginnings of coinage

Of the flashiest source of this luxury that fascinated the Greeks, the gold bullion and coin of Croesus, we have recently had a tantalizing glimpse. Not the treasury, not the royal jeweller's atelier, or even the mint has come to light. The Harvard–Cornell Sardis Expedition has, however, found a yet more fascinating place: a workshop where the grains and dust of native gold were processed to produce the raw material for those fabulous treasures. In many ways, this offers more of a challenge than admiring the superb finish or awesome weight of the worked products, because one can now wonder at the imagination required to bring the unlikely-looking raw material to a usable form, and at the skill required to produce gold of an almost perfect fineness. This fineness is all the more remarkable when one considers that the original raw material, alluvial gold, contained substantial amounts of silver and sometimes copper.[5]

The goldworking facilities had a finite life, even though the installations we found may have been reused several times. The product seems to have been made in small batches, to judge by the size of the furnaces. But small as the operation must have been, it supplies us with precious information about ancient gold-refining processes that has been found nowhere else.

It has been assumed that what was retrieved from the Pactolus was a natural alloy of gold and silver that contains up to 40% silver. But the analyses presented here, as well as earlier studies, indicate that the composition was variable, some of the alloy having a much higher gold content.

Fig. 1.3 Gold earring in the form of a lamb: length 10 mm, height 10 mm. It was found near the gold refinery at Sardis.

We should allow for variation in the natural conditions that cause concentrations to build up in the gravels. Bolin,[6] who claimed that the Lydian kings were debasing their currency, made an unwarranted assumption: that all the gold alloy found in the Pactolus was of the same composition. If this were so, then his argument that the Lydians were purposely debasing the coinage might stand. But evidence would suggest that there was variation in the raw material from the Pactolus (and presumably the other streams), and that silver had to be added to create an artificial alloy, electrum, of constant purity. Thus, Bolin's argument becomes unnecessary.

It is generally agreed that the Lydians invented coinage (as we understand it) for the Mediterranean world, but it is a matter of vigorous debate exactly when this happened. At some point in the seventh century BC – the range is from the early years to the very end of the century – they decided that creating and marking a series of small lumps of electrum at a consistent weight would be useful. The purpose for it or the advantage to be gained is perhaps less easily discovered than the date of its inception. A serious drawback to this system is that one cannot readily assess the amount of gold within a lump, even though the colour is a rough guide. An even more important innovation was to institute a bimetallic series of coins, presumably using gold and silver separated from alluvial gold in the first instance, augmented by pure silver from other sources (Fig. 1.4, p. 129). Knowledge of this momentous change is laconically preserved for us by Herodotus, who reports simply that 'they [the Lydians] were the first people that we know to employ minted coins of gold and silver...'.[7] The recent discoveries offer evidence that may shed light, at least indirectly, on this question.

By now the ultimate use of these lumps of metal as coins has been declared, even if that was not the original purpose of their creation. The use of agreed weights of precious metals as payment or as expressions of wealth goes back in the Near East at least 2000 years before Croesus. The use of coins, however, and the implication of even a partial monetary economy, is much more recent. The Lydians seem to have started this progression to a monetary economy with the electrum issues, and in essence perfected it with the new coins of gold and of silver.

A difficulty for numismatists lies in finding an appropriate time for initiating the electrum series, and King Alyattes, the father of Croesus, is a frequent candidate for this honour.[8] His hostile dealings with the Medes, whose territory was directly to the east of Lydia, and the need to pay his own army, might have suggested coinage as a solution. Paying mercenaries has been suggested frequently as a reason for creating coinage. The logic is that even at a very low wage the lump sums owed by the month or at the end of a campaign would become considerable, and even the small electrum pieces represented a notable store of value in the seventh- and sixth-century economy. The next question is where to place the introduction of the bimetallic system referred to by Herodotus.

The first coins were made from natural alluvial gold to which silver was added to create an electrum containing approximately 55% gold, 45% silver and 1% or 2% copper.[9] As Herodotus says, the Lydians created a coinage of pure gold and pure silver. This change is usually said (but without any explicit proof) to have taken place in the reign of Croesus, that is to say between 561 and 547 BC. There are strong contrary opinions, however, that put the initiation of Lydian gold and silver coins after the fall of Sardis and into the Persian era.[10]

The same types of Lydian coinage were continued under the Persians at Sardis, but the actual change from electrum to separate coins of gold and of silver has been attributed by scholars to Croesus and an independent Lydia.[11] The gold-working installations and the chronological indicators now discovered at Sardis make it clear that the parting of gold and silver, and thus the capability for issuing a coinage using the separate metals, was actually practised at the appropriate time for Croesus to have been the instigator. It is much more awkward to have the Persians, who had no experience with currency, suddenly take advantage of the Lydian innovation to issue a series of coins with Lydian devices before inventing their own standard type. None of the Persian standard types is in fact like the early Lydian or the disputed type. Thus we can confirm the hypothesis that the Lydians under Croesus initiated the bimetallic system of coinage, and even propose that the Sardis refining installations made its introduction possible.

Herodotus' description of the offerings of Croesus at Delphi makes it clear that the parting of gold and silver from natural gold must have been familiar to the Lydians and to the Greeks, as the use of the words *leukos* (white) and *apephthos* (literally, boiled down), for natural and refined gold, indicates.[12] The evidence of the processes used in the refining establishment and its probable date make the attribution of the change in the composition of the coinage to Croesus more likely than ever.

Even in the early seventh century BC the Lydians must have had access to gold, whether tribute or locally produced, both because of the specific reference of Archilochus, above, and because of more prosaic references to the gold mixing bowls of Gyges that could be viewed in the treasury of Cypselus at the sanctuary of Apollo at Delphi.[13]

Generally speaking, we suppose the gold was separated from the sands and gravels in which it lay by the traditional method of washing. It was in the form of 'dust'

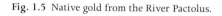
Fig. 1.5 Native gold from the River Pactolus.

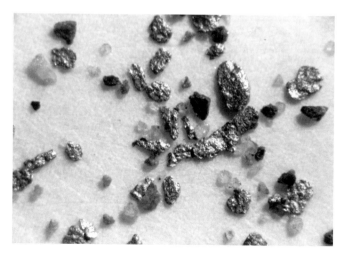

Fig. 1.6 Temple of Artemis at Sardis, looking east towards the Acropolis. Excavation house at left.

or quite small particles (Fig. 1.5). Nuggets are the exception, which is why they have commanded so much attention over the years. Howard Crosby Butler, who excavated at Sardis between 1910 and 1914, even mentions that his team occasionally found bits of gold when they were digging pits in the bed of the Pactolus.[14] They dug over 1000 tombs, most of which had already been looted. Nevertheless, in the few that were undisturbed, they found many pieces of gold jewellery.[15]

Sources for the Lydian gold

The source of the gold at Sardis is near the Palaeozoic massif of Mount Tmolus, ancient metamorphic rocks and intrusive veins of quartz, located to the south of Sardis, in a more recent conglomerate and its associated alluvium.[16] In earlier times, the action of the River Pactolus could have concentrated the sparsely disseminated gold from the alluvium (derived from the conglomerate) by actively reworking large quantities of the gravels. The present stream is sluggish, but tectonic movements such as the major

earthquake of AD 17, which partially buried the Temple of Artemis (Fig. 1.6), situated by the side of the stream, may have also altered the old channel, thus permanently changing the gradient.[17]

The ancient authors generally refer to the gold-bearing sands of the Pactolus or the Hermus as the principal source of the gold. The tale of Midas' release from his 'golden touch' by washing at the source of the Pactolus tends to corroborate this idea.[18] Curiously enough, these sources do not make much mention of gold in connection with the Phrygians, in their heyday during the eighth century BC, and finds of precious metals at the city mound of Gordion and in the tombs there have been meagre.[19] There are, however, ancient accounts of actual mining in the area of Mount Sipylus and we have a personal report of ancient workings in the foothills of Tmolus, immediately to the south of Sardis.[20] For the time being, we have not been able to identify the actual spots referred to. Certainly all the modern research to ascertain whether any gold remains in the area has concentrated on the conglomerate and the alluvium derived from it. These surveys have shown that the conglomerate forming the Necropolis hill on the west bank of the Pactolus at Sardis still contains gold, even if at very low concentrations. We can assume that the same is true for the matching hill used as the acropolis of ancient Sardis, and in that case the Lydians and Croesus were quite literally sitting on a gold mine.

There is no direct evidence of the organisation or whereabouts of other production centres. One cannot tell, for example, whether the conversion of locally mined gold dust was the main activity or whether refining 'scrap' coins was important too. Both forms of metal were certainly processed at the Pactolus North refinery. It is known that all the streams issuing into the Hermus valley from the conglomerate lying up against Mount Tmolus continue to carry gold (see Fig. 1.2).[21] We may infer that the quantity from each of these sources in ancient times matched that of the Pactolus, for which we have such vivid descriptions. There are scattered accounts of the actual mining of gold in some areas of Lydian-controlled territory, notably in the Troad near the town of Astyra on the Hellespont:[22]

[Gold mines] are now scant, being used up, like those on Mt. Tmolus in the neighbourhood of the Pactolus River.

Strabo XIII.1.23, translated by H.L. Jones

Another passage lists the sources of various rulers' wealth:

The wealth of Tantalus and the Pelopidae arose from the mines round Phrygia and Sipylus; that of Cadmus from those round Thrace and Mt. Pangaeus; that of Priam from the gold mines at Astyra near Abydus (of which still to-day there are small remains; here the amount of earth thrown out is considerable, and the excavations are signs of mining in olden times); and that of Midas from those round Mt. Bermius; and that of Gyges and Alyattes and Croesus from those in Lydia and from the region between Atarneus and Pergamum, where is still a small deserted town, whose lands have been exhausted of ore.

Strabo XIV.5.28, translated by H.L. Jones

Some of this text is corrupt, but it is interesting to note that the final area mentioned is not far from Adramyttion, where Croesus was 'governor' as a young man. Another snippet seems to confirm the Lydians' hold over the Hellespont:

Abydus [near Astyra] was founded by Milesians, being founded by permission of Gyges, king of the Lydians; for this district and the whole of the Troad were under his sway ....

Strabo XIII.1.22, translated by H.L. Jones

We may assume also that the Greek cities conquered by Croesus paid tribute in precious metal (presumably silver), but we cannot tell whether it would have come as bullion or coin.

Value and quantity of gold

The statements of Herodotus, if taken at face value, list more than 5 tonnes of gold given to the Oracle of Apollo at Delphi in the form of bricks or statues of a declared weight. Some have doubted that Herodotus was right or that Croesus actually did part with that much gold, on the grounds that the various measurements are irreconcilable with prevailing weight standards and the natural weight of the materials. We should not, however, regard the Lydian kings or Croesus as confidence tricksters, as some have suggested.[23] On the contrary, we maintain that Croesus was initiating a higher standard that would facilitate commercial exchange with the neighbouring Ionians.

After the elaborate list of offerings at Delphi (some of which may have been discovered in the French excavations of 1939),[24] Herodotus gives his account of the sack of Sardis and Cyrus' treatment of Croesus. There are several different accounts of their relations, both upon the first capture of Croesus and subsequently. The story of Croesus on the pyre is attested very early in the fifth century BC by the painting signed by Myson on a red-figured amphora in the Louvre (Fig. 1.7). It beautifully matches Herodotus' story, where Croesus is going to be burned to death by Cyrus because he was the enemy king. However, when Cyrus changed his mind, but could not put the fire out, Croesus prayed to Apollo, who sent a miraculous rainstorm that did. Bacchylides has the most coherent story, and Pindar mentions it too – all before Herodotus' own account.[25]

Herodotus also tells of gifts that Croesus made to various oracular sanctuaries of Apollo, and laconically observes that the dedications to Milesian Branchidai (i.e. the oracle of Apollo at Didyma) are equal in weight and similar to those at Delphi.[26] We should also remember that, in addition, Croesus contributed to a particularly expensive part of the Temple of Artemis at Ephesus – the column drums carved with figures, of which many broken fragments are now in the British Museum (Fig. 1.8).

Lydian riches were not confined to royal wealth, because Nicolas of Damascus recounts that when Croesus was young he tried to borrow a substantial sum of money from a private citizen by the name of Sadyattes.[27] Given his

except for various contributions in kind and 360 talents of gold dust from India valued at 4680 Euboeic talents of silver.

Herodotus tells a marvellous story about an Athenian aristocrat, Alcmaeon, who made a fool of himself by putting on special clothes to get all the gold that he could carry away from Croesus' storerooms. It goes like this:

> Alcmaeon, son of Megacles, gave all the assistance in his power to the Lydians who came from Croesus at Sardis to consult the oracle at Delphi; and Croesus, when the Lydians told him of the good service he had rendered, invited him to Sardis and offered him, as a reward, as much gold as he could carry on his person at one time. Alcmaeon thought of a fine way of taking advantage of this unusual offer: he put on a large tunic, very loose and baggy in front, and a pair of the widest top-boots that he could find, and thus clad, entered the treasury to which the king's servants conducted him. Here he attacked a heap of gold dust; he crammed into his boots, all up his legs, as much as they would hold, filled the baggy front of his tunic full, sprinkled the dust all over his hair, stuffed some more into his mouth, and then staggered out, scarcely able to drag one foot after another and looking, with his bulging cheeks and swollen figure, like anything rather than a man. When Croesus saw him he burst out laughing, and gave him all the gold he was carrying, and as much again in addition ....

Herodotus VI.123, translated by A. de Sélincourt

This story indicates that much of Croesus' store of gold was in the form of dust, which is not so easily used but it is more easily divided and apportioned than large ingots, especially when a considerable amount is to be allotted to the coinage. It is possible, also, for the parting to be done before the dust has been melted into lumps. The introduction of reliable gold and silver coins allowed for easier provision of small but measured amounts of precious metal in a form that did not require a second weighing or melting. This would have been much more economical but it is hard to prove, given that most of the metallic evidence from Sardis consists of thin pieces of gold foil of varying purity. They must have been melted once already before they could be hammered into foil.

Herodotus says that Gyges donated six golden mixing bowls, weighing a total of 30 talents, to the oracle at Delphi, as well as an indeterminate but larger weight of silver and other gold items.[30] These items are described as 'the Gygads', but elsewhere the Gygads are taken to mean the coins of Gyges, just as Croeseids are supposed to be the coins of Croesus or Darics those of Darius.[31] This argument

Fig. 1.7 Attic red-figured amphora by Myson, showing Croesus on the funeral pyre, *c.* 500 BC: height 58.5 cm.

name, which is the same as that of Croesus' grandfather, he was probably within the aristocratic circle, but his access to precious metal, even if it were not coined, adds to the impression that it was not closely held by the royal purse. Croesus had an unnamed citizen put to death for conspiring to elevate Croesus' half-brother to the throne and confiscated the citizen's considerable private resources for his own dedications at Didyma.[28]

The Pactolus North refinery is likely to have been capable of processing several hundred kilograms of gold each year, which would have been sufficient for tens of thousands of coins. Even so, it is likely that other facilities lay elsewhere, probably at the edge of town and in other valleys, since the fumes of lead oxide generated during the cupellation process, used to recover the silver, were extremely hazardous.

Storage and use

We have a quite explicit list of the tribute brought to Cyrus the Great by the various peoples of the Persian empire.[29] Almost all the assessments are in talents of silver,

Fig. 1.8 The face of a woman from one of the lower column drums of the Archaic Temple of Artemis at Ephesus, *c.* 530 BC.

is difficult to sustain if coins had not been invented by this time, as most scholars claim. It is not necessary that all this metal came from local sources; it could well have been tribute from the subjects of the Lydians' widening empire. Gold, however, was less accessible than silver to contemporary Greeks. Given the prominence of the Hermus and the Pactolus in the ancient texts, we may assume that some of the gold was local and that there was a big find in the time of Croesus or a more concerted effort to exploit the ore deposits. Certainly Sardis itself rather than Lydia as a whole is coupled with Bactria as the source for the gold used in decorating the palace of Darius at Susa.[32]

There is still a possibility that a much bigger area of goldworking lies in the unexcavated area to the north. There are, however, two important questions here: Where was the river in Lydian times? Would the Lydians have had the goldworks right in the middle of town? It is hard to imagine Herodotus not mentioning it in the section of his narrative where he described the burning of the city during the revolt of the Greek cities of Ionia against the Persians; there he explained that the *agora* or marketplace was divided by the river.[33]

Digging at Sardis

The Harvard–Cornell team is not the first group to excavate at Sardis. In 1910, a team directed by Professor Howard Crosby Butler of Princeton University, but financed by the American Society for the Excavation of Sardis, began work in the area of the Temple of Artemis. The clearing of a huge Hellenistic temple was not Butler's primary aim but he started at a spot known to have ancient remains, because of its two standing columns. It was also near the river, which was said to have flowed through the *agora* or marketplace of the Lydian city and was, therefore, preferable to beginning with any of the several large Roman buildings, whose remains were still visible above ground.[34] The physical situation of the city at the crossroads of a traditional east–west route to the interior and down to the Aegean Sea and of a north–south route, as well as its mythological reputation, made it a prime target for archaeological investigation. Unfortunately, the First World War interrupted the plans and an attempt to reopen the excavations in 1922 was foiled by the hostilities of the Turkish War of Independence. Even though Butler's team cleared the temple and investigated a great number of Lydian graves, the residential areas of the Lydian city continued to be elusive. It was left to the Harvard–Cornell team to look once more for the Lydian city. There is, in fact, a tenuous but real connection between the Harvard–Cornell expedition and Butler's

team, because in 1914 Professor George Chase of Harvard joined them to study the pottery found in the previous four seasons. In 1938, citing the pressure of administrative duties, he asked Professor George M.A. Hanfmann to collaborate with him. In 1948, Hanfmann visited Sardis and determined that most of the pottery and 'small finds' had been lost or destroyed and that there was not sufficient material for a substantial publication as had been planned. He does say that this trip caused him to decide that 'a new excavation at Sardis was needed'.[35]

The new excavations, the Archaeological Exploration of Sardis, were set up by Professor G.M.A. Hanfmann (Fogg Art Museum, Harvard University) and Professor A.H. Detweiler (College of Architecture, Cornell University). The sponsoring institution was the American Schools of Oriental Research, where Detweiler was president. 'The prime objective of any research at Sardis is the Lydian city of Croesus and his predecessors ... which remains completely unknown.'[36] Hanfmann emphasised the Lydian era because, even though the city was prosperous and influential under the Persians, the Hellenistic kings and the Romans, it never again matched the 'Golden Sardis' of the Lydian kings.

When the current excavations were finally started, in 1958, the professed aim was to find and explore the remains of the Lydian city, which are, after all, the unique aspect of Sardis in its 1300 year history as an urban centre. This had been Butler's aim also but first the clearing of the temple and then the war sidetracked his plans.

The original aims of the expedition have been achieved in a spectacular way and several important finds from the Roman era should be included in the list of discoveries. Detailed knowledge of just a few areas of the Lydian city has only shown how much more there is to be learned about its growth and status from the very beginning of the Lydian empire. The general urban development is poorly known because, unfortunately, the monumentality and extent of the later buildings at Sardis itself meant that frequently the Lydian remains were obscured, or even altered or destroyed, by Roman complexes. These have required their own careful investigation and in several cases could not be removed just to facilitate exposing the Lydian levels. An example of this situation can be observed on the overall plan of the excavation sector Pactolus North, where the presence of a substantial late Roman townhouse prevented excavation of domestic units of the Lydian period (Fig. 1.9). We can assume the continuity of the Lydian levels beneath the Roman building because of the rich remains of houses, and of some kind of public building, under the excavated Roman street, just to the south. We cannot easily reconstruct these Lydian domestic remains because the encroachment of the river at the west side has removed much of the evidence. The goldworks we have explored lies at the north-west corner of the excavated area known as Pactolus North (PN) (Figs 1.10 and 1.11).[37] The main industrial spaces represent only a small proportion of the area exposed at PN: roughly 900 out of a total of 2500 square metres, not including the purely Roman or Byzantine remains just referred to. Most of the excavating was done in the summers of 1968 and 1969, although small tests and a limited amount of analytical work were continued thereafter. What we have uncovered must have been quite a small operation. It is unlikely that this one area represents the complete gold-refining operations of Croesus or any other Lydian king.

Some of the precious metal produced at the PN refinery still exists in the form of coins and jewellery dug up over the years by farmers and grave robbers; much of it is now dispersed in many museums around the world. In addition, untold amounts must still remain underground in Lydia itself, as the recent publication of Lydian treasure from looted tombs has shown.[38] Finding the goldworks of Croesus was an unexpected and welcome bonus but much too specific as a research strategy. The field director might as well have announced that the expedition was looking for any one of the famous areas of the site as a specific purpose. In a sense, the approach from the very beginning could be said to have included looking for any of those desirable but elusive prizes.

Fig. 1.9 Overall plan of Pactolus North. The Lydian remains are delineated with bold 'stone-wall' lines, the Roman and later remains in outline.

Fig. 1.10 *(left)* Pactolus North, looking south-east. The goldworks are in the foreground, the Acropolis in the left background.

Fig. 1.11 *(below)* Pactolus North, looking north-west. The River Pactolus flows in the background, in front of the poplars and other greenery.

Notes

1 A. Ramage and N.H. Ramage (1971, pp. 143–60).
2 Archilochus 15 (Pedley 1972, no. 40).
3 Herodotus I.15–16; Strabo I.3.21, XIII.4.8, XIV.1.40 (Pedley 1972, nos 52, 49, 50, 51, with additional references).
4 Herodotus I.8f. King Kandaules was so proud of his wife's beauty that he persuaded his bodyguard, Gyges, to observe her undressing. She saw him and made it clear privately that he must either kill the king and marry her or she would have him executed. He took the hint and founded the Mermnad dynasty.
5 Topkaya (1984, p. 116).
6 Bolin (1958, pp. 22–37).
7 Herodotus I.94.
8 Weidauer (1975) discusses this at length in Chapter 2.
9 Cowell et al. (1998); Bolin (1958). See also pp. 169–73.
10 See p. 97, note 15.
11 For example, Kraay (1976, p. 30); Seltman (1955, p. 62). Main sources of evidence: Herodotus I.94; Pollux, *Onomasticon* IX.83 (Pedley 1972, no. 80).
12 Herodotus I.94.
13 Herodotus I.13–14 (Pedley 1972, no. 41).
14 Butler (1922, p. 16, footnote 1).
15 Published by C. Densmore Curtis (1925), and now in the Archaeological Museum, Istanbul.
16 For the general regional geology, see Brunn et al. (1971, pp. 225–55).
17 Topkaya (1984).
18 Ovid, *Metamorphoses* XI.142.
19 Young (1981) and Kohler (1995). But we should be wary of absolute judgments in the light of the hoard of Lydian electrum coins found in 1963 and published by Bellinger (1968), and the fact that a final report on the buildings and finds from the city mound has not yet appeared.
20 For Mount Sipylus: Pausanias VII.24.7. For Mount Tmolus: Letter from Clarence A. Wendell (US Minerals Attaché in Ankara) to G.M.A. Hanfmann, 11 August 1968: '… the other [photograph borrowed from Dr Birgi] is a photograph of an ancient mine opening on Mount Tmolus. The ancients followed a gold-bearing, quartz–arsenopyrite vein in mica schist rock country. This confirms my conclusion that I had seen old mine dumps in this area when I took my stroll with Professor Detweiler in 1967.'
21 Birgi (1944).
22 Attempts to reopen the mines were made in the late nineteenth century but were unsuccessful.
23 Bolin (1958, p. 23).
24 Amandry (1939 and 1977).
25 Bacchylides III; Pindar, *Pythians* I.184. Herodotus' account is in Book I.86.
26 Herodotus I.92.
27 Nicolas of Damascus, Jacoby (1923), 90 F 65 (Pedley 1972, no. 64).
28 Herodotus I.92.
29 Herodotus III.90–97.
30 Herodotus I.14. Thirty talents weigh 1135 kg.
31 Pollux, *Onomasticon* III.87: 'Highly prized is the Gygaean gold, as are the staters of Croesus.' (Pedley 1972, no. 80).
32 Dsf (Darius, Susa F) 35–55 (Pedley 1972, no. 303, with additional bibliography and explanations).
33 Herodotus V.101f: '[The Ionians] were prevented from sacking the place …. The Pactolus is the river which brings the gold dust down from Tmolus. It flows through the market at Sardis, and then joins the Hermus, which, in its turn, flows into the sea.' (Translated by A. de Sélincourt.)
34 Butler (1922, p. 3).
35 Hanfmann (1972, p. 10).
36 Hanfmann (1972, p. 11) quoting from his grant proposal to the Bollingen Foundation.
37 This slightly odd name is meant to distinguish it from Pactolus Cliff, a smaller sector about 200 m to the south of Pactolus North.
38 Özgen and Öztürk (1996).

CHAPTER 2

Historical Survey of Gold Refining

1 SURFACE TREATMENTS AND REFINING WORLDWIDE, AND IN EUROPE PRIOR TO AD 1500

P.T. Craddock

There are no detailed accounts of gold-refining processes extant until several centuries after gold production had ceased at Sardis. Even the one surviving ancient account, made in the second century BC, which does contain some practical detail,[1,2] is very possibly conflated (see pp. 34–5). Otherwise, the descriptions from classical antiquity are sparse in the extreme, such that in some instances it is not even certain that gold refining is being described.[3] Most of the literature considered here relating to the ancient processes originates from Europe or the Middle East, but there are other important early sources, notably from India, that are sometimes overlooked by classical scholars.[4]

However, through the Middle Ages and the Renaissance there are descriptions of the gold-refining processes in ever greater detail.[5] From a careful study of these and the remains excavated from the refinery at Pactolus North (PN) in Sardis, it seems that the procedures changed but little over the centuries and across continents. Thus the written descriptions of the methods, together with the rationale for carrying them out given by the more recent technical authors, are invaluable for our understanding of the ancient processes.

This chapter concentrates on the historical evidence for the refining of gold by cementation prior to the Renaissance. Other related chemical processes in the separation, assaying and refining of gold are dealt with separately in Appendices 3 to 5.

The parting processes left the purified gold as metal, but the silver which it contained as an impurity was left in the form of silver salts from which metallic silver had to be recovered. At the PN refinery, this was carried out by the process of cupellation, which was the standard method used throughout antiquity to recover silver both from base metal and its ores. The production of silver was not the prime concern of the PN refinery, and because of this and because there is an enormous literature on cupellation methods, including several detailed surveys,[6] the subject will not be considered further in this chapter.

By bringing together the literary evidence from around the world, some features which might individually be ignored or judged as erroneous, gain significance when they are found to appear in a number of independent accounts. These include such varied subjects as the introduction of lead and other base metals or metal salts at early stages in the processes, and an awareness of the platinum element group inclusions in gold (Appendix 4, p. 238). These subjects, taken in conjunction with the finds from the PN refinery, suggest new interpretations that have transformed our knowledge of the early methods of collecting and refining gold.

Surface enhancement of gold artefacts: the precursor to gold parting

The chemistry of the processes employed to enhance the surface of base gold are closely related to those used in the gold cementation processes, and thus it is appropriate to discuss them as a preamble, especially as the techniques considerably predate the first true refining processes.

Although complete separation of silver from gold was probably not practised in antiquity prior to the Lydian period, there is some evidence for at least the partial removal of silver from gold alloys right back almost to the inception of the use of gold. Thus Shalev[7] believes that the cast and hammered gold rings found at the Nahal Qanah Cave, dated to the fourth millennium BC, were deliberately treated to enhance the gold content at the surface. Specific gravity studies suggested that the metal of the rings usually contained between 65% and 75% by weight of gold but at the surface this had typically risen to between 80% and 95%, the remainder being of silver. This might be thought of as nothing more than the usual surface enhancement that occurs during prolonged burial, but the detailed study of the surface topography by scanning electron microscopy suggested to Shalev that the light hammering (in fact it seems to have been more of a planishing operation than true hammering) had occurred *after* the surface enhancement. He postulated that the rings:

> ... could have been passed through a surface oxidizing process, perhaps in an open fire on salty sand with NaCl. After the surface oxidation, the objects could be cleaned in natural organic acid to obtain the desired golden yellow colour, and their surfaces hammered in order to achieve their final shape and to regain their bright appearance.

Further evidence for the deliberate adulteration and surface treatment of gold has now been discovered, dating from the third millennium BC. The three famous gold chisels excavated from the royal cemetery at Ur[8] have recently been examined after it was noticed that the golden surface was peeling away in some areas, revealing much more coppery metal beneath.[9] Analysis of the larger chisel, Registration No. WAA 121349, showed the body metal at a depth of 1 mm typically to contain about 45% gold, 10% silver and 45% copper. By contrast, the surface examined had a gold content of 83%, with 9% silver and 8% copper. Clearly, both the copper *and* the silver had been preferentially removed from the surface layers. The surface was compact and heavily burnished, showing that the depletion had taken place *before* the final polishing of the gold and not whilst the metal was buried in the ground. Thus the chisels are a good and very early example of the process known as depletion gilding.

These are very similar to the better known processes of Central and South America, including depletion gilding of the *tumbaga* alloys.[10] In some of these treatments, copper alone was removed from the surface using organic acids derived from various fruit juices, but in other processes copper *and* silver were removed, which required much more active chemicals. According to descriptions made by the Spanish priest Fray Bernardino de Sahagún in sixteenth-century Mexico,[11] this was done by a combination of the application of alum and of pastes comprising a yellow earth and common salt, heat, and the burnishing of the base gold:

> And when it was cast, whatsoever kind of necklace ... then it was burnished with a pebble. And when it had been burnished, it was additionally treated with alum; the alum with which the gold was washed [and] was

rubbed was ground. And a second time the piece entered the fire; it was heated over it. And when it came forth, once more, for the second time it was at once washed, rubbed, with what was called 'gold medicine'. It was just like yellow earth mixed with a little salt; with this the gold was perfected; with this it became very yellow. And later it was polished; it was made like flint, to finish it off, so that at last it glistened, it shone, it sent forth rays.

In this treatment, it is likely that the alum was potassium alum, and the yellow earth was a hydrated ferric sulphate, which are all very corrosive. Before the distillation of mineral acids in the Middle Ages, corrosive sulphates or nitrates were widely used in the Old World as well as in the New. In the classical technical and alchemic literature, impure ferric sulphates were often referred to as *misy* (see p. 36).[12]

Bergsøe[13] carried out replication experiments based on the methods described by Sahagún, using alloys of gold and copper. He obtained a satisfactory surface with alloys containing 50% and even 25% gold, but the enriched surface peeled off an alloy containing only 13% gold.

Lechtmann[14] also carried out some experimental treatments to replicate the early American treatments, using an artificial alloy containing 60% gold, 30% silver and 10% copper, an alloy not far removed from the Sardis alluvial gold apart from the rather high copper content. The metal was hammered out and in the first experiments treated with a cement of brick dust and common salt in a sealed container, heated to about 350 °C. After only 10 minutes, a surface layer of almost pure gold had formed, and if the process was continued for much longer there was a tendency for the surface to peel off. Similarly, the gold-refining experiments of Notton[15] using common salt alone as the reactant demonstrated how easy it was to substantially remove the silver at quite moderate temperatures. Lechtmann thought that the cement in her first experiments had been too strong. Indeed, it was more reminiscent of the traditional European cementation salt parting processes rather than the Mexican process where the cement was just smeared on the surface of the metal, and with no mention either of a container or of heat in the second stage. One suspects that Lechtmann did not appreciate beforehand just how easily the silver could be removed. Accordingly, in the next experiment, an alloy containing 40% gold, 48% silver and 12% copper was coated with pastes of either ferric sulphate and common salt or ferric sulphate, common salt and ferric oxide, and left at room temperature for two days. After this, the paste was washed off, revealing a surface blackened by copper and silver oxides. These were removed by washing in hot brine, exposing a reddish-brown, rather porous surface which could be consolidated by burnishing, giving a smooth compact yellow surface. Lechtmann also suggested that heating to about 270 °C achieved the same effect, but it is not clear why this should be so, indeed the heat would be more likely to cause diffusion of silver from the interior to the surface, effectively reversing the whole treatment.

This method was much less rigorous and did not totally remove the silver except in the very surface, and would seem to have replicated the surfaces on the Ur chisels much more closely than the first experiments had done.

There are several descriptions of surface enhancement processes in some of the late Roman technical manuscripts such as the well-known Leyden papyrus, which was found at Thebes in Egypt:[16]

> **Recipe 14** For treating gold, or for thoroughly purifying it and making it brilliant. *Misy* 4 parts; alum 4 parts; salt 4 parts. Grind with water and having coated the gold with it place it in an earthenware vessel put in a furnace and luted with clay and heat until these substances have become molten, then withdraw it and scour carefully.

These methods of surface enhancement were often used on coins. Thus the thirteenth-century Islamic treatise of Manṣūr ibn-Baʿra describes how the gold flans were chemically treated before being struck:[17]

> After the gold flans are made round, they are placed in a pot and heated to redness. Pulverized Indian salt in a little fresh water is added to it. Then a fire is kindled under it with such intensity that the salt dissolves (melts?) as lead does. It [the molten salt] is decanted into a mould; then the dinars are removed to be washed with cold water, then fine sand. They are dried in a pot above a gentle fire and then stamped.

The Indian salt referred to is likely to have been saltpetre, principally potassium nitrate (KNO_3), together with a little sodium chloride and ammonium salts.[18] It would melt at quite low temperatures and remove some of the silver from the surface by the formation of silver nitrate and oxide.[19] It is very likely that the 'Indian salt' is to be identified with the Indus earth of the *Arthaśāstra* (described on p. 34), which was also used in gold treatments at a much earlier date.

Some of the traditional Japanese methods for surface enhancement of gold coins were observed by Gowland during his work at the Imperial mint in the later nineteenth century, and his descriptions are valuable for giving an insight into the ancient processes:[20]

> Owing to the large proportion of silver which the coinage alloys contained, the coins were still nearly white in colour, and it was necessary to give them a surface of gold. This was effected not by any ordinary gilding process, but by dissolving out the silver from the upper layers of the alloys (Fig. 2.1). The coins were first painted with a mixture composed of iron and copper sulphates, potassium nitrate, calcined sodium chloride, and resin made into a paste with water. They were then carefully heated to redness on a grating fitted over a charcoal fire (to convert the surface silver to silver chloride). After this they were immersed in a strong solution of common salt (to dissolve the silver chloride), washed with water, and dried. Their surfaces now consisted of a layer of pure gold. This process was followed by the old workers in gold, and with trifling modifications is still practised at the present day.

During his travels in West Africa in the early nineteenth century, the explorer Mungo Park observed gold dust being melted under the ashes of corn cobs, which were alkaline. This was done to give a better surface appearance to the metal when it had solidified.[21]

Following the introduction of mineral acids in the Middle Ages, surface treatments with mineral salts were only gradually replaced by acid treatments to remove copper and silver from the surface, and jewellers continued to use the older methods until comparatively recently, even in Europe. Thus, for example, Howard,[22] in the late eighteenth century, described gold 'cleaning' as follows:

> Some boil the pale (i.e. silver-rich) *gold* in a solution of these salts (verdigris, i.e. copper acetate, sal-ammoniac with vitriol, i.e. iron sulphate, or alum) made in vinegar; others moisten the salts with vinegar into the consistence of a paste, which is spread upon the *gold*, and the metal laid on burning coals till the mixture is burnt off. The sprinkling of the salts in powder upon the moistened *gold*, practised by some, does not answer so well.

In the early nineteenth century, Gill[23] stated that:

> It is a very curious circumstance, that the best workmen in this branch of jewellery have at this day no other menstruum for giving the last high finish in colour to their beautiful articles, than the compound salts of alum, nitre and common salt.

These tended to be traditional recipes of family firms and as such were closely guarded secrets that were often lost on the demise of the firms. Bury[24] has reported one of the very few that was published. The gold artefacts were boiled in an aqueous solution containing equal amounts of alum and common salt and twice the amount of saltpetre.

Fig. 2.1 Colouring gold coins in the Japanese Imperial Mint, mid nineteenth century. (From W. Gowland, *Trans. Proc. Jap. Soc.* 13 (1915), facing p. 32)

The resulting surface depended on the duration of the treatment, but afterwards the surfaces were treated by scratch-brushing or burnishing to give the required finish. Much the same effect could be achieved more easily with mineral acids but it was claimed that these had a 'diminished lustre'.[25]

The technical author Alfred Hiorns was much concerned that many traditional processes of metal finishing treatments were in danger of being lost through not being published. As head of the Metallurgy Department of the Birmingham Municipal Technical School around the turn of the nineteenth century, he was well placed to record the treatments. In his *Mixed Metals* he describes the treatment of gold alloys both with acidic solutions of salts, such as potassium nitrate and common salt, and with anhydrous salts used in a molten state.[26] The mixtures of salts specified include two parts potassium nitrate to one part each of common salt and alum, and another recipe containing equal parts of potassium nitrate, sal-ammoniac and borax. For the former mixture Hiorns states that:

These substances are ground to a fine powder, well mixed and placed in a previously heated blacklead (graphite) 'colour'-pot .... It is well to get the pot nearly red-hot before placing the 'colour' in it. The mixture must then be constantly stirred with an iron rod. It will first boil up as a greenish liquid, then solidify, and afterwards boil up a second time and

become thoroughly fused, having a brownish-yellow colour. At this stage the work, which has been previously annealed and dipped in dilute aquafortis (nitric acid), is dipped in the 'colour', being suspended on a silver or platinum wire, the latter being preferred, and kept in motion for about a minute and a half, then immersed in boiling water containing a little aquafortis. The immersion and swilling are again repeated, when the articles possess a beautiful colour. They are then washed in hot water containing a little potash, and finally dried in warm boxwood saw-dust.

Surface enhancement processes continued to be described through the twentieth century. Thus Fishlock's standard book on metal finishing[27] describes cements of alum, potassium nitrate, zinc sulphate and sodium chloride. These were mixed to a slurry in which the objects to be treated were dipped or it was applied as a paste. The coated objects were then heated over a charcoal or coke fire and plunged into water. The use of zinc sulphate is interesting, and recalls its inclusion in some parting mixtures listed by Agricola and Ercker,[28] and possibly in some of the Islamic sources.[29] Alternatively, the objects could be immersed in the molten reagents until the desired degree of surface enrichment had taken place. Another treatment was with a boiling aqueous solution of potassium nitrate, alum and common salt, in which the objects to be treated were immersed, for periods of between 10 and 20 minutes.

These recent treatments would have been familiar to craftsmen treating base gold millennia before.

The inception of gold parting

The evidence from remote antiquity for the deliberate surface treatment of gold artefacts which removed both silver and copper, coupled with the comparative ease with which silver could have been totally removed from gold using either common salt or a salt/corrosive iron sulphate mixture, suggests that gold refining was well within the technical capabilities of the ancients long before the Sardis refinery. Thus the possibility must be considered that true gold refining has a much longer history.

First, it must be understood that there was no *a priori* reason for the ancients to refine gold or to consider what came out of the ground as impure. Our idea of pure gold as a single and precisely defined element is based on the relatively modern scientific concepts of the nature of elements, and in particular on the Law of Constant Composition, by which each element has precise, invariant and unchangeable properties. To us, this seems no more than stating the obvious, but the ancients did not have such a concept of an ultimate, pure elemental material.[30] Thus metals such as gold, coming from various sources, could have widely differing properties but still be gold. Given the widely held belief that metals 'grew' in the ground, it would seem only logical to expect the properties of the metal to depend on their environment. For example, a light-coloured natural alloy from a given locality would be regarded as the gold of that place, rather than necessarily a natural alloy of rather high silver content. Refining would be a long and expensive process to improve the colour but reduce the weight of the gold, whereas surface treatment would be much easier and enhance the colour whilst preserving the weight.

The discovery of the application of depletion gilding to the Ur chisels shows that, even in the third millennium BC, the technology existed to remove at least some of the silver from the surface of gold–silver alloys. Clearly, the whole object of the treatment was to make the metal look golden at the surface whilst preserving as much of the weight as possible.

There seems to have been genuine confusion in antiquity between the status of 'real' native gold alloys and adulterated golds or electrums, even though they could be physically and chemically identical. This confusion persisted right through into the post-medieval period and beyond, as exemplified by Ercker's comments on the improvement or graduation of gold (see p. 64). As discussed in Appendix 5, p. 247, the principal method of assay was by the touchstone, which was essentially a surface method. The method of specific gravity measurements, which seems much more obvious and infallible to us, was apparently little employed, long after Archimedes' famous, if apocryphal, discovery of the principle of the method, until balances of sufficient accuracy were developed during the Renaissance in Europe. Thus although the processes of gold refining were likely to have been known from remote antiquity, little or no natural gold alloys were treated until the introduction of coinage created the need for pure metal or at least metal of consistent quality. In short, this was a technology awaiting a use.

Direct evidence of gold refining

Perhaps the most direct evidence for the refining of gold in antiquity is the purity of the surviving metal. In general, ancient goldwork contains appreciable silver compatible with unrefined natural gold, but Nicolini,[31] for example, has claimed that the high purity of some ancient gold artefacts indicated they had been made of refined metal. However, these high-purity pieces are not part of a consistent series but rather isolated examples amongst otherwise impure metal. There is the additional complication that the analyses themselves may not be representative of the true composition of the metal.[32]

From such analyses of gold antiquities as have been performed, there is no evidence prior to the introduction of coinage for the use of a true cementation process that would have totally removed silver and any other metals from throughout the gold. Without coinage, there was, of course, little incentive to refine and thereby reduce the weight of metal. In an interesting paper, Ogden[33] has shown that there are marked differences in the composition of gold artefacts made by those who produced a gold coinage and those who did not. It would seem that before coinage was adopted the unrefined gold was employed with little or no treatment, but where coinage was produced then, although the gold used in artefacts might still be base, it was usually deliberately debased by the addition of regulated quantities of silver and/or copper to refined gold. Particularly striking is the comparison between the composition of the gold artefacts of the Greeks, who had gold and silver coins, and that of the gold artefacts of their contemporaries the Etruscans, who did not. Similarly, the pre-Colombian civilisations, expert in the surface treatment of quite base gold alloys (see p. 28), certainly had the technical knowledge to purify gold, yet this was never done. Without the concept of coinage, there was no motive.

Indirect evidence of gold refining

The material excavated at Sardis is the earliest surviving physical evidence for the parting of gold and silver. Other, more indirect archaeological and literary evidence has been put forward to suggest that gold parting was practised even earlier. These claims will now be discussed.

There are many ancient texts from both Mesopotamia and Egypt referring to the refining of impure gold.[34] Unfortunately, they are accounting records of quantities rather than technologies, thus usually they state no more than the weight of the metal before refining and the weight loss after refining. Without exception, no details are given of either the metal with which the gold was debased or of the refining process, beyond that it was often performed by fire in the furnace. Sometimes the gold was subjected to more than one refining operation, as in this example of the first millennium BC from the archives of Erech:[35]

> Five minas of gold, five weight measures of revenue of the king were put into the fire. At the first firing two thirds mina, five shekels of gold disappeared. It was reduced to four minas fifteen shekels. In the second firing, half a mina, two shekels of gold were lost to yield three and two thirds minas, three shekels of gold.

It might be thought that as gold was naturally alloyed with silver and that generally ancient goldwork only contains small amounts of copper and other metals, it is most likely that these refining operations refer to some form of parting process by which the silver was removed. However, gold alloyed with copper was quite widely used in both Mesopotamia and Egypt at various times,[36] and copper or lead would be the obvious metals with which to debase gold fraudulently. There is also some evidence, discussed in Appendix 3, p. 235, that lead may have been used in antiquity to extract very finely divided gold from the ore, and the resulting gold could still have contained some lead. Thus, in all probability, the early records of refining operations refer to cupellation, which would remove base metals such as copper and lead but not affect the silver. It is noticeable that there is no reference to the recovery of the metal removed from the gold, rather suggesting that it was not silver. Even in Roman times, Pliny could still note that gold was refined by roasting with lead, that is a cupellation process,[37] although elsewhere he describes the true parting processes (see pp. 35–6). The general consensus of the literary sources would seem to be that gold–silver parting was not practised before the mid first millennium BC – that is, the time of the Sardis refinery.[38]

A different approach has been suggested by the requirements of gold used for the production of gold leaf.

It has been stated that in order to produce true gold leaf[39] gold of a high purity was required, and thus the occurrence of very thin gold leaf could be taken as an indication that gold refining was being practised. Thus Oddy[40] states that:

> The invention of gold leaf was impossible before the perfection of methods for the purification of gold as only pure gold, or gold rich alloys, free from certain impurities can be beaten out to produce the thinnest leaf. The introduction of gold refining is generally put at around 2000 BC in Mesopotamia, but it is quite possible that a better estimate could be made from a study of the dates of surviving objects which are covered in {traces of} gold leaf.

Gold was beaten into thin sheets or foils from the very beginning of its use, and by the second millennium BC in Egypt true gold leaf was being produced.[41] However, although ductile metal substantially free of lead or copper would have been a prerequisite and could have been produced by cupellation, it is quite possible to produce acceptable gold leaf from gold containing small amounts of copper and appreciable quantities of silver,[42] and some of the gold leaf from antiquity is decidedly impure.[43] Thus the introduction of gold leaf cannot be taken as an indication that gold parting was being practised.

Literary evidence for cementation parting

Unequivocal literary references to, or descriptions of, the parting processes all post-date the Sardis refinery, but do show that the technology was in use around much of the Old World well before the end of the first millennium BC, usually where a gold coinage was also in production, the apparent exception being India.[44]

None of the descriptions is very detailed, and sometimes they are difficult to interpret or fully understand, due variously to their brevity, style or translation. In the main, they seem to describe the treatment with common salt and/or corrosive sulphates,[45] often in the presence of added base metal. Modern commentators, who have tended to be working on individual texts in isolation rather than taking all the texts together, have suggested that the use of mixed reagents and especially the presence of additional base metals is likely to be an error. However, instructions listing these ingredients recur again and again, including the much later reliable and detailed accounts from the European Renaissance (given on pp. 58–9 and 63–4). Taken with the evidence for the presence of both lead and copper, apparently at early stages in the process at the Sardis refinery, these early references now require careful reappraisal.

Herodotus (I.14 and 50)[46] describes in some detail the donations of precious metal made by both Gyges and Croesus to various temples and sanctuaries:

Gyges offered a lot of silver, but also much gold.
(I.14.1)

Croesus offered half-ingots, six palms long and three palms wide and one palm thick. (I.50.1)

Their number was 117 and four of these were of purified gold, each of the weight of two and a half talents while the other half-ingots of white gold had a weight of two talents. (I.50.2)

It is maybe significant that whereas the offering of Gyges are just described as being of gold, Herodotus specifically states that some of the donations made by Croesus were either of white gold (which could be a reference to the alloy containing about 45% silver found in the electrum coins) or of refined gold. Taken together, these perhaps suggest that the refining process began with Croesus.

There are several early descriptions relating to the purification of gold from India, although all are rather confused. A lost work, *Indika* by Ktesias, written about 400 BC, apparently contained descriptions of both gold refining and the production of crucible steel, but these seem to have got conflated in the surviving abridgement made by Photios, a Byzantine bishop, some 1200 years later.[47] The section on gold reads:

Concerning the Spring which is filled every year with alluvial gold, from which 100 earthen pitchers are drawn up annually. The pitchers must be earthen since the gold when withdrawn is in a solid state, and it is necessary to break them to extract it. The spring is of a square shape with a perimeter of 16 cubits [i.e. 24 ft] and a fathom [i.e. 6 ft] in depth. Each pitcher weighs a talent [i.e. 57 lb].

Commentators have long considered the so-called spring should be regarded as a furnace, but suggested that it should be iron that was in the pots, as 'storage (of gold) in pitchers would not affect its physical condition'.[48] However, if the spring was indeed a furnace, then treatment of alluvial gold in the solid state in sealed earthenware pots sounds very like the later descriptions of the cementation parting of gold. The furnace dimensions are rather large, but the square shape is similar both to the excavated PN furnaces and later descriptions given below and in Chapter 3.

Several descriptions of the treatment of gold, including refining processes, are to be found in the *Arthaśāstra* in the sections outlining the duties of a mine manager.[49] There are several references to the purification of gold, including the following at 2.13.5–8:

Of the best [varieties], the pale-yellow and the white are impure. He should cause that because of which it is impure to be removed by means of lead four times that quantity. If it becomes brittle by the admixture of lead, he should cause it to be smelted with dried lumps of cow-dung. If it is brittle because of [its own] roughness he should cause it to be infused in sesamum-oil and cow-dung.

This should be a reference to the cupellation of gold to remove base metals such as copper, but there is a problem in that gold normally contains silver and it does specifically state that the impure gold to be so treated was pale or white, which strongly suggests that silver was indeed the impurity, and of course cupellation would not remove it. However, the whole description is similar to the much later passage in the *Mappae Clavicula,* where it is implied that it was the removal of the silvery-white platinum group element inclusions that is described.[50] The quantities would be appropriate to form the gold–lead intermetallic compound, Pb_3Au, and of course cupellation would remove the PGE inclusions.

The very next section of the *Arthaśāstra* at 2.13.9 contains a direct reference to the refining of mined gold:

[Gold] produced from the mines, becoming brittle by the admixture of lead, he should turn into leaves by heating and cause them to be pounded on wooden anvils, or should cause it to be infused in the pulp of the bulbous roots of the *kadalī* and the *vajra* plants.

Once again, mine gold should contain silver not lead, unless the lead was being deliberately added to the finely ground gold ores to gather up the precious metals in a process akin to the later liquation process, only operating on the powdered ore instead of on molten copper, as discussed in Appendix 3, p. 235. Hammering out the impure gold into leaves could be significant in that it forms one of the stages in parting processes, but this does not seem to be a reference to a cementation process.

The most convincing reference to true cementation is at 2.13.47 in a section on gold working rather than gold production, and would be the earliest description of the cementation process if the *Arthaśāstra*, or at least this passage of it, is indeed of the fourth century BC:

Ornamental gold of the best kind, possessed of excellent colour, passed through an equal amount of lead, turned into leaves by heating [and] made bright with Indus-earth, becomes the base of blue, yellow, white, green and parrot-feather colours.

Indus earth is normally taken to be the saline soils from the great river valleys of India that were latterly collected at specific times of the year and used as a reagent to patinate the zinc-rich bidri wares,[51] or as a feedstock for the production of gunpowder.[52] The earth is typically rich in common salt, nitre and ammonium salts, and would form an ideal cement for the parting process (see p. 29 for the use of 'Indian salt' in medieval Islamic gold treatments). The further reference to the prior treatment with lead, for whatever purpose, is interesting. Thus the sequence would seem to be cupellation, hammering into foils, followed by parting with saltpetre. This sequence is strikingly similar to the stages one believes took place at Sardis to refine scrap gold: cupellation, hammering into foils, followed by cementation, but using common salt instead of saltpetre.

Returning to the classical world, reference is made to gold refining by Plato in the *Politicus*. However, it is only given as a metaphor, and from this the impression is conveyed that the process was difficult.[53] No technical detail is given, although copper, silver and even the PGE inclusions were apparently to be removed.

The best known and certainly the most detailed description of the cementation process as practised in antiquity is that quoted by Diodorus Siculus[54] in the first century BC, from a more detailed but now lost work, *On the Erythraean Sea*, by the grammarian and geographer Agatharchides of Cnidus, which was written in the second century BC, and apparently based on first-hand observation.[55]

The whole quotation is on gold production in Egypt. The first part is a quite detailed description of the mining of the gold using firesetting to weaken the host rock. Firesetting was only used against hard homogeneous rock and thus it was the mining and processing of gold from a primary quartz deposit that was described here rather than the working of secondary placer deposits such as those from the River Pactolus. This could be of significance in the choice of processing technique adopted, particularly those to separate the gold from the gangue, as the gold in the primary deposits would tend to be much more finely dispersed than in the secondary placers. Note also that gold from a primary source would have no PGE inclusions (see Appendix 4, p. 238). The description continues with the beneficiation and purification processes:

> In the last steps the skilled workmen receive the stone which has been ground to a powder and take it off for its complete and final working; for they rub the marble (i.e. the mine gangue) which has been worked down upon a broad board[56] which is slightly inclined, pouring water over it all the while; whereupon the earthy matter in it, melted (washed) away by the action of the water, runs down the inclined board, while that which contains the gold remains on the wood because of its weight. And repeating this a number of times, they first of all rub it gently with their hands, and then lightly pressing it with sponges of loose texture they remove in this way whatever is porous and earthy, until there remains only the pure gold-dust.
>
> Then at last other skilled workmen take away what has been recovered and put it by fixed measure and weight into earthen jars, mixing with it a lump of lead proportionate to the mass, lumps of salt and a little tin, and adding thereto barley bran; thereupon they put on it a close-fitting lid, and smearing it over carefully with mud they bake it in a kiln for five successive days and as many nights; and at the end of this period, when they have let the jars cool off, of the other matter they find no remains in the jars, but the gold they recover in pure form, there being but little waste.

This is clearly a description of the salt cementation process,[57] and equally clearly it is related to the process practised at the PN refinery some 400 years earlier. It is noticeable that the otherwise complete description omits any mention of turning the mined gold into foils prior to parting, strongly suggesting that the gold dust was treated in that condition. There is in fact only one ancient western source that states that the gold dust should be made into foils for the parting process.[58] The remains from Sardis, as shown in subsequent chapters, also suggest that the freshly mined gold was treated directly without first being converted to foils.

The quite unambiguous additions of controlled quantities of lead and also of some tin are noteworthy, especially in view of similar statements in some other early descriptions and the presence of lead on the surfaces of the parting-vessel sherds from the PN refinery, and of tin on some of the gold globules and a foil (see pp. 112 and 186).

The function of the lead and tin in the cement has always proved problematic, and hence the accuracy of the whole quotation has been questioned. Convincing modern parallels are hard to find. Percy records a process practised up to the nineteenth century at Oker, near Goslar in the Harz mountains (Germany), for the recovery of trace amounts of gold from silver using litharge and sulphur.[59] However, the function of the lead in that process was specifically to remove the overwhelming quantities of silver as economically as possible and the process could not work at all effectively on metal containing more than a few per cent of gold. Some scholars, notably Healy,[60] suggested that the two processes of parting and of cupellation, to recover silver from the spent cement, had been conflated, but unless the account is totally mixed up, it is quite clear that weighed quantities of lead and a little tin were part of the contents sealed in the parting vessel. Halleux[61] suggested a double process of cementation and cupellation proceeding simultaneously. This is ingenious but chemically impossible, as the whole point of the cementation is to turn metallic silver *into* a salt and cupellation recovers metallic silver *from* its salts. Burstein suggests that the lead was added to recover precious metal from the residuum of the ground ore after washing, but before cementation. Whilst this interpretation is technically quite feasible it does require a gross conflation in the section dealing with the recovery of the impure gold. Some of the practical manuals published in Germany in the early sixteenth century[62] describe the addition of lead to the molten impure gold in order to facilitate its granulation when poured into cold water prior to parting. In Agatharchides' description, however, the washed gold seems to have been poured straight into the cementation vessels. Several later sources do describe lead (zinc?) or copper *salts* as components of the parting cement.[63]

Notton,[64] who carried out gold-parting replication experiments based on Agatharchides' description, found that the addition of lead metal actually impaired the removal of silver from the alloy of base gold with 55% silver and 7.5% copper that he used. He suggested that if lead really was added then its function could have been to remove any silicious gangue material from the gold. However, overall, Notton seems to have doubted whether lead and tin were really added to the charge.

The addition of barley bran is not mentioned in other early descriptions.[65] It has been suggested that the organic material was necessary as a source of hydrogen ions to create hydrogen chloride as the active reagent in the process.[66] However, the process works perfectly well with common salt alone, as Notton demonstrated, and the addition of any organic material, by making the atmosphere inside the vessel more reducing, would actually inhibit the reaction.

It is possible that the bran or straw was the external fuel rather than an internal reactant, as both Pliny and Strabo in their descriptions (given below) of gold refining state that a fire of chaff should be used rather than charcoal. The use of straw as a fuel in the gold refining is also mentioned in the Babylonian Talmud, which states that a clay container with openings was used to let in the flame of the barley straw fire.[67] However, straw or chaff does not seem to be a very convenient fuel for a process lasting for five days and nights, and once again it would require a gross mistake in Agatharchides' text as it now stands.

Thus it does appear that the ingredients as listed in the surviving accounts are impracticable. As this was a real industrial process, it would appear that there is something wrong with these descriptions. The most likely fault is conflation and it is possible that, as Burstein suggests, lead was used to recover further gold from the crushed ore after washing, followed by cupellation and cementation. Alternatively, the refining process itself was twofold: first the gold was cupelled with the lead and tin and barley bran, followed by parting with common salt. Traditional cupellation or fire assay often uses lead oxide with flour as a reducing agent.[68] This explanation does at least account for all the ingredients, but it should not have been necessary to have cupelled the raw gold prior to parting, and the order in which the ingredients are listed in the existing accounts gives no support for a putative twofold process with the metal and bran first, followed by the common salt.

It is important to note that there is no reference to brick dust, clay or any other inert carrier, and at the termination of the process the jars contained pure gold and very little else. Noting this absence, Notton carried out one experiment with no brick dust and found that the process worked perfectly well but the silver was almost entirely lost from the sealed parting vessel, and onto the furnace walls, reminiscent of the walls of the Sardis furnaces.

The specification of a closely fitting and luted lid is important, as there is no direct evidence that the Sardis vessels were lidded. All subsequent detailed descriptions specify that the parting vessel should be sealed, and on the literary evidence alone it is likely that the Sardis vessels were also lidded.

Pliny refers to the purification of gold several times in his *Natural History*.[69] The first reference is at 33.60:

> The first proof of quality in gold is however its being affected by fire with extreme difficulty; besides that, it is remarkable that though invincible to live coal(s) made of the hardest wood it is very quickly made red hot by a fire of chaff, and that for the purposes of purifying it it is roasted with lead.

This is in a passage on assaying, and the process is one of cupellation that by itself would not remove any silver in the gold.

The second description is at 33.69 in sections on gold mining, where it states that:

> **The extracted material is pounded, washed, burnt and ground. The powder from the mortar is called *scudes*; the silver, which comes out of the furnace, is called sweat (*sudor*). The dirt, which is thrown**

out of the furnace, is called *scoria* in the case of all metals. The one in the case of gold is pounded and melted again. The bowls are made of *tasconium*, this is a white earth similar to clay and no other (earth) can stand the bellows and the fire and the red-hot material.

No details of reagents are given but the treatment of gold, possibly from a primary context, which produced a silver scum might suggest a parting process. However, no active salts are mentioned, and it is more likely that Pliny had in mind the smelting and recovery of gold from the ore rather than a parting process. The reference to the reworking of the slags (probably with lead although this is not stated) to recover more of the gold is extremely interesting and recalls the Roman gold-smelting process excavated at Três Minas in northern Portugal.[70]

There are more detailed descriptions of gold parting in sections dealing with the medicinal uses of gold (33.84), where it states that:

> It (gold) is also roasted with a double weight of salt and three times weight of *misy* and again with two portions of salt and one of the stone which is called *schiston*. So it draws poison (*virus*) from the things which are burnt with it in an earthen vessel (*vas*), itself being pure and uncorrupted. The ash left is preserved in an earthen pot (*olla*), mixed with water and smeared on the face (cures) eruptions, but it is better to wash it away with *lomentum*.

Pliny was describing here a double process using mixed reagents, in this instance first common salt and iron sulphate, followed by common salt with talc (an inert hydrated mixed silica/magnesia, SiO_2/MgO, mineral). Experiments by d'Elhuyar in the late eighteenth century,[71] using a variety of reagents either singly or in combination, showed that cementation could be made to work at a much lower temperature using a mixture of common salt and sulphates, either in the form of gypsum (calcium sulphate) or barytes (barium sulphate), than when the same reagents were used singly.

The translation given in the Loeb edition has a number of inaccuracies. *Misy* is translated there without comment as copper pyrites, a mixed iron/copper sulphide,[72] whereas in fact it is correctly translated as a mixture of corrosive sulphates, with ferric sulphate predominating. *Schiston* is just translated as 'splittable'; in fact from other contexts it is likely to be talc, although Healy[73] suggests alum. The translation also suggests that the vessel in which the reaction took place was specifically an earthenware cooking pot, that is an *olla*, whereas the word used is *vas*, which is

the word for an ordinary pot; only the residue was to be kept in an *olla*.

Halleux[74] believed that Pliny was working from an original Greek text and that *virus* should be read as the equivalent of the Greek *ios*: that is, rust rather than poison. *Lomentum* is translated by Rackham as 'lotion' but in fact has the specific meaning of bean meal.[75]

In the third description, which occurs at 34.121, the preparation and uses of *misy* are described in detail.[76] The short reference states that:

> A mixture of *misy* is employed in the magical purification of gold.

At 35.183, in sections dealing with the preparation and properties of alum, a mixed potassium aluminium sulphate, Pliny notes that black alum (most likely to be a mixed iron potassium sulphate) was used in the dyeing of dark cloth and that it could be used to purify gold.

Note that in the last two references parting is referred to only in the specific context of the uses of the *misy* or alum and thus the presence of other salts (such as common salt) in the process is not excluded.

Strabo, in his *Geography*,[77] describes gold refining by the Galatae, who inhabited the Cevennes and Pyrenees, as follows:

> The product of smelting the gold and refining it with a kind of 'styptic' earth is 'electrum'. They further smelt this mixture of gold and silver: the silver is burned away and the gold remains. This type of alloy (that is the electrum) is easily fused and hard like stone. It is for this reason that the gold is preferably melted by a fire fuelled by chaff since the flame is gentle and suitable for an alloy which yields and fuses easily; a charcoal fire, on the other hand, consumes a lot since it over-melts the gold and volatilizes it because of its intensity.

Strabo's Greek term στυπτηριωσης γη, styptic earth, is likely to have been a corrosive sulphate, such as alum or ferric sulphate, the Greek equivalent of the Latin *misy*.[78] Note that the styptic earth is apparently used only in the first part of the process to produce refined electrum, although in fact it only makes sense as part of the parting cement. The explanation why chaff was to be preferred is also somewhat misunderstood; a low temperature was necessary in order not to melt the finely divided electrum, thereby reducing the surface area exposed to the cement. As already noted (p. 35), chaff seems an unlikely fuel for a process of long duration.

Recipes for the refining of gold featured in some of the

Alexandrian alchemic literature of late antiquity have been collected and discussed by Halleux.[79] The following unusually clear and concise recipe is taken from the *Pseudo-Democritus*, which has its origins in the first centuries AD, but only survives in a later Syriac version, from which a translation was made for Berthelot:[80]

> Du sel et de l'alun, chacun une once, du vitriol noir de Chypre, 2 grammes; broie ensemble et mets dans un creuset. Plonges-y des lames d'or. Dispose par couches de la chalcite [?] entre les lames. Mets sur des charbons ardents et souffle. Quand le produit sera fondu, tu verras quel produit tu obtiendras.

Once again, this is a mixed chloride/sulphate process, this time using both alum and ferruginous sulphates. Note the reference to the gold being in the form of sheets (leaves?). This is apparently the first such specification in western gold-refining literature.

The *Mappae Clavicula*[81] has a typically rather confused account of salt cementation in recipes 2 and 3, described as methods of making or 'extending' gold:

Again making gold

Melt one ounce of silver, half an ounce of copper and one ounce of gold. Again, take sand and press it onto a level place. Cool until it is dry, and mix again some salt and roast it in the furnace for a day and a night. Afterwards take it out and wash it until the salt runs off; and again dry it and knead it in vinegar and set it aside for a little until it absorbs it and dries out. Then again put into the furnace a piece that has not been washed and do this once and again; knead it in vinegar every time you put it into the furnace. Now you ought to put it into the furnace four or five times until it becomes almost as if it is all cooked away; and when you take it out use a silver withdrawing tool, which in Greek is called *elquison*. Take a weight equal to the former amount, mix it all together and grind it.

Then melt separately the two kinds of material that you have concocted [that is the residual gold and the salt-containing cement which now contains silver] and sprinkle them gradually [onto a lead bath?] until it is used up. Then cool it and you will find that hard lead has been made. Melt this together with *cepsonium*, that is, kneaded ashes. As is shown according to the Key, *psomion* is ash kneaded with water, which you lay underneath in the furnace to the thickness of a finger.

The next chapter, 3, continues:

Again

Take only a little for experiment when you do it once until you learn it thoroughly. Take one ounce of reddish Cyprian copper in solution [? *posios*], one ounce of quite good silver, and melt it with chaff, until when hammered out it does not make a noise, and then melt it together with one ounce of gold and the same amount of natron. Then turn face to face two little bowls, that is two hollow earthenware pots and put inside them the widened [that is, hammered out] melt that has been prepared, and mix in *antisma*. What had been a little bead of copper is now turned into a little bead of four ounces of silver. In the bead we find more than an equal amount of gold. [Take] one part of Pontic sinopia (a ferruginous clay containing hematite), two parts common salt, grind them all together, lay the sheets on the bottom and sprinkle [the sinopia and salt mixture] over them and coat them with pot clay so that they cannot breathe. Put fire under them until you feel it is all right. Take them out and you will have the very best gold.

At first reading, these two descriptions would seem to be conflations of the 'extension' of gold by the addition of silver and copper in the ratio of two parts to one respectively, and of the refining of gold by parting followed by cupellation.

However, these descriptions could be the earliest reference to graduation, followed by refining. That is, the practice of adding more of the metal to be removed in order to allow an open porous structure to develop throughout the metal grains during the refining process, ensuring that all the silver or base metals in the gold were exposed to the parting agent.[82]

Smith and Hawthorne believed recipe 3 to be just a surface treatment to give the impression that the metal was pure gold throughout. However, salt cementation was usually associated with complete parting, especially when the metal was in the form of foils. It would anyway have been useless to produce surface treated foils as the surface enhancement would have been lost when they were melted up for use. Treatments that merely enhance the surface are only of use on the finished artefact.

Thus there are three possibilities:

- A conflation of gold extending and refining. Note: Chapters 2 and 3 *are* entitled respectively 'Again making gold' and 'Again', following 'Making the most gold', which is the title of Chapter 1.

- Surface treatment, but the products are respectively molten or foils.

- Graduation.

Of these, the third seems most reasonable and requires least alteration of the text as it has come down to us.

The comment on sealing the top of the mixture with clay is interesting in view of the apparent absence of lids at the PN refinery.

At Chapter 246a, there is a description of the production of gold leaf that includes the earliest description of the parting furnace. It states:

Making gold leaf

How the leaf should be made. Mix together into an ingot [? *clavum*] one ounce of Byzantine gold, and one ounce of clean silver. Purge it by means of lead and afterwards cast it. Then mix and beat out a sheet, and when it has been beaten thin, cut it [into square pieces each] weighing five Byzantine *tremis* [1²/₃ solidi]. If one is long or short, equal it out in breadth and in length with a hammer. Eight leaves should result from the [initial] two ounces, after they have been made equal in size. Heat them on the hearth, beat them, holding them in iron pincers, and while you beat them, they should be spread from inside to the outside, so that they appear thinned in the middle. When they have increased by a half, cut them with a small knife three times by measure to give four equal pieces. Fold them edge to edge equally, extend them and cut them with shears [to give 32 pieces]. The pieces should be placed on each other edge to edge on the hairs [of a skin] pressed lightly by hand and put in oil. 64 leaves have now been made from the above eight. Then make a pouch out of [sheet] copper and always beat them in it; and place more copper for the beating, one leaf above and one below. And as you beat with a flat-headed hammer, strike as many blows on one side as on the other. And when they have increased a half, cut them and lay one on top of another. Next put them in oil and always fold copper [between them] and fit them together and beat them long enough. Repeat this until out of the [original] eight leaves 1028 (actually 1024!) are made. Trim them with the shears and wrap the trimmed pieces in a linen cloth, since they ought to be heated in the furnace where the gold leaf is put.

The furnace itself should be two feet high from the ground and there should be placed on its wall [as a cover] a perforated tile with three holes on one side and three on the other and one in the center. And another tile should be put half a foot above the ground, with a hole in the center. And at ground level a hole should be made [through one of the walls] through which the wood is put, and [higher up] in front, [another hole] through which the gold is put. And you should clean the gold well [by heating it] with ashes of cow dung mixed equally with salt, burnt

and ground like the ashes. In the first [heating] you should put in old ashes, in the second, new ashes, and in the third ashes similarly sieved.

It is instructive to compare this description both with the remains excavated at Sardis and the later medieval descriptions of parting furnaces given by Theophilus and other European and Islamic writers, to which one must now turn.

Medieval European descriptions

The account given by Theophilus, probably written in the twelfth century, is the clearest and most complete medieval European description of the parting process and is quoted here in full, following the translation of Hawthorne and Smith.[83]

Chapter 33. Cementing Gold

Take any kind of gold and hammer it until it becomes a thin flat sheet, three fingers wide and as long as you can make it. Then cut pieces as long as they are wide and lay them together exactly and make a hole **all over** them with a thin **chaser**. After this, take two fire-tested earthenware dishes (*testa*) of such a size that the gold can lie in them; then break into tiny pieces a tile or a piece of burnt and reddened furnace-clay and when it is powdered, divide it into two equal parts by weight and add to it a third part of salt of the same weight. It should then be lightly sprinkled with urine and mixed so that it does not stick together but is just moistened. Put a little of it into one of the dishes [to cover] approximately the area of the gold, then a piece of gold, then the composition again, and gold a second time. Gold should always be covered with the composition so that it is not touched by other gold. In this way fill the dish up to the top and cover it with the second dish. Lute the dishes carefully all around with mixed and kneaded clay and put them at the fire to dry.

Meanwhile build a furnace of stones and clay, two feet high and one and a half feet wide, broad at the bottom but narrowing at the top. Here there should be a hole in the middle. Inside the furnace there will project three rather long and hard stones which can endure flame for a long time. On these stones place the dishes with the gold and cover them freely with other dishes. Then put the fire and wood below and see that a plentiful fire is not lacking for the space of a day and a night. In the morning, however, take out the gold, melt it again, hammer it, and put it into the furnace as before. After another day and night take it out again, mix a little red copper with it, melt as before, and put it back into the furnace. And when you have taken it out for third time, wash it and carefully dry it. Weigh it, when dried, and see how much has been lost, then fold it up and keep it.

One is immediately struck by the overall similarity of shape and size of the furnaces described here with the surviving remains from Sardis (see pp. 83–6), and this description suggests that the latter may not have been much taller than the preserved height.

The use of the word *testa* to describe the parting vessels deserve some comment. The word often meant a sherd, and in Chapter 23, in a description of fire assaying, Theophilus states:

> Sift some ashes, mix them with water, and take a fire-tested earthenware dish *(testa ollae)* of such size that you believe the silver which is to be refined can be melted in it without running over.

*Testa olla*e means literally the sherds of cooking pots, but Hawthorne and Smith believed that in this context it should be interpreted as a shallow dish and that the prefix *testa* was to distinguish it from the ordinary *ollae,* which were round. However, broken sherds of earthenware cooking pots would describe the utilised sherds found at Sardis quite well. For the actual cementation, an intact vessel capable of being sealed had to be used, the *olla* rather than the *testa* as stated, but the use of cooking pots, as specified by Pliny 1000 years before (see p. 36),[84] both whole and as sherds, is significant for the interpretation of the utilised sherds of cooking vessels found at Sardis.

The addition of copper to the metal towards the end of the process is the earliest certain reference to the practice of graduation: that is, adding an excess of the metal to be removed.[85]

In Chapters 69 and 70, Theophilus describes the use of sulphur to quickly refine small quantities of gold (see p. 67).

European alchemists

The medieval Islamic and European alchemists do refer to the processes. This exemplar is from the anonymous *Sur la très précieuse et célèbre orfèvrerie*:[86]

> Sur l'affinage de l'or. Prenez du sel marin, mettez dans un vaisseau solide, lutez-le par-dessus et mettez au feu jusqu'à a ce qu'il brûle, et mettez deux mesures en poids de sel tamisé, un tiers de brique tamisée, mettez dans deux vaisseaux une couche de sel et une couche d'or. Qu'il soit martelé en lames et que le vase soit luté tout autour par le lut de sapience. Ensuite mettez-le à cuire au fourneau. Voici le four: prenez une cruche, percez les côtes au milieu en croix, passez-y deux tiges de fer, mettez les vases avec l'or au centre de la croix, et faites un trou au fond de la cruche pour la sortie de la cendre. Alors emplissez de charbon et efforcez-vous de cuire l'or; si l'or est le centre [?]. Le lendemain, amollissez la brique avec le sel et laissez de nouveau cuire quelques heures.

The pseudo-Aristotelian *De Perfecto Magisterio*[86] has a somewhat similar description:

> *On the Separation of Gold*
> **Make out of it sheets similar to your fingernails and cement them with this powder. Take two parts of purified (lit. separated, different, particular) common salt, one part of an old brick found in the bank of a river or on the shore of the sea (i.e. already impregnated with salt), which is better, triturate (all this) very well and pass through a sieve of close texture, then leave the said sheets cemented in this way in a medium-sized *athanor* on a tripod for one day and one night. And if you fill the *athanor*, which is a small oven (furnace), with live coals and when they diminish you always add other (coals,) then after having removed the crucible from the fire and cooled it, open it and you will find the gold perfectly purified.**

The descriptions of gold parting often occur where the alchemists are discussing the properties of matter, particularly metals, and the descriptions are made to exemplify some of the reactions. The comments of the alchemists are important as their theoretical basis was firmly grounded in Aristotle, and even earlier philosophical Greek thinking, and thus offer some insight into ancient ideas on what was taking place during the reactions of gold parting.

The *Summa Perfectionis*,[88] which was probably compiled in its present form in the thirteenth century in Europe from Islamic originals dating back several centuries previously, and traditionally ascribed to the great Islamic scholar Geber, includes a section attempting to explain the reactivity of various metals in terms of their behaviour in the gold parting process:

> <84> Particular Discourse on the Cement
> We say therefore that since some bodies are more burnt up by the method of calcining fire, and others less, that those which contain a greater quantity of sulfur are more combustible, and those which contain less, are less. Since sol (gold), therefore, has a lesser quantity of sulfur than the other bodies, it is therefore burnt up least among all the mineral bodies by the inflammation of fire. But after sol, luna (silver) participates least of

all the remaining bodies in sulfur, though it has more than sol. Hence it can stand a shorter space of time at this firing of flame than sol, and it can less tolerate things burning it on account of their similar nature. Because venus (copper) is of more sulfur and greater earthiness than sol and luna, it therefore tolerates less inflammation than they. But jupiter (tin) participates more in sulfureity and earthiness than sol and luna, though less than venus. Hence it is burned less through inflammation than venus, though more than sol and luna. Saturn (lead), however, has preserved more earthiness and sulfureity by nature in its com-mixture than the foresaid bodies; therefore it is inflamed quicker and easier than all the bodies just mentioned, and attacked quicker by inflammation; it is not, however, consumed more quickly on account of this – that it has an especially well-conjoined sulfur and more fixed than that of jupiter. Mars (iron), on the other hand, resists being burnt up not *per se* but *per accidens*. For when it be mixed to bodies of much humidity, it absorbs them on account of its lack of humidity, and thus having been conjoined, it is not inflamed nor burnt up if the bodies conjoined to it are neither flammable nor combustible; but if the bodies commixed with it be combustible, it is necessary that mars be burnt up and inflamed according to the nature of its combustion. Therefore since the cement is com-posed of flammable things, the necessary cause of its discovery is manifest, and it is namely that all com-bustibles be burnt up. So, since there is only one non-combustible body (i.e. gold), only that or another prepared according to its nature is preserved in the cement. Yet some last more and some less in the cement, though which more and which less are known, with their said causes. Hence luna remains more, but mars less, still less jupiter, and less still venus, but least of all saturn. Let us therefore relate the method of the cement, since in the test of perfection it is highly necessary for our cognition thereof. We therefore say that the composition of this is from flam-mable things: the things of this genus are all blacken-ing, fleeing, penetrating and burning, such as vitriol (hydrated iron sulphates), sal ammoniac, flower of brass (iron or copper sulphates ?), old ground potter's stone, a very small quantity of sulfur or none at all, male urine, and similar sharp and penetrating things. So let all these be cemented with male urine upon thin tablets of that metal on whose account the test is to be performed. Then let the tablets be spread upon an iron grate in a sealed pottery vessel so that one of them not touch the other, with the result that the virtue of the fire freely penetrate each of them equally, and thus let the pot be kept in strong fire for three days. Let caution then be applied so that the table(t)s be fired but not melted. After the third day, you will find the tablets clean of all impurity if the body of them has been in perfection (i.e. if there has been some gold there in the first place); if not, you will find them wholly corrupt and burnt up in calcination. But certain men put the tables in a fire without cement, and they are purified also, if they are made of a body of perfection, but if not, they are wholly burnt up. How-ever, in this final test which is perfected with the burning of fire alone, they need longer space of com-bustion than those which are tested by the judgement of cement. But since luna is not far from the difference of sol's nature, with a little treatment it rests with that in the judgement (i.e. this latter treatment does not separate gold and silver because they are too similar).

It is interesting that common salt is not amongst the possi-ble ingredients of the cement, but copper salts apparently are included.[89] The reference to an iron grate inside the pottery vessel seems very strange. It also puzzled Percy,[90] and is almost certainly an error.

However, the account of cementation given by Alber-tus Magnus in the fourth volume of his *Book of Minerals*,[91] compiled in the thirteenth century, singles out common salt as the most important ingredient of the cement:

The substances that purify gold are sharp and extremely dry, such as salt, especially sea salt, and the soot of substances that are unctuous but dry, and brick dust. When gold is to be purified, an earthenware pot is made in the shape of a *cucurbita* or *scutella*, and over this a similar one is placed, and they are cemented together with the stiff clay the alchemists call 'lute of wisdom' [*lutum sapientiae*]. In the upper one there are many holes by which the vapour and smoke can escape. And next the gold is beaten out in thin, short sheets and arranged in the vessel in such a way that each layer of [gold] sheets has above and below it [a layer of] powder made of soot, salt, and finely ground brick mixed together. And it is cooked in a hot fire until it is extremely pure and the ignoble substances in it are consumed. The 'lute of wisdom' of which the pots are made is composed of ground-up pottery, remixed and baked; for such a vessel, when placed in the fire, does not shrink perceptibly in the fire. There are other ways of preparing 'lute of wisdom' in alchemy, but let this, which is used by the goldworkers, be sufficient. This, then, is the method of purifying gold, and nothing is burnt away in it except ignoble material.

And for this reason Hermes aptly says in his *Alchemy*: 'Sulphur itself, because of a certain affinity by which all metals are closely related to it, burns and reduces them all to ash, except only gold: for the pores [of gold] are tightly closed and cannot be opened'.

The reference to the shape of the parting vessels given in this description is very important for the study of the Sardis material. The vessel is described as being like a *cucurbita* or *scutella*. The first would indicate a gourd-shaped vessel, and although the *scutella* was a flat round tray in Roman times, it had come to mean a much deeper vessel by the medieval period. Thus both descriptions suggest the vessels to be used in the cementation were not too dissimilar from the Lydian coarseware cooking pots that it is postulated were employed at Sardis.

The comments on the 'lute of wisdom' are of interest, if a little confused. It is first stated that the lute used to join the two earthenware vessels was composed of a stiff clay. It then goes on, apparently, to describe another 'lute of wisdom', to be used for the fabric of the vessels themselves. This lute contained ground-up pottery as an inert filler, to prevent shrinkage (cracking ?) in the fire. This is the only place where a special composition for the vessel is suggested, most other descriptions are content with specifying an earthenware vessel.

Another important medieval European account is given in the *Tractatus Novae Monetae*, which occurs in *The Red Book of the Exchequer*, compiled in 1381.[92] It states:

> In this way an assay of gold ought to be made: in the fire for three days and three nights in a pot with a cement made of a powder of good red tile and *sal de Peyto* [saltpeter ?] added to it in four times the quantity of the gold to be purified. In other words, to a pound of gold for purifying, containing 23 carat three and a half grains in pure gold as above, there should be added four pounds of cement, of which one seventh or one eighth part should be of saltpeter and the whole remainder of the cement of powdered good red tile. Thus, that is, when the gold contains more of alloy a greater quantity of the salt is placed with it, and if less, a smaller quantity.

Sal de Peyto was translated by Smith as saltpetre (KNO_3), and this seems entirely reasonable. Saltpetre was apparently unknown in classical antiquity, although the active agents in the 'Indus earth' of the *Arthaśāstra*[93] (see p. 34) and 'Indian salt' in some of the early Islamic recipes[94] (see p. 29) almost certainly included saltpetre. The earliest description of the production of the purified mineral is from China in the mid first millennium AD,[95] but no earlier than the twelfth century in the West.[96] Smith claims this

reference to *sal de Peyto* was the earliest mention of the use of saltpetre in parting, but there are the much earlier references in the *Arthaśāstra* and the Islamic sources noted above, and thus it should not be entirely left out of the discussion of the ancient processes. India was the major producer of saltpetre and supplied the Islamic world and possibly Europe in the medieval period. Parting with saltpetre seems to have been more prevalent in India, which is considered next.

Indian descriptions

The earliest Indian references to gold refining are contained in the *Arthaśāstra* of Kauṭilīya and possibly in the *Indika*, which have already been discussed (see pp. 33–4). Some medieval descriptions have been published.[97] The most detailed are contained in the *Rasaratnasamuchchaya*, which was probably compiled between the thirteenth and fifteenth centuries. In Book V, it states that:

> Gold is to be purified and killed, as otherwise [if taken internally] it robs one of strength, virility and happiness, and brings a series of maladies.   11

> Gold-leaf of the weight of one *karsha* is to be smeared with salt and placed between two earthen saucers and heated on a charcoal fire for an hour and a half, when its true colours will come out.   12

This was clearly a salt cementation, but there then follow descriptions of treatments using mercury sulphide:

> The best method of killing all the metals is with the aid of the ashes of mercury [generally sulphide of mercury]. The next best is through the agency of roots, whereas *killing* with sulphur is least to be recommended.   13

Killing, which Rây sometimes italicises, is usually taken to mean converting a metal to a mineral, often the sulphide or oxide, although in the case of gold, amalgams or purified gold seem to be meant. The stress on the use of mercury is no more than part of the Tantric tradition to which these works belong, and which used mercury-based preparations extensively. The description continues:

> When a metal is *killed* with *ariloha* [meaning not clear], it is injurious. Gold-leaves, pierced with holes and coated with a paste of lemon-juice and the ashes of mercury, and roasted ten times, are thereby killed.   14

> Project into melted gold its own weight of the ash of mercury; [when cooled] powder it and rub with lemon juice and cinnabar, and roast it in a covered crucible

twelve times. The gold thus acquires the colour of saffron.   15–16

Gold-leaf is *killed* by being rubbed with one-fourth of its own weight of *killed* mercury and acid of any kind, and roasted eight times.   17

Clearly, in these treatments the impure gold on heating would have been exposed to both sulphur and mercury as the mercury sulphide decomposed. The sulphur would have formed sulphides with all the impurities, most of the mercury would have evaporated but some could have formed an amalgam, at least fleetingly, and thereby facilitated exposure of the gold to the sulphidic materials. The product, as Râmaya concluded, is likely to have been purified gold.

A detailed description of the parting of silver and gold using saltpetre (KNO₃), as practised at the Delhi mint, is given in the *Ā-īn-i Akbarī*, compiled in the sixteenth century by Abū al-Faẓl Allāmī, a chief minister to the Mogul emperor Akbar:[98]

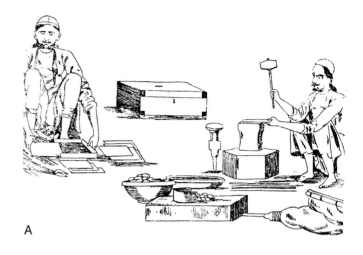

A

> The melter of the ore makes small and large furrows in a slab of clay, which he smears with grease, and pours into them the melted gold and silver, and leaves them to solidify (Fig. 2.2). For copper, instead of using grease he sprinkles the moulds with ashes. The plate-maker makes the impure gold into plates of the weight of 6 or 7 máshahs each , and 6 fingers square [i.e. about 4.5 inches] (Fig. 2.2): these he carries to the assay-master, who measures them in a mould made of copper, and stamps such as are found correct, in order to prevent alterations, and to show the work done. When the plates have been stamped, the owner of the gold, for the weight of every 100 jaláli goldmuhurs, must furnish 4 sérs of saltpetre, and 4 sérs of unburnt brick dust.
>
> The plates, having been washed with clean water, are coated with the above mixture, and put one above the other, and the whole is surrounded with dry cow-dung from the fields. They then set fire to it, and let it burn gently, till the cow-dung is reduced to ashes, when they leave it to cool; then these ashes, being removed from the sides, are preserved in order that the silver which they contain may be extracted. These ashes are called in Persian *khák i khalác*, and in Hindi *salóni*. The plates and the ashes below them are left as they are. This process of setting fire to the dung, and removing the ashes at the sides, is twice repeated. When three fires have been applied, they call the plates *sitái* [which means three times treated]. They are then again washed with clean water, and stratified (?)

B

C

Fig. 2.2 Gold refining in sixteenth-century India.
A Impure gold hammered into regular plates and checked in a gauge ready for cementation.
B The plates mixed with parting cement and heated under a mound of cow-dung fuel.
C The pure gold melted and cast into bars in open clay moulds. (From *Ā-īn-i Akbarī*, D.C. Phillott, 1927, Plates II and III)

three times with the above mixture, the ashes being removed from the sides.

Note the cementation was apparently not carried out within any vessel but just buried under the cow dung ashes, although here the salt formed would have been predominantly of silver nitrate or oxide, rather than the more volatile chloride. The description continues:

> This operation must be repeated, till six mixtures and eighteen fires have been applied, when the plates are again washed. Then the assay-master breaks one of them; and if there come out a soft and mild sound, it is a sign of its being sufficiently pure; but if the sound is harsh, the plates must be coated with the mixture once again, and passed through three more fires. Then from each of the plates is taken 1 máshah, of which aggregate a plate is made and tried on the touchstone; and if it is not sufficiently fine, the gold has again to pass through one or two fires. In most cases, however, the desired effect is obtained by three or four fires.
>
> The following method of testing the quality of the gold is also used. They take 2 tólahs of pure gold and two tólahs of the gold which has passed through the fires, and make 20 plates of each, of equal weight. They then spread the above mixture evenly, and apply the fire, wash them, and weigh them with an exact balance. If both kinds are found to be equal in weight, it is a proof of pureness. The gold thus refined is melted and cast into ingots.
>
> Treatment of the ashes. – The ashes [khák] are washed, 2 sérs at a time; and whatever intermixed gold they may contain, will, from its weight, settle to the bottom. The impure sediment is triturated with quicksilver, in the proportion of 6 máshahs per sér. The quicksilver from its friendly attraction draws the gold to itself, and this [i.e. the amalgam] is put into a glass vessel, and the gold separated by means of fire.

There then follows a very detailed account of the complex treatments carried out to recover the gold and silver in the ashes. The presence of gold in the spent cement is initially somewhat surprising, but if the saltpetre also contained common salt (which is very likely considering its source)[99] there is a possibility of the mixed nitrate–chloride ions dissolving the gold itself. Thus Lewis[100] in his description of saltpetre parting states that nitrates and chlorides must never be used in conjunction or else they would dissolve the gold itself, and this warning is also given by Howard.[101] Auric chloride is not stable at high temperatures, but in this predominantly low-temperature process accompanied by frequent washings, it is possible that it could form and deposit gold in the ashes.[102] Note, however, that many of the cements described in some detail by Agricola and in one instance by Ercker[103] contain common salt and saltpetre together, which rather suggests the problem was not as serious as Lewis and Howard believed. See pp. 59 and 66 for further discussion of this point.

A process to refine gold using sulphur and copper is also described in the *Ā-īn-i Akbarī*, and this is summarised on p. 68.

Traditional parting processes as carried on in the Punjab were described both by Baden-Powell and in some detail by Maclagan[104] at the end of the nineteenth century:

> The mass of gold to be purified is melted down in a small crucible. This is accelerated by adding a little *sohága* or borax. On melting, the gold is poured into a *reza* in which it becomes a small thin bar about 9 inches in length and about 1.5 inches in circumference. This bar is made longer by beating and is then cut up into small pieces weighing about 4 *máshás* each. These pieces are beaten into small squarish plates or *patras*, in size about 3.5 by 4 inches and of the thinness of stout brown paper. The *sunár* then places an *opla* or dried cake of cow-dung on a heap of ashes in an *angíthi*, which for such purposes is the lower half of a broken *gharra* or earthen pitcher. On the *opla* is spread a layer of powder composed of two parts of *surki* (brick dust) and one part of common salt. The *sunár* now seizes one of the plates of gold with his *chimta* or tongs and immerses it in water. When sufficiently wetted, the *patra* is taken out and the same *massála* or powder is sprinkled on it until a good coating adheres to the plate, which is then placed on the layer of powder spread on the *opla*. This process is repeated with all the plates, and they are placed on the top of the other. Over the bank of *patras* a miniature dome with small apertures at intervals is erected of small pieces of cow-dung. A *batti* or wick made of twisted cloth is soaked in oil and lighted. This is thrust into the dome through one of the apertures and thus the cow-dung is fired, and allowed to smoulder for three hours, by which time it is, as a rule reduced to ashes. Three hours are allowed for cooling. All this time the heated powder has been acting on the *patrás* and absorbing impurities. If, after assaying, it is found that the gold has not been thoroughly purified, the same process is repeated except that no powder is sprinkled on the plates. The *angíthi* is always put out of the way of draughts. If a draught was allowed to play on it, an amount of heat would be generated which would have the effect of melting the plates at the edges. Great care must also be taken in preparing the powder. If it is not properly

prepared, it will either not act on the plates or it will act harmfully.

With a fine strong gold, which is alloyed merely with silver, will be purified in 4 to 6 hours: otherwise it is put in two or three times, and will take as long as 12 hours. It is customary in Pindi and Muzaffargarh to put it in three times and in Kohát three or four times. It is usual in Delhi to make the leaves 8 *máshás* in weight and about 6 inches square; and the mixture of brick dust and salt is there known as *saloni*. The leaves weigh 3 *máshás* in Jhang and 8 *máshás* in Amritsar, in Pindi they are 3 inches by 4. In Muzaffargarh, Umbálla and Guidáspur the *saloni* is mixed with water and smeared on to the leaves instead of being sprinkled over the moistened leaves as noted in the above quoted description. In Lahore the powder applied consists of one part of common salt and one part of *opla* ashes; in Gurgáon of one part salt and two parts of brick powder; in Amritsar of equal parts of alum, salt and *opla* ashes, 3 *tolás* of the mixture being required for 20 *tolás* of gold. In Jhelum {where the process is called *aga*} it is oil, burnt clay and salt; in Pindi, Jhang and Dera Ismail Khan it is water, salt and *bhitor* or earth from the *chúlha*: in Jullundur it is *opla* ashes salt and water; and in Kohát two thirds of *opla* ashes, and one third of salt mixed with water.

A process is used in Peshwar which is very similar but more expeditious. The gold is plastered with a mixture of black salt and borax and put into a cake of burning cow-dung with other cakes heaped round about; the alloy is said to be removed in five minutes. In Jhang a mixture of one *rati* of bluestone to 4 *ratís* of sal ammoniac mixed with water, and *geru* is applied and the gold again put in the fire. In Hoshiárpura a process of purification by salt, alum and pip (?) is mentioned. In Gurgáon the goldsmith knows when the gold *pannas* are pure by the sound they give when moved and also by the powder ceasing to stick to them.

The sort of process above described has the disadvantage that the metal forming the alloy is practically lost. Accordingly recourse is very commonly had to the following practice. The alloyed gold is mixed with silver either in equal quantities or 1 *tolá* of gold to 2 of silver. The mixture is either cut into small pieces and placed in water or is melted and thrown into water where it sinks to the bottom in separate small particles. These pieces or particles are then put into a porcelain bowl containing acid and pure gold falls back in black grains to the bottom. Water is added and the liquid is carefully poured away, leaving the pure gold which

can then be melted into a nugget at pleasure. The acid used is nitric or nitromuriatic acid. In Jullundur the ingredients are powdered nitre and *kahi mitti* {an earth containing iron as a sulphate}; and in Kohát they are alum nitre {*qalmi shora*} and *kahi mitti*. In Dera Ismail Khan the ingredients of the acid there known as *faruki tezáb* are given as nitre and *kasis* or *kahi mitti*. This process is that employed by the *niáryás*: it is specially used in Dera Ismail Khan for extracting pure gold from old ornaments and the gold thus extracted is known as *tezábi*. The silver in this process is recovered from the acid by putting in copper plates to which it adheres in powder; the powder is shaken off and collected in a cloth.

A third process is used in Amritsar and applied in Jhelum to gold obtained from the sand known as Gorhi gold, by which the gold is fused with borax and purified by application of calomel (mercurous chloride, $HgCl_2$) and corrosive sublimate (mercuric chloride, $HgCl_3$) {*ras-kapúr*}.

Note the cementation process was essentially unenclosed, even though in this case the relatively volatile silver chloride would have been produced. Note also the range of regional variation in the gold-refining processes.

A somewhat similar account is given of the procedure carried out at the Hyderabad mint in the south of India by Viccajee for the assay of gold, but the same procedure could be used to refine the metal:[105]

About half a *tolá* (86.125 grains) of the gold to be assayed is taken and beaten into four very thin leaves, each weighing approximately 21.5 grains. The weight of each gold leaf is recorded before and after treatment to cementation.

The cement consists of two parts of finely-ground red brick dust and one part of finely crushed common salt, with sufficient Tamarind water or rice gruel, to make the whole into a paste.

A layer of this cement, about $1/4$ inch thick, is evenly spread on a piece of broken earthenware pot about 6 in. × 4 in. × $3/16$ in. thick.

The four leaves are then smeared all over with cement and placed on the piece of earthenware, one on top of the other and arranged like slates on a roof so that the leaves overlap and leave about $1/4$ inch of each leaf exposed. The leaves are now covered with dry cement and put aside. Four leaves, thus prepared, constitute an 'assay' and any number of such assays may be got ready as required. Forty cow-dung cakes, about three inches in diameter, are now arranged in a circle about 18 inches in diameter on the floor, and in the centre an 'assay' is placed and covered up with sixty other cow-dung cakes in a conical heap, care being taken not to disturb the cement.

When all the required number of assays are thus made ready, the cakes are lighted and the fire allowed to burn down. The assays are allowed to remain in the ashes until the earthen pot pieces are perfectly cool, which usually takes from 10 to 12 hours. For convenience this work is got ready in the evening and allowed to burn down all night. The room during the night is locked. The next morning the assays are carefully removed from the ashes, the leaves cleared of the adhering dust, and then weighed.

If the purified leaf has a beautiful pale-gold colour, and is soft and does not emit a sharp metallic sound on being rung on a hard surface, but a dull sound, the assay is considered to be complete; invariably, however the touchstone is used to verify this. If these tests are not satisfactory, the leaves are again subjected to the same process. When carefully carried out, the refined gold by this method assayed from 998.6 to 999.0 fine.

It is instructive to compare these accounts with that of Abū al-Faẓl Allāmī made some three centuries previously. Much of the description on the preparation of the plates of impure gold and of the open-firing arrangements is very similar but the nineteenth-century cement seems to be common salt or mixtures of salt and sulphatic material such as alum or 'black salt' (i.e. copperas hydrated ferrous sulphate), rather than the saltpetre used previously. However, the cow-dung cakes would contain both nitrates and ammonium salts and these were in direct contact with the salt cement and indirectly with the metal. Thus in reality the impure gold would have been exposed to a combined chloride, sulphate and nitrate attack, and there was a possibility of some gold loss.

Maclagan makes the important comment that the other metals with the gold are lost in the salt cementation processes which he described as taking place in an open environment, revealing again the problems containing the volatile silver salts if it was intended to recover them.

Traditional gold purification methods are still employed to some extent by jewellers in India. R. Skelton (personal communication) made a video recording of these processes at both Varanasi and Jaipur in the early 1990s, and their procedures are obviously related to those observed a century earlier by Maclagan and Viccajee. The gold that is to be treated is in the form of wires which are beaten out into long thin ribbons which are then wrapped around another strip to form a packet which is then beaten out still further until it is about 40 μm thick.[106] This is coated with ashes, cow dung and common salt, made into a paste with mustard oil. These are much the same basic ingredients as listed in previous recipes, although only common salt is specified as the active agent and there is no mention of brick dust. The coated gold is then placed in an open ceramic pot and covered with cakes of dung to which a little kerosene is added and the whole is set alight and allowed to burn for about two hours. The gold is then boiled in lime juice to remove the tarry deposits formed from the partial combustion of the mustard oil, and washed in water. There is a further stage where the gold is beaten again and then given a short heat treatment on a charcoal fire. The gold is then washed again and is ready for use. The purpose of the second beating and heating was probably to consolidate the porous structure of the treated gold.[107] Note that the jewellers still rely on the touchstone to judge the purity.

Islamic descriptions

There are several important Arabic accounts of the process, including that of al-Hamdānī, compiled in the Yemen during the tenth century (see Appendix 3, p. 236, note 6). There it is stated that the gold recovered by amalgamation with mercury from the finely ground ore was first melted and a scum (of fine rock dust?) skimmed off. The gold was then hammered out and cut into strips. These were placed in layers in pots packed in a mixture of brick dust, common salt and zāj. Several varieties of zāj[108] are specified including al-zāj al-abyad al-martokī, literally white lead sulphate. This is also the earliest unequivocal reference to the use of brick dust. The vessel was apparently then placed on top of a small square furnace with entrances, rather similar to those already described on p. 38 (Fig. 2.3). Al-Hamdānī's description suggests that this was a double process: first a smoking followed by a firing of the gold, although the exact meaning of this is not clear.

Here again, common salt and sulphates were to be used together, although the addition of lead sulphate is puzzling. Perhaps this could be understood if the low-temperature 'smoking' stage was the cementation followed by the high-temperature 'firing', that is the cupellation to recover the silver, in which case the addition of lead salts is more explicable. Allan noted that 'it is clear that al-Hamdānī's recipe is a combination of two different processes for extracting silver from gold, the salt and the sulphur (sic) process'. However, this survey of the early processes suggests that mixtures of reagents, particularly of chlorides and sulphates, used together was quite common, and the addition of base metals was also widespread.

The instruction to place the vessel on top of the furnace does seem most strange, as all the energy for the

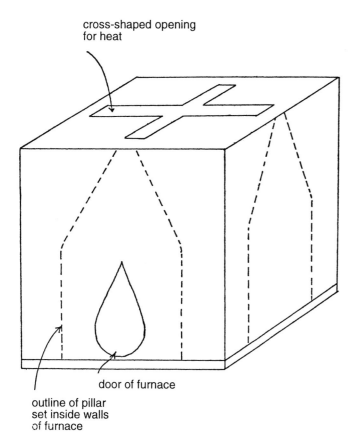

cross-shaped opening
for heat

door of furnace

outline of pillar
set inside walls
of furnace

Fig. 2.3 Reconstruction of a gold-refining furnace based on the description of al-Hamdānī (fol. 30r, v), made in the tenth century. (From *Persian Metal Technology*, J.W. Allan, 1979)

erable detail on both the refining and testing of gold and silver. The first and fourth chapters of the work are devoted to gold refining and parts of Levey's translation are given here:

> The Emir gathers equal parts of the three types of gold; he melts them, makes them into leaf, cuts them as large as the clove, and then brings them together in the furnace of the mint in the customary way for a night. The result of these three will come about from the state of the complete mixture. A strong acacia charcoal fire is lit to melt what is in the body of the gold and silver which the exalted Allah created. Nature is incomplete in its ripeness until the gold is again obtained; the gold is, therefore, not melted but kept on the fire to perfect its nature and to complete it [in maturity]. However, it loses in weight a measure of silver that is separated in the refining of the fire, thus diminishing its body for its ripeness was insufficient. That silver is extracted from the ore (spent charge?) of the refining mixture with the propriety that its recollection brings. Its weight is made true and written on the silver.
>
> First, refine it one night; then refine the gold again a second night. Work on the extraction of the silver from the ore (spent charge?). The reckoning depends upon all the silver extracted according to the continued [process], as on the first, second, third and fourth nights until the impurity is removed from the ore in the process of refining.
>
> Then, one returns to the silver to refine it first one night. Silver, without any gold, is placed on it, equal in weight to it.

The description continues with detailed instructions to refine to constant weight to ensure that all the silver had been removed and methods of assay to test the quality of the gold; practical but vital steps in the production of metal for coinage purposes.

The three types of gold, described in more detail elsewhere,[113] are probably to be equated with gold from the primary deposit, secondary gold from the placer river gravels and sands, and, less certainly, pyritic gold. The constant reference to refining and other processes performed over periods expressed in nights reflects the general practice of carrying out metal smelting and refining at night, when not only would it be cooler, but it would be easier to judge the conditions inside the furnace from the glow within and colour of the flame, etc.

The fourth chapter is a detailed description of the parting furnace and of the process:

> A cupola[114] is built, circular inside and square outside. Its cross-section is of four spans by four spans. The

process must be supplied from the fire and thus, in order for the process to proceed uniformly, it is necessary for the source of heat to surround the vessel. If the vessel was basically spherical or cylindrical, as most early parting vessels seem to have been, then the process could not really have been satisfactorily carried out. Possibly this instruction has been mistranslated or at least misunderstood. Another possibility is that the parting vessels were flat and shallow such as some of those of the Roman and Saxon periods excavated in Britain,[109] and also of the early Edo period in Japan.[110] If this is so, then possibly the process could work satisfactorily heated only on the underside, but even here it would have been better to place the vessel inside the furnace.

Al-Kāshānī, writing in the thirteenth century,[111] gives a recipe for a parting cement containing one part common salt to two parts brick dust to four parts white clay, to be used in a process lasting for three days.

Manṣūr ibn-Baʿra was specifically concerned with the production of refined metals for the Cairo mint, also in the thirteenth century,[112] and thus his treatise goes into consid-

exterior of its walls is made of pure clay and salt when it is built; its interior is fashioned with clay and salt to the edge of the cupola. It is then sealed. It is difficult as with a delicate piece of pottery. It is opened to kindle the fire; there is a door for it like the door of an oven. It has a clay grating which is punctuated with designated separations in the construction. The bottom of the cupola is raised from the earth the distance of a brick.

Some new, thin red brick is well pulverised and then sieved. For every 2 kails (a kail was equal to about 16.7 litres) of brick, 1 kail of salt is admixed. It is moistened with a little water.

Description of the refining of the gold: Some of this mixture is put into a red clay cementation pot into which is also placed pounded gold cut up as thin as finger nails. The thin gold and brick mixture fill the pot. Then another pot is inverted above it. A union is made strong with a seal carefully made of clay as required. It is put into the middle of the furnace above another brick. The cementation pots are overturned so that the two both contain gold mixed with the other. To make it permanent, the pot of the good gold above must be struck by the fire whose heat is maintained. The pot of the lower quality will be underneath. It is treated gently; the fire is cut off from it a little. A cover of broken-up acacia is placed between the walls of the cupola with the cementation pots still in the middle of the furnace. It is ignited until it burns properly. The opening is closed with a block from the first [part] of the night until the second [part] of the morning. The cupola is then opened to remove what is in it. The seal is broken from the pots and what is in them is sieved. Under it is placed a small clay water dish to weaken the heat. It is kept with the earth to extract the silver in it. The loss of the gold in that fire is determined correctly with the balance. It is returned to the refining until it is known that the proper condition is at hand. At that time, the surface of the good part is rubbed.

Levey describes this as a combined cementation/cupellation process, but there is only a passing mention that the silver is to be recovered, right at the end, otherwise this is a detailed description of cementation. Clearly, gold of two different standards was being treated together, possibly gold that had already been through one treatment, was reprocessed at the same time as the raw untreated gold. The almost pure gold would have a higher melting point than the alloy and this could explain the instruction to expose it to the full heat of the fire whilst dealing more gently with the base gold beneath. The instruction to include salt in the clay used to make the walls of the

furnace is apparently unique, but interesting. If the wall was sufficiently hot, it would help to maintain an atmosphere of ferric chloride around the cementation vessels. Note that the parting vessels stood on a central brick as at Sardis.

'Alī ibn-Yúsuf, writing in the mid fourteenth century in Morocco,[115] also gives detailed descriptions of the refining of gold, using both sulphur and common salt:

As to the working of gold from silver and copper, work in two steps. Wash (refine) it from copper as silver was washed from copper and lead according to the previous example (that is by cupellation). Sulfur may be added and the copper is burned leaving pure gold.

As to washing gold from lead, then, the example of the washing of silver from lead is the same. As to the washing of gold from silver, the work is in two operations. One is with the ores, and the other with the mixtures. In the case of the ores, the gold mixed with silver is pulverized until it is very fine. It is then spread out on a bed of powdered baked bricks, limestone, and salt .... A fire is kindled for it in the furnace known as the 'oven of soot'. The silver goes into the pores of that earth. The metal is pure. This work has been performed with a blaze and salt in the same way. Gold has thus been worked from silver as was copper from it .... It is all melted and yellow sulfur is evident. So the gold is purified from the silver and the pure one remains. The first is the best.

These descriptions very much parallel those of Theophilus, where the natural gold alloys were parted with common salt but gold was recovered from gilding with sulphur.[116]

'Alī ibn-Yúsuf also describes the recovery of gold from the furnace walls:[117]

The gold is in two kinds, gold particles and others which are gilded. The gold particles are pulverised in a mortar, then sieved. The sieve holds the 'acceptable' ['ashūr]. Then what comes out from it is rubbed with mercury. Only the gold amalgam is retained to obtain the gold. This is mixed with the 'ashūr, weighed, and fired with borax. Then it is emptied into the marāṭ, a vessel in which gold cools. Then the gold is weighed.

Levey then paraphrases the remainder of the process:

The next step is to make gold into leaf and put it into a soot crucible [shaḥīra] with a powder of new red brick and salt stone. It is spread in earthenware and heated unsealed. A cover is put on it of that powder. It is then sealed. Essentially, the minter carries out a combined cementation–cupellation procedure to obtain the pure gold.

Note that, although the spent salt cement probably was cupelled to recover the silver, there is no mention of it in the description to warrant Levey's last comment; only cementation is described there.

Parting in the Far East

China

Gold refining was practised in southern China, where gold coinage was in use on a considerable scale in the Chu state over 2000 years ago.[118] There are detailed records of cementation processes involving common salt from the mid first millennium AD, including this description, from Needham's translation of *Huang Ti Chiu Ting Shen Tan Ching Chueh*,[119] concerning the use of gold in the preparation of elixirs:

> Although one may have obtained some gold [it may not be very pure]. So one should beat it into thin plates, as is often done among the common people, and then bake it with salt [in a furnace] for a day and a night. Remove it [from the crucible], melt it [again], beat it, and bake it again [with salt]. Continue the operation until there is no loss of weight, then stop.

Further on more detailed instructions are given:

> Method of refining crude gold taken from streams [placer gold]. Use clean earthy clay [*kan thu*] to make a crucible, and dry it by baking over a fire. Use pinewood charcoal and place the crucible in the furnace, and then put the crude gold inside it. Work the furnace bellows to blow up the fire, and when the metal has melted put powdered salt in with it, stirring well. Watch until full fusion has taken place and then use a thornwood stick to remove the slag [*o wu*]. Continue to add powdered salt, and go on stirring and removing slag. When the process is complete pour the metal out into a mould making sure that no cracks or fissure appear (presumably in the solidified metal, not the mould). If they do, then mix equal amounts of iron filings ground to powder, ash of burnt cow-dung, and powdered salt, warming them over a cow-dung fire until the mixture becomes uniformly granular then add it to the re-heated gold. [After some time] take a look, and if it has become soft, remove it and beat it into thin sheets. Mix equal amounts of yellow alum and poplar-balsam and liquefy them [by heating] with mud; then smear the mixture over the gold leaves and heat them over a charcoal fire. Stop when red heat is reached. Repeat the process four or five times, and you will get the best quality red gold.

This detailed description has both similarities with the Western methods and features which are apparently completely different. This was a two-stage process in which the impure gold was first treated with common salt in the molten state and only afterwards beaten into thin sheets which were then treated with alum. The addition of ground iron filings and common salt if the metal was still impure is interesting, and apparently unique.[120] Presumably, it would have reacted with the salt to form ferric chloride, which is intensely reactive. However, there is another possible explanation. The final product is claimed to be red gold, which is usually taken to mean pure gold, but the addition of very small quantities of iron to gold can also impart a rosy hue to the metal, and this was certainly known and practised in ancient Egypt.[121]

Japan

Quite sophisticated metallurgical processes to smelt and refine gold and silver from a variety of ores intensified

Fig. 2.4 The salt cementation process as practised in nineteenth-century Japan. *Background:* Loading the cementation hearth. *Foreground:* Making the free-standing fustrums of the cement containing the powdered gold. Note: there was no attempt here to restrict the volatile silver chlorides but see Fig. 2.6. (From W. Gowland, *Archaeologia*, 69 (1917–18))

after contact with the Portuguese and Spanish in the sixteenth century. Some processes, such as liquation to recover silver from copper, were almost certainly learnt from the Portuguese. There is little contemporary Japanese documentary evidence on these processes, but archaeological research at some of the smelting sites is revealing interesting information. Excavations at the smelting site of Emaki, near the gold mines on Sado Island, off the northwest coast of Honshu, the main island of Japan, have unearthed evidence of sixteenth- and seventeenth-century gold refining that is very different from that described and illustrated in the nineteenth century (Figs 2.4, 2.5 and 2.6, p. 129).[122] Large numbers of shallow oval trays, about 30 by 15 cm and about 5 cm deep have been recovered. They are of a very rough ceramic and have the distinctive purple coloration of clay that has been exposed to salt cementation. It is suggested that these vessels, which were apparently unlidded, were the parting vessels. There is no evidence for separate lids; possibly the vessels were stacked in the furnace with the upper vessel forming the

lid of the vessel below, as described by Ercker in contemporary Europe.[123] Furthermore, analysis shows several per cent of silver in the fabric, which confirms their use as parting vessels. Fragments of clay rods were also found in quantity with the vessels and it is suggested that these formed the bars of the grate on which the vessels sat. In the vicinity of the ceramics were a number of small square cut pits that showed signs of fire. It is suggested that these were the bottoms of the cementation furnaces.

Some quite detailed records of traditional Japanese metallurgical processes were made in the second half of the nineteenth century by Western metallurgists who had gone there to bring Japanese technology into the modern world after Japan's long isolation. Amongst them was J.G.H. Godfrey, whose account was recorded in Percy.[124] At Sado, gold was obtained both from surface placer deposits, as at Sardis, and from primary deposits of native silver, silver ores, the sulphide argentite, and argentiferous copper. Clearly, the silver frequently contained enough gold to be worth extracting. In the first stage, the auriferous silver was repeatedly melted with sulphur. This gave rise to a mixture of sulphides and unreacted metal. The former contained most of the silver and a little gold, which was then treated with iron to release the silver. The unreacted metal contained most of the gold with a little silver. In order to remove the silver from the gold and also from impure gold dug from the placer deposits, the gold was then granulated and repeatedly cemented with:

> common sea-salt in clay dishes, each cementation lasting about 24 hours. For the treatment of 0.32 lb. of argentiferous gold are required 62.5 lbs. of salt and 167 lbs. of charcoal; one foreman and 2 labourers being employed in this operation. The yield of cemented gold from the quantity of argentiferous gold above stated is 0.095 lb. The cementing mixture is washed with water, and the residue which contains the silver as chloride, weighs 4.2 lbs. From this residue 0.174 lb. of silver is obtained by fusion with the addition of 1.7 lb. of metallic lead, and cupelling the resulting argentiferous lead, the consumption of charcoal in both operations amounting to 41.6 lbs.

It is significant here that the impure gold was apparently treated in clay dishes rather than the usual closed parting vessels and there is no mention of lids, yet the silver was to be recovered. It is however significant that, based on Godfrey's figures, about 22% of the silver was lost, assuming that silver was the only metal removed. Other descriptions of contemporary Japanese cementation processes show clearly that they were done on an open hearth with no enclosing vessel at all.

Fig. 2.5 The salt cementation process as practised in ninteenth-century Japan. Removing the fustrums of spent cement containing the purified gold after firing. (From W. Gowland, *Archaeologia*, 69 (1917–18))

Gowland[125] recorded the refining of gold in Japan in 1872 whilst he was engaged in establishing a modern mint for the Japanese government (Figs 2.4 and 2.5). He states that:

> The gold was first reduced to a coarse powder by heating it near to its melting-point, and rubbing on an iron plate with a stone or iron rubber. The coarsely powdered gold was mixed with common salt and a certain proportion of clay, and piled up in the form of a cone on an earthen dish. The whole was then placed in a furnace containing charcoal as fuel and was kept at a red heat (normally about 900 °C), insufficient to melt the gold, for at least twelve hours, by which means the silver was converted into chloride, which was adsorbed by the clay. The dish with its contents was then removed and the gold separated by washing with water. The silver was extracted from the residue by smelting with litharge in a simple furnace.

These Japanese processes differed significantly from most of the other recorded processes, with the exception of the traditional Indian gold-refining operations,[126] in that no parting vessel was used. At red heat the loss of silver as the chloride may have been considerable (cf. Godfrey's figures quoted above), and it certainly demonstrates beyond all doubt that a sealed vessel is not necessary for the process to function satisfactorily. As the cement was not surrounded by an earthenware container, the clay in the cement was chemically essential,[127] and the cement of the first description must almost certainly have contained some clay or brick dust as a source of ferruginous clay.

The cementation process is illustrated on the well-known Sado mining scrolls,[128] including the copy preserved in the British Library (Fig. 2.6, p. 129).[129] This and the other copies, all compiled during the late eighteenth and early nineteenth centuries, clearly show the truncated cones of the cement mixture being cast from a former very much in the shape and manner of sandpies onto a dish. A strange feature shown on all of the scrolls is a lid being added to the top of each cone before it was put into the bed of the furnace, and being removed from the cone after treatment. The function of a lid without a vessel is not at all clear.

Notes

1 In the passages quoted throughout Chapters 2 and 3 and Appendices 3 to 5, the present author's comments are contained in parentheses ( ), those of the translators/editors are in brackets [ ] and the very few original parenthetical items are contained in braces { }. Where the translation differs from that published, the text is in bold. The present author is grateful to A. Giumlia Mair for checking and retranslating some of the passages in the light of new scientific knowledge.

2 Agatharchides of Cnidus, *On the Erythraean Sea*. The original, which contains descriptions of the mining and refining of gold in Egypt, is lost but is preserved in part in Diodorus Siculus III.12–14 (Oldfather 1935, pp. 115–23) and in the works of Photios (Müller 1855). Both sources are discussed and compared in Burstein (1989).

3 The classical and early medieval literature on the assaying and refining of precious metals is surveyed in general by Moesta and Franke (1995) and Moorey (1994, pp. 218–21), and in more detail by Halleux (1985). However, in the latter work, some of the related processes from excavated sites seem to be misunderstood, most noticeably those from Sardis itself, taken from the reports of Ramage (1970 and 1978a). Thus although the foils are described as cemented, they seem to be considered as just surface enhanced (p. 45), and more seriously the cupellation hearths are interpreted as remains of a preliminary process to remove copper from the native gold (p. 50). Similarly, Halleux suggests that the lead imported into Rio Tinto was to treat a putative auriferous copper, whereas in fact it was to collect the silver from the 'dry' jarosite silver ores (Craddock et al. 1985). There are other important general studies relevant to the processes at Sardis by Nicolini (1990), Healy (1978, 1980a and 1989) and Szabó (1975). Previous publications by the authors of this study are given in Ramage (1970, 1978a and 1987), Meeks et al. (1996), Cowell et al. (1998), Craddock (1995, pp. 115–17) and Craddock et al. (1998a).

4 *The Kauṭilīya Arthaśāstra* (Kangle 1972). The *Arthaśāstra* has been traditionally ascribed to Kauṭilīya, the chief minister of the Mauryean Emperor Chandragupta, who ruled in North India in the fourth century BC. However, it is by no means clear how much if any of the work in its present form is of that date. The lost *Indika* of Ktesias (Wilson 1836, and see p. 33), which was broadly contemporary, also described gold cementation.

5 Pre-eminent amongst these in English translation are the following: the anonymous *Probierbüchlein* (Sisco and Smith 1949); the *Pirotechnia* of Biringuccio (Smith and Gnudi 1942); the *De Re Metallica* of Agricola (Hoover and Hoover 1912); Ercker's *Beschreibung aller*

fürnemisten mineralischen Erzt- und Berckwerksarten (Treatise on Ores and Assaying) (Sisco and Smith 1951). For more bibliographic details, see the volumes themselves and note 1 on p. 70.
The detailed discussion of the old parting processes given by Percy (1880) in the volume of his great *Metallurgy* devoted to silver is of inestimable value to the modern scholar because of the historical perspective of his treatment of the metallurgical processes. C.W. Ammen's (1997) book on the refining of metals, intended for the small operator, is also of great interest because it describes many of the traditional operations of assaying and refining as current living processes.

6 See especially Gale and Stos-Gale (1981) and Craddock (1995, pp. 221–31) for the early history of cupellation, and Percy (1870) for a more general coverage. See also the sources listed in note 5.

7 Shalev (1993).

8 Woolley (1934, p. 303, Plate 158). See also Plenderleith in the same volume.

9 La Niece (1995).

10 Bray (1978 and 1993); Emmerich (1965, pp. 177–83).

11 Fray Bernadino de Sahagún, *Florentine Codex* (Dibble and Anderson 1959).

12 Footnote in *De Re Metallica* (Hoover and Hoover 1912, pp. 573–4).

13 Bergsøe (1938).

14 Lechtmann (1971 and 1973). The detailed experiments are described in the 1973 article.

15 Notton (1971 and 1974). See also pp. 35 and 178–9.

16 Caley (1926); Hunt (1976); Halleux (1981). Recipes 14 (Hunt 1976, p. 87), 24 (p. 91) and 74 (p. 101) all seem to deal with the surface enrichment of base gold alloys.

17 Levey (1971, pp. 35, 66).

18 The traditional method of making saltpetre in India from evaporated salts collected from the ground in the dry season is described in Marshall (1915). Analyses by Leather and Mukerji (1911) showed that the sodium chloride contents of between 1% and 3% in some refined products were not uncommon.

19 Darling and Healy (1971) and Healy (1980a) tried a similar treatment with saltpetre on some base gold but believed that they were unsuccessful because all that happened was that the coins went black. This was due to the formation of silver and copper oxides, which could have been removed with a strong brine solution, as the South Americans had done, or possibly with weak organic acids, revealing the enhanced surface.

20 Gowland (1915).

21 Park (1907, pp. 218ff., 229).

22 Howard (n.d. [1788], p. 1069).

23 Gill (1822).

24 Bury (1991, pp. 346–7).

25 Gee (1892, pp. 42–5).

26 Hiorns (1912, pp. 388–9).

27 Fishlock (1962, pp. 294–5).

28 Hoover and Hoover (1912, p. 453); Sisco and Smith (1951, p. 185). See also pp. 59 and 63.

29 Allan (1979, p. 8). See also pp. 45 and 53, note 107.

30 Empedocles envisaged materials, including gold, as composed of varying degrees of the four elements, fire, earth, air and water. This was incorporated in Aristotle's philosophy and as such remained the prevalent concept in the West for almost two millennia. Not until the post-medieval period did the modern idea of the chemical elements emerge, as exemplified by the unalterable 'corpuscles of gold' propounded by Robert Boyle in 1661 in *The Sceptical Chymist*. Pliny's rather confused and contradictory views on the reality or otherwise of different varieties of metal, as expressed at various places in his *Natural History* – 34.143–4 on iron and 34.94–5 on copper (Craddock 1980) – are typical of ancient thought.

31 Nicolini (1990).

32 The analyses cited by Nicolini (1990, pp. 39–40, note 124) and others are in general taken from the huge analytical survey of prehistoric goldwork carried out by Axel Hartmann (1982). These were performed by emission spectography on scrapings taken from the very surface of the goldwork, as observed by the present author. These samples are subject to the problems of potential surface enrichment, made especially serious now that it seems that deliberate surface enhancement was prevalent almost from the inception of gold usage (as discussed on pp. 27–9).

33 Ogden (1993).

34 Fossey (1935); Levey (1959a, b); Forbes (1971, pp. 173–6).

35 Levey (1959b, p. 191).

36 Lucas (1962, pp. 229, 490); Ogden (1993).

37 *Pliny: Natural History*, 33.60 (Rackham 1952, pp. 48–9)

38 Ogden (1992, p. 40), Levey (1959b, p. 192), Joannès (1993), Moorey (1994, p. 219), Lucas (1962, p. 229) and Rose (1915–16). De Jesus (1980, p. 89) believed that there was no early refining of silver from gold in Anatolia. Halleux (1974) believed gold refining probably began in Mesopotamia in the second quarter of the first millennium BC, which is possibly a little early. Joannès (1993) believed that gold refining was practised by the neo-Babylonians in the mid first millennium BC. As Lydia and Mesopotamia were by then both parts of the Persian Empire, it is very likely that knowledge of the process should have already spread out from Sardis.

39 The term 'gold leaf' has been variously applied to very thin gold sheet with quite a wide range of thickness. Untracht (1982, p. 662), for example, defines leaf metal as being between 0.000 08 and 0.0001 mm thick. Perhaps the best practical definition is that it is gold sheet that is too thin to support its own weight.

40 Oddy (1981), followed by Scheel (1989).

41 Scheel (1989); Farag (1981).

42 Modern gold leaf made by traditional methods regularly contains varying amounts of silver, up to 50%, and several per cent of copper. It is elements such as iron and aluminium that seriously embrittle gold and must be absent (Untracht 1982, p. 663; Hatchfield and Newman 1991; Lins 1991).

43 First studied and analysed by Berthelot (1901a and 1906). These and other analyses collected in Hatchfield and Newman (1991).

44 The earliest gold coinage in India dates from the beginning of our era, but gold refining is described in the *Arthaśāstra* of Kauṭilīya. See note 4 on the dating of the *Arthaśāstra*.

45 Unfortunately, several modern scholars commenting on the ancient classical and Islamic references to what are quite clearly sulphate processes have described them as sulphur or sulphide processes (Healy 1978, p. 155; Rackham 1952, p. 67; Allan 1979, p. 8; see also pp. 36, 45 and 52, notes 72, 76–78), which are chemically very different, and for which there is no evidence prior to the Middle Ages (see pp. 67–8).

46 Godley (1920, p. 57). See also Pedley (1972), and p. 20.

47 Wilson (1836); Crindle (1882). The quotation is taken from Hulme (1940–41).

48 Hulme (1940–41). Beck (1884, p. 29) was the first to suggest that the 'spring' was, in fact, more likely to be a furnace.

49 Kangle (1972). See also p. 50, note 4.

50 Smith and Hawthorne (1974, pp. 45–6). See also Appendix 4, p. 240.

51 Craddock in Stronge (1993).

52 Marshall (1915). See also note 18, above.

53 *Plato: The Statesman*, 303e (Fowler 1925, pp. 166–7). See also Appendix 4, p. 241.

54 Diodorus Siculus III.12–14 (Oldfather 1935, pp. 115–23). There is another very similar version in *De Mari Erythraeo*, written by Photios (Müller 1855). Halleux (1985, p. 59) suggests that the similarity of the two texts confirms the accuracy of Diodorus' transcription, but as Photios was so much later than both Agatharchides and Diodorus he could equally well have used the Diodorus text rather than the original. The two texts were translated and published together by Burstein (1989).

55 Nicolini (1990, p. 41) suggests that the greater complexity in the process described by Agatharchides, notably the salt treatment, over that by Plato two centuries previously represents technical progress. In reality, it would have been impossible to remove the silver, let alone the PGE inclusions (see Appendix 4, p. 238), from the gold by heat alone. The real difference is that Agatharchides was describing a technical process, whereas Plato was making a political metaphor.

56 Burstein suggests that the reference to a board or wooden table is incorrect as the surviving tables at the goldworks in the Eastern Desert are of stone. However, these were probably used for the primary sorting of the ore from the coarse rock. For the final separation of the gold from the finely ground ore, a wooden surface would be preferable. The washing tables described by Agricola (Hoover and Hoover 1912, Book 8), for example, are all of wood. The modern equivalent, known as vanning tables, use rubber surfaces to retain the gold.

57 Although Healy (1979, footnote 122) suggests that the salt could have been alum.

58 This is a medieval copy of the *Pseudo-Democritus*; see p. 37 and note 79.

59 Percy (1880, p. 361).

60 Healy (1978, p. 154).

61 Morrisson et al. (1985, p. 50).

62 Collectively known as *Probierbüchlein* (Sisco and Smith 1949, recipes 133–5, 142). See also pp. 54–7.

63 al-Hamdānī (Allan 1979, p. 8); and see p. 45. See also Ercker (Sisco and Smith 1951, pp. 182–90); and p. 63, for example.

64 Notton (1971 and 1974) and p. 178. See also Healy (1974); Moesta (1986, pp. 142–5); Moesta and Franke (1995).

65 Pliny (*Natural History* 33.85) associates the residue and spent cement from gold refining with *lomentum*, which Rackham (footnote p. 66) translates as 'barley meal mixed with rice'. However, there is no justification for this translation: bean meal is the correct meaning, and thus of no relevance to the reagents listed by Agatharchides.

66 Percy (1880, pp. 395–7) and Liddell (1926, p. 286) thought the active agent was hydrogen chloride, generated from the reaction of the salt present in the cement with the damp oxidising atmosphere of the parting vessel. But see pp. 180–2 for more on the chemistry of gold parting.

67 Levey (1959b, p. 192).

68 Haffty et al. (1977).

69 *Pliny*: *Natural History*, 33.60, 33.69, 33.84, 34.121 and 35.183 (Rackham 1952). See also the work of the Projektgruppe Plinius (1993).

70 Bachmann (1993). See also p. 235.

71 d'Elhuyar (1790, p. 200), quoted in Percy (1880, p. 394).

72 Rackham (1952, p. 67), as does Forbes (1971, pp. 180–1), both misquoting Bailey (1929, p. 204), who describes *misy* as being 'more or less oxidised pyrites, containing sulphates of iron and copper'. An unequivocal description of *misy* is to be found in Pliny's *Natural History* at 34.121 (see note 76).

73 Healy (1981).

74 Morrisson et al. (1985, p. 60).

75 In a footnote (p. 66), Rackham goes into more detail and inexplicably translates *lomentum* as barley meal mixed with rice. See also note 65.

76 And where it is made plain that *misy* is to be translated as impure hydrated ferruginous sulphates *not* pyrites (ferric sulphide), as given in the Loeb translation. A similar mistake was made by Allan over the ingredients specified by al-Hamdānī (see p. 45). As stated elsewhere, there is no evidence for the use of either elemental sulphur or metal sulphides in the parting processes of classical antiquity.

77 Strabo, *Geography* 3.2.8 (Healy 1978, p. 155). For the complete Greek text see Jones (1923, p. 41).

78 Healy (1978, p. 155) mistakenly describes this as a sulphur cementation, although correctly identifying styptic earth as being based on sulphates. Elsewhere, Healy (1980b and 1989) defines styptic earth as natron containing more than 50% salt.

79 Halleux (1985).

80 Berthelot (1893, p. 67).

81 Smith and Hawthorne (1974, pp. 29–30).

82 See Ercker's detailed description on p. 65, and Appendix 5, p. 250, note 15, for a description of the modern process of estimating silver in gold where silver is normally added.

83 Hawthorne and Smith (1963, pp. 108–9).

84 *Pliny: Natural History*, 33.84 (Rackham 1952).

85 See p. 37 for the earliest references to graduation; p. 65 for Ercker's detailed description; and Appendix 5, p. 250, note 15, for details of modern procedure.

86 Translation by Halleux (1985, pp. 67–8). The surviving manuscript preserved in the Bibliothèque Nationale (BN gr 2327, fol. 280 r, v) is in medieval low Greek, copied in 1478, but the language seems to be about eleventh century. Several different sections can be distinguished; the oldest would seem to be of Egyptian origin.

87 *De Perfecto Magisterio* is on p. 65 of Halleux (1985).

88 Newman (1991, pp. 774–6).

89 The 'flower of brass' is probably to be identified with Galen's *chalcanthos*, literally 'flower of copper' (Walsh 1929). This is usually interpreted as hydrated copper sulphate (Koucky and Steinberg 1982), but from the context in which Galen describes it, may also mean a form of copperas (hydrated iron sulphates) or even alum (mixed aluminium potassium sulphate), all of which would have been effective parting agents. Dioscorides' description of *chalkanthos* in his *Materia Medica* V.114 (Gunther 1934, p. 639) would indicate copperas. Pliny also uses the same term in his *Natural History* at 34.114, where *chalcanthum* is identified with *atramentum*

sutorium, literally shoemaker's black. At 34.123–7, the production of this is described in detail, but it is still unclear whether the product was green or blue vitriol: that is, the hydrated sulphate of copper or iron. Note that, as the copperas was prepared from the mineral-rich waters collected in copper mines, it almost certainly contained some copper even if the main metal was iron.

90 Percy (1880, p. 386).

91 Wyckoff (1967, p. 230).

92 *The Red Book of the Exchequer* (Hall 1896); also quoted in Smith and Gnudi (1942, p. 203).

93 *Arthaśāstra* 2.13.47 (Kangle 1972, p. 114).

94 Levey (1971, pp. 36–7).

95 Needham (1980, p. 190).

96 Stillman (1924, pp. 197–8), for example. Needham (1980, p. 195) believed the absence of saltpetre was the limiting factor in the development of gunpowder in medieval Europe. See also Partington (1960).

97 Rây (1956, pp. 180–1).

98 Translation by Blochmann, quoted from Percy (1880, pp. 379–81). See also the revised edition of Blochmann, edited by Phillott (1927, pp. 21–2), republished 1989.

99 See p. 51, note 18, for analyses of Indian saltpetres at the beginning of the twentieth century.

100 Lewis (1763), *Commercium Philosophico-Technicum*, quoted and discussed in Percy (1880, p. 382).

101 Howard (n.d. [1788], p. 1069). See also p. 66.

102 Note the gold content of approximately 0.064% in the litharge cake 44393V, which also contained 13.5% silver. (The numbers with a letter suffix used thoughout this book to identify the finds were assigned by the British Museum's Department of Scientific Research. See Appendix 6, p. 251, for a concordance.)

103 Hoover and Hoover (1912, pp. 453–4); Sisco and Smith (1951, pp. 187–8). See also pp. 59 and 64.

104 Baden-Powell (1868–72); Maclagan (1890, pp. 21–2).

105 Viccajee (1908, pp. 40–3), reprinted in Smith (1923–4).

106 The Sardis foils are about 60 μm thick. The symbol μm is the abbreviation for a micrometre (0.000 001 m or 0.001 mm).

107 See p. 185 for the condition of the Sardis foils after cementation.

108 Allan (1979, p. 8) translates *zāj* as pyrites, i.e. iron sulphides, but as the specific variety of *zāj* described slightly later on is translated as lead sulphate, it seems likely that sulphates rather than sulphides were intended throughout. It is also possible that zinc sulphate was meant. (See Ercker's description of a cement, p. 63.) These interpretations would also conform much better to known ancient practice using *misy* or alums.

109 For example, Bayley (1991a, b); Bayley and Barclay (1990).

110 See p. 49.

111 al-Kāshānī (Abū al-Qāsim Kāshānī) lived in Tabriz about AD 1300, and wrote a comprehensive book of metallurgical practices. See al-Kāshānī (n.d.); see also Allan (1979, p. 8) for an English summary.

112 Levey (1971, pp. 54–5 and 63–4), dealing with gold refining.

113 Levey (1971, p. 53).

114 In current English terminology, a cupola is specifically a shaft furnace used for melting metal (McCombe n.d.). Probably 'furnace' would have been a better translation here.

115 'Alī ibn-Yúsuf, *Al-dawḥa al-mosht'abika fī ḍawābiṭ dār al-sikka*, quoted in Levey (1971, p. 23) from Husain Mu'nis (1960).

116 Theophilus, Chapters 69–70 (Hawthorne and Smith 1963, p. 147). See also pp. 38–9.

117 Levey (1971, p. 123). See note 114.

118 Needham (1974, pp. 47–54) gives a wide ranging discussion on the development of gold usage in early China, and the evidence for gold refining. See Ivotchkina (1993) for a discussion of use of gold plates as a currency from the fifth century BC.

119 Taken from Needham (1974, p. 58).

120 Needham claims that the addition of iron was also usual occidental practice. In fact, iron was added to release the silver from the silver sulphides formed during the sulphur parting of gold (see p. 67), but not from salt cementation.

121 Frantz and Schorsch (1990). However, Needham (1974, pp. 257–68) has suggested the special 'red' or 'purple sheen' golds cited in the early Chinese literature contemporary with these descriptions were alloys of copper with only small quantities of gold, inlaid and patinated, analogous to the more recent Japanese *irogane* alloy, *shakudo*, or equivalent Chinese *wu-tong* alloys (Wayman and Craddock 1993). See Craddock and Giumlia Mair (1993) for a general discussion of the possible connections between the oriental purple-black patinated metals and similar but much earlier occidental metals.

122 The present author is most grateful to the archaeologists working at Emaki for the opportunity to examine at first hand this unpublished material. The interpretation given here is based on informal discussions and should not be taken as final.

123 Ercker (Sisco and Smith 1951, p. 186).

124 Percy (1880, pp. 551–5).

125 Gowland (1917–18, p. 137).

126 See p. 42.

127 See p. 180.

128 Nihon Gakushiin (1958, p. 212).

129 Todd (1998).

CHAPTER 3

Historical Survey of Gold Refining

2 POST-MEDIEVAL EUROPE

P.T. Craddock

Renaissance Europe: *Probierbüchlein*, Biringuccio, Agricola and Ercker

The accounts given by these four Renaissance sources are perhaps the most important for the study of the ancient process of salt cementation.[1] They are pivotal to the whole study, being the first truly comprehensible, detailed and, above all else, reliable accounts we have, but made at a time when the process was clearly little changed from that practised in the Middle Ages and even antiquity. Above all, they were made by writers familiar with the processes they describe, and not, in the memorable phrase of Cramer, 'written like one who never blackened his fingers or singed his beard in metallic operations'.[2] Thus these accounts published in the sixteenth century are obviously the same processes as those mentioned with much less detail and clarity in previous centuries. Furthermore, they are in the same tradition as the process at Sardis, which operated over 2000 years before. Indeed, much of the evidence found during the scientific examination of the process debris from Sardis can only be satisfactorily explained by reference to these works of the European Renaissance. At that time, cementation with common salt, and to a lesser extent with nitre, was still usual for the refining of impure gold. The newer processes based on sulphur or sulphides (notably antimony sulphide), or mineral acids (see pp. 67–9), were still reserved for other more specialised separations.

The *Probierbüchlein*

The various works which go under the name *Probierbüchlein – Little Book on Assaying* – contain several different recipes for the cementation process as well as for graduating gold.[3] It is obvious that the more famous technical authors of the sixteenth century who also feature in this section borrowed heavily from the various anonymous editions of these works, which were probably first compiled in the early sixteenth century.[4] They do, however, contain some features which are not included in the later works, including the process of granulation.

Granulation

The gold to be cemented was sometimes granulated to produce a large surface area rather than hammered into foils. The granulation was facilitated by the addition of antimony or lead to the gold. This would also have had the effect of graduating the gold, although copper or silver were the metals usually added.[5] The granulation procedure is described in detail:

133. *How to Refine Gold*

For one mark of completely roasted gold take two marks of lead.

134. *How to Cast 12-Carat Gold*

Use for one mark of gold two marks of antimony and one and a half quarter-marks of lead.

135. *How to Cast 9-Carat Gold*

Use for one mark of gold one mark of antimony and no lead.

The next section describes how the gold is to be melted in a crucible containing ashes, and then:

When it is thoroughly molten, stir the gold well and take an assay sample. This assay will enable you to make your computations [concerning the necessary additions]. Inspect the crucible frequently after the gold is molten to be sure that it has not been damaged. After you have taken the sample, put in the addition and at once cover the crucible again with a lid to prevent the addition from smoking away or running over. Let the gold with the addition stand for the same length of time that it would take you to say 12 paternosters. Then take a small crucible, skim the gold, and granulate it in a tub or barrel of cold fresh water {the colder the better}. Hold the crucible high over the barrel by means of tongs and pour slowly so that the granules will not become too fat. The hollower and thinner the granules are the better they are.

This addition of lead early in the process is of obvious significance to the procedure described by Agatharchides (see pp. 34–5), where, perhaps significantly, there is no mention of gold foils, and also to the presence of lead on the surface of one of the Sardis cementation-vessel sherds.[6]

The following recipe (137) recommends using a sherd from a cementing pot to skim the dross off the surface of the molten gold, recalling the apparent use of potsherds at Sardis. The sherd is recommended to be about a span in length, and kept hot in the fire ready for use, presumably to minimise the thermal shock on contact with the metal.

Recipe 142 contains similar information on the alloying of gold prior to granulation:

142. How to Melt and Alloy Gold
[Prior to Granulation]
If you have 12- to 16-carat gold, use four *lot*[7] of lead to one *lot* of copper; in the case of 16- to 22-carat gold, use three *lot* of lead to one *lot* of copper. When you have prepared the lead–copper mixtures according to the carat, that is, different ones for gold of different carat contents, allot of this copper plus lead as much per mark as described hereafter. And when you wish to add this matter to the gold, be sure that the gold is first molten.

Then, when you wish to add some of the matter described above, take for one mark of 6- to 12-carat gold two and a half *lot*; for 18- to 21-carat gold, one and a half *lot* is adequate. Melt the gold for the duration of 12 paternosters and let it react thoroughly with the matter; then granulate it through a piece of split wood into a tub.

Cementation
Many recipes are given in the *Probierbüchlein* for cements, including sulphur or antimony, but only those of common salt are given here:

144. A Real True Cement from Kremnitz in Hungary
Take two parts of brick and one part of salt, grind them very fine, and mix them with each other. When you wish to prepare the powder for use, moisten it

with vinegar like cupel ashes. Finally, the pot must [not] be cracked when you want to charge it.

Only one edition of the *Probierbüchlein* has the 'not'; in the others the instruction was apparently to use a cracked pot. This seems improbable, but recipe 154 (see p. 56) states that the cement and gold are to be placed in a divided (*zerspalten*) vessel. Sisco and Smith believed this could mean the pot was split, possibly to allow air to penetrate.

Recipe 145 goes through the whole cementation process from the granulation stage:

145. How to Granulate and Cement Gold
Take a piece of wood that is split half way into four parts and have the water into which you are going to granulate stirred vigorously therewith so that it runs around in rings. {Do that every time you part gold in a crucible so that your granules do not become elongated.} When the gold is granulated, transfer it to a jug, decant the water from it, and flush it into a trough. Leave a little water on the gold to keep it moist and then cement the gold [as follows]: Use a clean new pot, cover its bottom with a layer of cement and level it with a cementing tool. Then put gold granules on top of that, laying them evenly beside each other in such a way that they do not touch and the cement can be effective everywhere. Then {depending on how wide the pot is} throw one or two handfuls of cement on top of the gold and distribute the cement evenly; repeat this over and over again, putting more gold on top of this, and continue until the pot is full.

When, now, the gold has been put neatly in the pot, place the pot with the gold in a cementation furnace, which previously should have been heated for 12 hours or a little longer, and let it stand there for 24 hours. Or, if you wish to place it in a cold furnace, leave it in for 36 hours. And clean the furnace when you are ready to put in the gold. After that lift out [the pot with] the gold, take a sample, and assay it on the touchstone. If the gold shows 23 carats, let it stand [for a while]. Then transfer the gold to a trough, letting water drip on it, thus quenching it carefully lest the silver go off in fumes. Or, if you have a kettle, quench it therein, the same way as you were told. If you have enough time, however, it would be so much better to let it cool by itself.

After it has cooled, tumble it with a bent tool, thus breaking up the cement gradually as long as it is lumpy. After then passing the cement through a sieve into a kettle, empty the gold [which has stayed on the sieve] into a pan, and tumble it once more to free it completely from all cement. Next put the cement that was passed through the sieve in a jug and carefully wash out over a large barrel the small pieces of gold that have passed through the sieve with the cement. Then pour warm water on this gold from a jug and stir it with a piece of wood or with a broom so that the cement will soak off. Next wash the gold in a jugful of warm water in a kettle having holes in it and after that pour it into an iron pan. Then wash it again with warm water in another wooden vessel, and when it has been washed clean, empty it into a trough. Finally put it once more in a pot with cement, alternating as before a layer of cement with a layer of gold, making the latter a little thicker and using a little less cement than the first time. Place it once more in the cementation furnace, putting as before one pot close to another so that their tops touch. Keep a sample of the gold that came out of the first fire, or the first cement, and rub it on the touchstone. If the sample shows 22 carats, leave the gold in the furnace for 12 hours. If it shows less than 22 carats, leave it in the furnace a little longer, depending on how much lighter you find it than 22 carats.

If, in the first fire, the gold attains more than 22 carats, say 22½ or 23, leave it in the second fire for a shorter time than 12 hours lest it become too perfect. If, after you have taken it out, it is good enough, leave it out; but if it does not quite come up [to the desired degree of purity] and does not compare satisfactorily with your assay needle, you must cement it once more and put it back in the furnace. Only when it lacks nothing toward its [desired] perfection, wash it and dry it, and when it is dry, melt it. For this purpose use a dry crucible that has never been wet; put in the gold piece by piece until all of it is in. Start by putting a little in the crucible and let it get thoroughly heated through. When it is very hot, use a strong blast and then put in [the rest of] the gold slowly. Be careful with the gold; do not let it touch the floor or let it become contaminated since this is fine gold. You must be careful not to touch it with tongs or pincers that have been used to handle white gold nor to use a tub or trough or crucible that has contained white gold.[8] If, on [remelting] fine gold, you wish to use a bottom [as protection against a leaking crucible] take an old used crucible that has contained fine gold. Unless you watch all this, the gold becomes brittle and is not suitable for working.

Recipe 152 contains a unique reference to the use of stones to control the fire around the parting vessel, using a natural-draught wind furnace (Ercker's 'lazy Heinz', see pp. 61–2 and Fig. 3.2):

152. A Precious Cement in Which to Make Fine Gold out of Rhenish and Other Gold

Take one part of powdered colored brick and one part of crushed salt, moisten them with wine vinegar like ashes for cupels and half fill a crucible therewith. Then put in the gulden (thin gold coins, see Ercker's account on p. 63) in such a way that they do not touch the crucible and finally tamp more of the above matter solidly on top of the gulden until the crucible is filled. After closing it tightly so that no fumes can escape, put the crucible in a wind furnace on top of a stone, one span high, being careful that there are no ashes on this piece of stone. Then surround the crucible with small stones until the whole furnace is filled with stones. Underneath it all light a blazing wood fire. In 24 hours the gold will be cemented satisfactorily and will be good gold.

Several other recipes are given, which although basically similar, do include some significant details.

154. A Common Way of Cementation

To employ it, prepare a powder [as follows]: Take one part of salt, which should be ground as fine as possible, {and be sure to use good, purified salt}, and two parts of good brick {which should not be sandy since sand steals and attacks strongly}, which must also be ground extremely fine. After they have been mixed well, the cement powder is ready. Next make the addition, that is, add to each mark, or half a pound [of gold] copper with its added lead, and melt all this in a crucible. Then granulate it in water and dry the granules. Following this, take a divided pot and put in it layer upon layer. {Note, in passing, that cement powders should be prepared, or sprinkled, with sharp vinegar in such a way that they will be neither too moist nor too dry, after which they are to be mixed once more.} Then charge the pot and close the vent of the cementation furnace with brick. The heat should be regulated in such a way that at first there is only a gentle fire and then a stronger one, fed by good, dry wood. And the longer it is maintained the better it is. It should be increased for 26 hours and then be made less strong for 10 hours, so that it will be effective for 36 hours; and it should be neither too cold nor too hot. After all this has been done, take some of the gold, melt and cast it into a small ingot, and assay it on a touchstone. If it has not yet reached the desired number of carats, take out the largest granules and put them on once more as before. After they have been acted on for another 24 hours, take them out again, remelt the gold in a melting crucible, and cast it into ingots. If the gold has become too high in carats, take malleable silver and add to the cement[ed gold] as much thereof per mark of gold as it can stand in accordance with your wishes. Assay it until it complies with your intentions.

155. Another Way of Cementation

Use two parts of brick to one part of common salt, and gold that is hammered into thin sheets. Put all this in a crucible *stratum super stratum*, that is, one layer upon another; but first moisten the cement with children's urine. Then cover the crucible with a pan of the type used for making tests; place it in a furnace among glowing charcoal, and keep it in a gentle fire for a whole natural day. Then, if you have two pounds of gold [for instance], take an equal amount of copper, hammer it into thin sheets and, after melting the gold in a crucible, take a copper sheet, put the copper in the gold, and stir. After melting them together, take a sample and assay it on a touchstone. Then take sulphur and throw it on the liquid gold until the sulphur has consumed all the copper in the gold.

The last part of this recipe seems to be a graduation process, removing the added copper with sulphur rather than with salt.

156. Cement Powder That Has Proved Effective

Take brick that is neither stony nor sandy, nor too strongly burnt, but only within reason. Then use a pounding iron to crush it on a hard stone – just an ordinary one. After that sift it once or twice through a small-meshed sieve. The grains that remain on the sieve must be crushed again so that the brick becomes fine – the finer the better. Then take salt and crush it, making it just as fine as the brick by grinding it in a hemp or mustard mill. Make a cement of the two; but take two parts of brick to a third part of salt so that there is twice as much of the brick as of the salt. If you mix this well and moisten it with children's urine, but not too much, the cement will be good and right.

This seems to be a standard cement, simple, but effective. The principal active agent was common salt, but note that the urine would have supplied some nitrates, although these would have soon dissipated in a 36-hour process.

Recovery of the silver from the spent cement

Finally, brief instructions are given on how to recover the silver from the spent cement by cupellation or by treatment with mercury. The latter is of some importance as the earliest reference to the use of mercury to recover silver minerals rather than just metallic silver.[9]

157. How to Smelt the Cement

Take the washed cement, mix it with granulated lead and litharge, and smelt all this together. This will yield the silver present in the cement.

158. Another Way

Take the powdered cement and pour quicksilver thereon. Then put the used pulverized cement with the quicksilver in an ore-washing pan and mix it in well by hand. Thus, the mercury will take into itself what has been retained by the powder. Collect the mercury, add it to lead in a test, and cupel it. That will give you the silver.

Biringuccio

The account given by Biringuccio in *Pirotechnia*[10] is the last and shortest chapter in Book 4, devoted to the recovery and purification of precious metals. It is quoted here in full:

The Method of Cementing Gold and of Bringing It to Its Ultimate Fineness

I have striven to demonstrate to you as well as possible the method of bringing the metals of your ores to their ultimate perfection. But one cannot always do with gold that is found in the ores what is done with other metals, for gold is not always accompanied by fine silver nor is it so accompanied that it can or should be treated by means of aqua fortis (nitric acid) without doing anything else. The practical philosophers therefore have found a method using the force of fire and certain attractive minerals with which they achieve a certain result which to me is surely miraculous, for they separate a mixture of two or three metals joined together, and one, uninjured in its form, remains deprived of the companions that it contained.

But let us now leave this discussion. I tell you that in order to do this you first make a little furnace like those that melt by wind. At the head where the grating would be are one or two thick irons, one *dito*[11] or more wide, reaching across the whole diameter of the bottom. This should be one and a half *braccia*[12] high from the ground or a little more. Level with the opening where the fire is introduced, make a little iron grating which holds the wood suspended in the air and separated from the coals which it makes in burning. This should be one-half *braccia* from the lower floor.

This done, take an unglazed pot (**earthenware pot**, *olla*) or crucible or **small** vessel of fire-resistant clay of the size that you think you need. Then take well-sifted old tiles or **well-sifted** brick dust, having made a fine powder of the quantity you wish, and one-third of the whole of finely ground common salt; both are very well sifted with a sieve and mixed together. There are some who add to this composition an eighth part of vitriol, but usually the said powder and the salt are enough, with the tiles.

After this, have the gold that you wish to cement beaten and all made into sheets as thin as paper. This done, a quantity of the said powder composition is put in the bottom of the vessel and spread out in a layer. Then a layer of pieces of your beaten gold is spread out above it, having been first moistened in vinegar or in urine in which some sal ammoniac has been dissolved, and it is then covered with the said powders. Proceed thus, layer upon layer, putting in one layer of gold and one of powders, until the vessel that you took is completely full, or until you lack gold to fill it. Then it is covered with a lid made for this purpose of a raw or baked tile or some similar clay, and the whole is luted and dressed with *lutum sapientiae* (i.e. the lute of wisdom, a common alchemic term for a strong air-tight sealant[13]) and dried out. The vessel prepared in this way is put in the aforesaid little furnace on the two little irons that you built in crosswise at the top. It is then covered over with a tile or flat bricks, well sealed in the furnace and luted, leaving only two or three air holes at the corners as vents for the fumes and breathing places for the flames. Fire is applied with **thin pieces of** soft wood, beginning with a little at first and then increasing it gradually. It is continued for twenty-four hours, always taking care, however, not to apply a fire so strong that it melts the gold and materials together {for they would then not act; indeed your labor would be increased} but let it be only enough to keep the vessel always red (i.e. 700–800 °C).

Then, at the end of this time, slow down the fire – nay, put it out completely – and open the little furnace from above. With tongs or pincers take out the heated vessel as hot as possible. When it is out, open it by lifting off the cover with tongs or some other method and put (**pour**) it all into urine or cool ordinary water. When it is very well quenched and can be handled, wash off the cement that is on the gold with a small brush or with the hand and make it clean. When this has been done, take some of the pieces of gold, and, rubbing them on a touchstone, compare them with the gold needle of the carat that you desire and see whether they correspond. In case the gold has not attained this, repeat, giving it another cementing or two with new powders in the way shown you above. When your gold has been brought to the desired point

after the first, second, or third cementing, melt it with a little borax or a little sal alkali, lime, or furnace ashes and cast it into bars or whatever form you wish.

This done, you will have your gold in its ultimate perfection and fineness, in that beautiful color which you desire, and in its true value, although it lacks that weight of silver or copper or whatever was its companion before. Even the silver is not lost because it remains absorbed in the powders. In order to recover it, it is put with washings and other remnants and finally is made into cakes and smelted in the blast furnace, with cupeling hearths or other sweepings, as I taught you in its place concerning the smelting of litharge. In this way you will recover from these all, or almost all, of the silver that was in the gold you cemented.

As with the descriptions given in the *Probierbüchlein*, the process was of many hours duration and the main active parting agent was common salt. Note also there that, although the cementation vessel was sealed, the lid was just a piece of tile or clay luted into place with the special clay.

Agricola

The description of salt parting given by Agricola is contained in Book 10 of *De Re Metallica*,[14] devoted to the purification of gold, and continues after the descriptions of the manufacture and use of mineral acids, and of the sulphur and antimony (*stibium*, modern stibnite Sb_2S_3) processes.

Cements

First, the various cements are described:

We use cements when, without *stibium*, we part silver or copper or both so ingeniously and admirably from **raw** gold. There are various cements. Some consist of half a *libra*[15] of brick dust, a quarter of a *libra* of salt, an *uncia* of saltpetre, half an *uncia* of sal-ammoniac, and half an *uncia* of rock salt. The bricks or tiles from which the dust is made must be composed of fatty clays, free from sand, grit, and small stones, and must be moderately burnt and very old **and this must be in all cases**.

Another cement is made of a *bes*[16] of brick dust, a third of rock salt, an *uncia* of saltpetre, and half an *uncia* of refined salt. Another cement is made of a *bes* of brick dust, a quarter of refined salt, one and a half *unciae* of saltpetre, an *uncia* of sal-ammoniac, and half an *uncia* of rock salt. Another has one *libra* of brick dust, and half a *libra* of rock salt, to which some add a sixth of a *libra* and a *sicilicus*[17] of vitriol. Another is made of half a *libra* of brick dust, a third of a *libra* of rock salt, an *uncia* and a half of vitriol, and one *uncia* of saltpetre. Another consists of a *bes* of brick dust, a third of refined salt, a sixth of white vitriol (hydrated zinc sulphate, $ZnSO_4.7H_2O$), half an *uncia* of verdigris, and likewise half an *uncia* of saltpetre. Another is made of one and a third *librae* of brick dust, a *bes* of rock salt, a sixth of a *libra* and half an *uncia* of sal-ammoniac, a sixth and half an *uncia* of vitriol, and a sixth of saltpetre. Another contains a *libra* of brick dust, a third of refined salt, and one and a half *unciae* of vitriol.

These cements were specifically intended to treat raw, that is freshly mined, gold. The treatment of scrap gold had been discussed in preceding sections, and for these mineral acids or sulphur was already preferred.

These are, of course, a whole series of recipes that Agricola had collected, probably rather uncritically, from various sources all over Central Europe. One is immediately struck by the complexity of the mixtures: the chlorides from the common salt are always used in conjunction with at least one other salt, either nitrates from the saltpetre and impure sal-ammoniac or sulphates from the vitriol, and sometimes all three, recalling the more condensed ancient and medieval descriptions that are often accused of conflating separate processes. The earliest specific European reference to the use of saltpetre in the parting process is contained in *The Red Book of the Exchequer*, published in 1381.[18] The combination of common salt and saltpetre in so many of the cements listed here is surprising. Later authorities[19] state very firmly that common salt and saltpetre were not to be used together because of the danger of the combined vapours of hydrochloric and nitric acids attacking the gold itself.[20] It is perhaps significant that none of the other major sources mentioned here deliberately mixed common salt and saltpetre in their cements, with the exception of one special recipe given by Ercker (see p. 64) specifically to produce very pure gold from the 23-carat gold which was normally judged sufficiently pure for most purposes. It is possible that Agricola was in error here, and later on (p. 457) notes, correctly, that gold and silver can be parted by common salt and brick dust alone. The identity of the white vitriol is not certain. Hoover and Hoover suggested it could be dehydrated ferrous sulphate, rejecting zinc sulphate; white lead sulphate is another possibility (cf. al-Hamdānī, p. 45). However, zinc sulphate was a component of later cements,[21] and zinc sulphate would certainly have been available as the naturally occurring white hydrated zinc sulphate mineral, goslarite.

Cementation vessel and its charge

Agricola's description continues with details of the preparation of the charge in the cementation vessel:

> Those ingredients above are peculiar to each cement, but what follows is common to all. Each of the ingredients is first separately crushed to powder; the bricks are placed on a hard rock or marble, and crushed with an iron implement; the other things are crushed in a mortar with a pestle; each is separately passed through a sieve. Then they are all mixed together, and are moistened with vinegar in which a little sal-ammoniac has been dissolved, if the cement does not contain any. But some workers, however, prefer to moisten the gold granules or gold-leaf instead.
>
> The cement should be placed, alternately with the gold, in new and clean pots in which no water has ever been poured. In the bottom the cement is levelled with an iron implement, and afterward the gold granules or leaves are placed one against the other, so that they may touch it on all sides; then, again, a handful of the cement, or more if the pots are large, is thrown in and levelled with an iron implement; the **white** granules and leaves are laid over this in the same manner, and this is repeated until the pot is filled. Then it is covered with a lid, and the place where they join is smeared over with lute,[22] and when this is dry the pots are placed in the furnace.

This description, accompanied by an illustration (Fig. 3.1) of the parting vessel, is very important, demonstrating that the coarseware cooking pots would have formed quite acceptable parting vessels in the PN cementation furnaces, which are similar in basic form to those described by Agricola:

The furnace

The furnace has three chambers (Fig. 3.1), the lowest of which is a foot high; into this lowest chamber the air penetrates through an opening, and into it the ashes fall from the burnt wood, which is supported by iron rods, arranged to form a grating. The middle chamber is two feet high, and the wood is pushed in through its mouth. The wood ought to be oak, holmoak, or turkey-oak, for from these the slow and lasting fire is made which is necessary for this operation. The upper chamber is open at the top so that the pots, for which it has the depth, may be put into it; the floor of this chamber consists of iron rods, so strong that they may bear the weight of the pots and the heat of the fire; they are sufficiently far apart that the fire may penetrate well and may heat the pots. The pots are narrow at the bottom, so that the fire entering into the space

Fig. 3.1 Cementation furnace for salt parting, built around an iron frame. Note the vase-shaped parting vessel in the foreground.
Key: *A* furnace, *B* pot, *C* lid, *D* air hole. (From *Georgius Agricola: De Re Metallica*, H.C. Hoover and L.H. Hoover, 1912, p. 455)

between them may heat them; at the top the pots are wide, so that they may touch and hold back the heat of the fire. The upper part of the furnace is closed in with bricks not very thick, or with tiles and lute, and two or three air-holes are left, through which the fumes and flames may escape.

This gives a good explanation of why the rounded form of pot was desirable. But note that the pots illustrated by Ercker (Fig. 3.2) are cylindrical.

Cementation process
The description of the process continues:

The gold granules or leaves and the cement, alternately placed in the pots, are heated by a gentle fire, gradually increasing for twenty-four hours, if the furnace was heated for two hours before the full pots were stood in it, and if this was not done, then for twenty-six hours. The fire should be increased in such a manner that the pieces of gold and the cement, in which is the potency to separate the silver and copper from the gold, may not melt, for in this case the labour and cost will be spent in vain; therefore, it is ample to have the fire hot enough that the pots always remain red. After so many hours all the burning wood should

be drawn out of the furnace. Then the refractory bricks or tiles are removed from the top of the furnace, and the glowing pots are taken out with the tongs. The lids are removed, and if there is time it is well to allow the gold to cool by itself, for then there is less loss (presumably of silver: see *Probierbüchlein* recipe 145 on p. 55, which gives a similar warning); but if time cannot be spared for that operation, the pieces of gold are immediately placed separately into a wooden or bronze vessel of water and gradually quenched, lest the cement which absorbs the silver should exhale it. The pieces of gold, and the cement adhering to them, when cooled or quenched, are rolled with a little mallet so as to crush the lumps and free the gold from the cement. Then they are sifted by a fine sieve, which is placed over a bronze vessel; in this manner the cement containing the silver or the copper or both, falls from the sieve into the bronze vessel, and the gold granules or leaves remain on it. The gold is placed in a vessel and again rolled with the little mallet, so that it may be cleansed from the cement which absorbs silver and copper.

The particles of cement, which have dropped through the holes of the sieve into the bronze vessel, are washed in a bowl, over a wooden tub, being shaken about with the hands, so that the minute particles of gold which have fallen through the sieve may be separated. These are again washed in a little vessel, with warm water, and scrubbed with a piece of wood or a twig broom, that the moistened cement may be detached. Afterward all the gold is again washed with warm water, and collected with a bristle brush, and should be washed in a copper full of holes, under which is placed a little vessel. Then it is necessary to put the gold on an iron plate, under which is a vessel, and to wash it with warm water. Finally, it is placed in a bowl (trough), and, when dry, the granules or leaves are rubbed against a touchstone at the same time as a touch-needle, and considered carefully as to whether they be pure or alloyed. If they are not pure enough, the granules or the leaves, together with the cement which attracts silver and copper, are arranged alternately in layers in the same manner, and again heated; this is done as often as is necessary, but the last time it is heated as many hours as are required to cleanse the gold.

Some people add another cement to the granules or leaves. This cement lacks the ingredients of metalliferous origin, such as verdigris and vitriol, for if these are in the cement, the gold usually takes up a little of the base metal; or if it does not do this, it is stained by them. For this reason some very rightly never make use of cements containing these things, because brick dust and salt alone, especially rock salt, are able to extract all the silver and copper from the gold and to attract it to themselves.

The last paragraph is a very interesting if somewhat convoluted comment on the practice of adding metallic salts to the cement, which clearly confused and puzzled the otherwise clear-headed Agricola – as it was to puzzle more recent experimenters such as Notton.[23] However, as this survey has shown, from antiquity there is some evidence for metal or metal salts as a component of the cements, as apparently recorded by Agricola himself and Agatharchides before him.

The most detailed sixteenth-century description of this and other aspects of gold refining is that of Ercker.

Ercker

His *Treatise on Ores and Assaying*[24] describes all the methods of gold refining in great detail, including salt cementation, which will now be quoted:[25]

On Cementation and What It Is
Cementation is an interesting and beautiful art by which silver, copper, brass, and other metals can be gnawed and eaten away from gold by means of a moistened powder, so that gold is separated from its alloy without being injured. However, it should be used only on those gold compositions that are more than half gold. When there is more silver and copper present than there is gold, the other methods of parting are to be preferred (that is sulphur, antimony or acid parting, see pp. 67–9), as they can be performed in less time and with less effort and expense. Note, moreover, that the cement must be prepared in a manner depending on how high- or low-grade the gold is.

Cementation powders consist of substances and ingredients that attack and consume silver and copper on account of their sharpness {such as sharp, salty species}; they are mixed with verdigris, *aes ustum*, and the like, because these give gold a beautiful color and serve to graduate it (see p. 64 and Appendix 5, p. 250, note 15), or with *lapis hematites, crocus martis* (yellow hydrated iron oxides), tutty (zinc oxide, prepared from calcined zinc carbonate, smithsonite or calamine), or calcined vitriol, all of which may be added to a cement according to individual taste as long as nothing is used that is adverse to gold or makes it impure or damages it.

The presence of copper in the salts would not really graduate the metal where the effect of the base metal *in the gold* is to

create extensive porosity when it is removed. In fact, the chemicals listed and the description of the beautiful surface make this opening account read more like a surface treatment than a true cementation.

The presence of verdigris (copper acetate) in the cements is especially interesting, given the high copper content of the litharge cakes and its presence on some of the parting vessel sherds found at Sardis. Experiments by Hall[26] showed that chlorine is likely to have been the active chemical attacking the silver in the gold. It is believed that the iron salts in the cement are also important in catalysing the production of chlorine but copper is much more efficient. Percy,[27] noting the then recently invented Deacon process where chlorine gas was manufactured by blowing hydrochloric acid vapour through pulverised brick impregnated with copper salts, suggested that the presence of copper in the cement could have promoted the production of chlorine. Gold refining by bubbling chlorine gas through the molten metal has been extensively used from the nineteenth century on.[28]

Ercker's account continues:

Powdered brick is used in cement because the brick dust absorbs everything that is gnawed loose and eaten out of the gold by the sharpness and moistness of the other ingredients. This includes the silver and copper, which otherwise would continue to adhere or to be attached to the gold. In this way the gold is made pure and clean. I have described hereafter several cements that I have used and found good and satisfactory. If it is to be performed on a large scale, this process of cementation requires a special furnace that holds a fire for a long time, and I shall therefore tell first how it should be constructed.

Note that salt cementation only requires a special furnace for large-scale operations, suggesting that smaller operations could be performed on ordinary hearths, for which the special ceramic containers studied by Bayley[29] from a variety of Roman and medieval metalworking contexts in Britain, would probably have been essential. In Ercker's detailed account of the parting process, he is careful always to describe the parting vessel as a pot, never a crucible.

Ercker's description continues with the furnace itself:

How to Build the Furnace for Cementation
If there is much to be cemented, there is no better way of doing it than by building a furnace like the one shown in the following figure (Fig. 3.2). Although there are other cementation furnaces, it is my belief that this one is the most efficient type, since it holds a steady fire for about twenty-four hours without

Fig. 3.2 A self-stoking furnace for cementation – the so-called lazy Heinz. **Key:** *A* lower opening or 'mouth hole', *B* upper opening or 'mouth hole', *C* position of earthenware plate on iron grate, *D* vents, *E* small vent, *F* vent plug, *G* scorifier, *H* and *K* cementation pots, *L* poker. (From *Lazarus Ercker's Treatise on Ores and Assaying*, A.G. Sisco and C.S. Smith, 1951, p. 185)

requiring much attention. It is made as follows: Build of brick a square structure that at the inside should be one ell wide and one ell high until you come to the break where the furnace begins to taper. (In the old German long measure, an ell was approximately 60 cm.) Then continue upward for half an ell from this break to the tower of the Heinz (a general term for a convenient or labor saving device, in this instance the lazy Heinz, a furnace that operated by natural draft rather than requiring a bellows), and make this tower one and a half to two ells high and one and a half [*sic*] ells wide at the inside, so that the whole furnace is three to three and a half ells high. The front of the furnace should have two mouth holes; the lower one should be one-third of an ell long and one-sixth of an ell high {through which the draft or the air can pass}, and the

upper one, one and a half quarter-ells wide and of the same height. At this mouth hole, inside the furnace, there should be an iron grate, which is indicated in the figure by the line *C*, and on the grate in front of the mouth hole, inside the furnace, an earthenware plate. A muffle, which is as high as the mouth hole, is put on this plate, and the prepared cement is then placed under this neatly and cleanly. Or you may omit the muffle and put on the plate a pot containing the prepared cement. There should also be four smoke vents, one in each side of the tapering part, as indicated by *D* in the figure.

When you wish to use this furnace for cementation, and the cement has been charged, burning coals are first poured from the top into the Heinz or tower, followed by charcoal. It is then closed with a cover or lid so that no air can escape, and only the mouth hole *A* and the air holes or vents *D* are left open so that the fire can start up and have some draft. Later, all the air holes and vents are closed very tightly except, high up in the furnace near the lid, the little hole called *E*, which leaves an opening wide enough for a little finger to be stuck in. In this way, the cement will not become too hot and no harm will be done.

In such a furnace a fire can be kept up for twenty-four hours, as said above, without the need of tending it much or of adding charcoal, since the charcoal moves along in the Heinz and feeds a fire of the right required intensity for this length of time. If the heat is to last for more hours, more charcoal can be poured into the Heinz during the twenty-four hours, and the fire will be kept up as long as needed.

I shall pass over the merits and demerits of the cementation furnaces that have been used by others before this. The difference will be self-evident when the two kinds are used competitively.

This last comment is almost certainly a reference to Agricola and the furnace he described, as Ercker specifically criticises *De Re Metallica* in his introduction. Ercker continues:

> If in an emergency such a furnace is not available for cementation, put the prepared cement in a clean charcoal fire between bricks and let it heat therein for as long as you will be told hereafter, without letting it melt. Cementation can be performed in this way; but it takes more watching and more charcoal.

Ercker then describes the operation of gold refining, using Rhenish gold. That is, gold originating from placer deposits such as the Pactolus gold, rather than mined primary gold, although here in the form of bars. Note, however, that none of the Central and Western European sources of gold contains the PGE inclusions.

How to Cement Rhenish Gold

If you have Rhenish gold, either as ingots or in little flat bars, and you wish to cement it, you must hammer it thin – the thinner, the better – and cut it into little pieces as big as a crown. If, instead, you have Rhenish gold guldens that are to be cemented and purified as quickly as possible, hammer them a little thinner; but if the case is not so urgent, leave them as they are. Such hammered gold or gold guldens should be cemented as follows: Take 16 *lot* (1 *lot* = 14.5 g[30]) of brick dust {obtained from an old, dry brick which was not fired too much and which should not be very sandy}, 8 *lot* of salt, and 4 *lot* of white vitriol (hydrated zinc sulphate, $ZnSO_4.7H_2O$). After all these ingredients together have been ground small and moistened with urine or sharp wine vinegar, just like cupel ashes, the powder for cementation is ready. Then take the gold that you wish to cement, and first anneal it in a fire and let it cool again. Then sift some of the powder into a scorifier or pot, making a layer as wide as the pot and one finger thick, and put the gold, after having previously moistened it in urine or wine vinegar, on the cement powder. Lay one little piece beside another over the whole width of the vessel. Follow this with more of the moistened cement powder, half a finger thick, and then with gold wetted in urine as before. In this way put in one layer after another until the scorifier or pot is full. The top layer should be a cement powder, a finger thick, so that no gold is visible. Cover it with another scorifier or a lid and seal the joint with lute so that the vapors or spirits which would like to come out cannot escape (this applies especially to the volatile silver and iron chlorides), and then place the scorifier or pot thus laden with gold and cement in a cementation furnace. See to it that it is left there in a uniform heat for twenty-four hours. It should glow with a dark-red color, but the gold in the cement should not melt, as that would do damage. This is because the silver and copper that were extracted by the cement would go back into the gold if the gold melted in the cement, and all the effort of laminating and cementing would be wasted.

This is not necessarily true, especially of the copper. The real problem is that the surface area would be very much reduced, thereby restricting the contact between the gold and the cement.

The scorifier would normally have been a shallow earthenware dish (Fig. 3.2), or even just a flat tile. It is interesting that a tray or open vessel was specified, stacked in the furnace, with the upper dish acting as the lid of the dish below. Scorification (or scourification) was used in the assay and refining of precious metals to reduce the bulk of lead prior to cupellation. This was done by putting the lead containing the precious metal on the scorifier and heating to 950–1000 °C, whereupon the bulk of the lead oxidises leaving the precious metal still within the lead.[31] The surviving detritus of the process can be difficult to distinguish from that of cupellation, but it seems that the process was certainly regularly employed as a distinct stage in medieval Europe,[32] and there are detailed descriptions of the process elsewhere in Ercker's book.[33]

The description continues:

> After the gold has been in the cement for twenty-four hours, close the furnace everywhere and let it cool. Then take it out, open the pot, and wash off the cement powder with warm water. The gold will be found to be almost 23 carats fine. If you wish to increase this fineness, put the gold in a second separate fresh cement, for which you should take 16 *lot* of brick dust, 8 *lot* of salt, 4 *lot* of white vitriol, 1 *lot* of saltpeter, and 1 *lot* of verdigris. Put this in the furnace again for twenty-four hours, as in the first cementation, and repeat this until the gold is entirely pure or has been made as high in assay as you wish. It can be done in fewer hours, but leave it in the furnace for twenty-four hours if you have the extra time, it is that much better and safer. This fact and how much each cement actually improves the gold are shown by assay.

This stronger-acting cement contains a combination of chloride and nitrate salts in the form of common salt and saltpetre, together with copper. Note this was not an ordinary cement but was only to be used to produce pure gold.

Ercker then gives more specific instructions for the treatment of low-grade gold and other gold:

Another Cement for Low-Grade Gold

The first cement used on low-grade gold should always consist of two parts of brick dust and one part of Hungarian or other salt. In making the second cement, equal parts of verdigris, *lapis hematites* (hematite), and calcined vitriol should be added to these two ingredients. When this is boiled with urine until it is completely dry and then crushed fine, it will make the gold high-grade and beautiful.

The second cement would indeed be powerful, containing sulphates and nitrates from the urine, with iron and copper salts to promote the production of chlorine and ferric chloride.

A Good Cement Suitable for all Kinds of Gold
Take 14 *lot* of brick dust, 4 *lot* of hematite, 1 *lot* of *crocus martis*, 1 *lot* of verdigris, 6 *lot* of white vitriol, and 3 *lot* of saltpeter, all of it ground fine. Use it after the first cementation to cement gold moistened with urine as before, and you will obtain beautiful gold. There are some who use in this and several other cements antimony (antimony sulphide is meant here not the metal, see pp. 67–9) and sal gemmae (rock salt). Everybody should do as he pleases; but it is logical to think that if gold can be cemented satisfactorily and as you desire by using two, three, or four ingredients {which I have checked to be true}, it is unnecessary to take seven or eight substances. When somebody wishes to do something special for the sake of graduating (a special process involving the purification and possibly colouring of gold with copper to give a redder colour, see below) [the gold], I can see his point, though one thing is certain: Any gold that is very pure and of high assay has in itself its correct gold graduation and its correct beautiful, natural gold color. It is, of course, possible to give gold an extra high color until it surpasses in this respect all other kinds of high-assay gold. In my opinion, however, such gold is less attractive than gold that has a naturally high beautiful color.

More Information on Cementation
After gold has been cemented and is almost pure, some people like to put this cemented gold into another cement that is made of four parts of brick dust, one part of sal ammoniac, one part of sal gemmae, and one part of salt, all of it mixed and ground small. After wetting it in urine, they put the gold in a cementation scorifier, seal it with lute [and place it in the furnace] for twelve hours until the gold is altogether pure. Concerning the addition of sal ammoniac, which ordinarily attacks gold [when in aqua regia], you should know that it does not do so raw, especially when it is mixed with watery salt, as in this case. On the contrary, it cleans it so that no trace will be left of another metal which might have remained in or on the gold, though gnawed loose by the other cements.

This is another example of a secondary, stronger cementation where pure gold is required, using a combination of chlorides and possibly nitrates coming this time from impure sal-ammoniac. Ercker seems to be aware of the potential danger of this, but the statements concerning solid state and watery common salt are confusing. Similarly, the cryptic comment of Sisco and Smith 'when in aqua regia' does not seem very relevant as sal-ammoniac

was used neither in the manufacture of *aqua regia* nor as a common additive to the acid. In fact, as Ercker notes, the combined reagent cleaned the gold much more effectively, with the clear implication that it was a more powerful cement.[34]

Ercker then went on to describe how low-grade gold, which would be difficult to hammer into thin foils, could be prepared in a finely divided form to give the required large surface area by granulation, in which the molten gold alloy is poured into water, reminiscent of the methods described in the *Probierbüchlein* (pp. 54–5), but without the addition of lead. Alternatively, the alloy could be cast in thin sheets using moulds made from or lined with canvas. Once the initial cementation had rendered the gold malleable, it could be hammered into foils in the usual way.

Finally, Ercker recounts how the silver could be recovered from the cement:

> You have to know further that when all the cementing is done and you have large amounts of used cement, which contains the silver and other alloy that was in the gold, the values can be recovered from this cement by putting it through the smelter together with other sweepings that contain no gold. In this way you will recover the silver that the cement has extracted from the gold. Gold is not absorbed by the cement.

In other words, the spent cement would be cupelled to recover the silver, as was the case at Sardis, although at Sardis much of the silver would have been lost as the volatile chloride (pp. 178 and 208). The final comment about gold not being absorbed was not really correct. The cements containing both chlorides and nitrates could well have absorbed some gold.[35]

Ercker also describes the production of a still more superior gold by adding copper to the refined gold and performing cementation again, in the process known as graduating:

How to Graduate Gold
To graduate gold means to give gold a still higher color, which surpasses its beautiful natural gold color and which shades more toward red than its own proper high color. It is done as follows: Take good pure gold and add to it an equal amount of fine or pure copper, melt and cast the two together, hammer the ingot thin, and purify it again by cementation so that the gold regains its original weight. Then add once more an equal amount of pure copper to the gold, melt and cast the two, again hammer the ingot thin, and cement a second time. Repeat this until you like the color. Some people think that, when this process of

melting gold with copper and then separating the two again by cementation is repeated about thirty times, gold can be given such a high color that it surpasses copper in the height of its color, provided that the following cement is used, which is supposed to be much better than ordinary cement. Namely: Take good, dry brick dust {which should have dried well in the sun} and ordinary salt {which should have been dissolved once, cleaned by being passed through felt, and ignited once}. Make of each one separately a fine powder and pass it through a hair sieve, and then add Roman vitriol (hydrated copper sulphate, $CuSO_4.5H_2O$), which should have been first rubified as follows: Take good red wine vinegar, which should have been distilled with an alembic. Dissolve the vitriol therein, make the solution nice and clear by passing it through felt, and put it in hot ashes to evaporate. It will then be right. After that, put it in a new pot in a bed of charcoal and stir it with a little piece of wood until it turns blood-red. Grind it small after it has cooled, and it will be rubified. Then take verdigris, also dissolve it in distilled vinegar, distil the solution *per filtrum*, let it evaporate and become red-hot, just as you did with the vitriol. Also dissolve an equal amount of sal ammoniac in red wine vinegar. Then take equal parts of all the above-mentioned powders, mix them well, sprinkle the mixture with the vinegar in which the sal ammoniac was dissolved, and the cement is ready.

This stronger-acting cement, once again for the production of pure gold, contained both copper salts and a combination of chlorides, nitrates and sulphates in the form of common salt, sal-ammoniac *and* vitriol.

This passage is of considerable interest as the most detailed early description of graduation,[36] together with an attempt to explain the theory behind the process and thereby justify it.[37]

Ercker again seems to be distancing himself from the merits of graduating. It is an operation that other people perform and the cement is only *supposed* to be much better. Before the modern idea of elements having immutable and constant properties, there was no reason why there should be only one natural gold or why it should not be improved by processes such as graduation and still be pure gold, although Ercker seems rather unconvinced. The problem was that there was still no certain concept of what constituted pure gold, or even that there was such a single entity.[38] A very similar description to that in Ercker was given in the *Probierbüchlein* some 50 years previously, and it is likely that the final stage of Theophilus' cementation process in which copper was added was also a graduation

process, performed for a similar motive (see pp. 38–9). It could even explain the presence of high copper contents in the gold-refining debris from Sardis. Ercker is the one early source that attempts to give the motivation for such an apparently quixotic procedure as adding one of the metals that were to be eliminated, after the metal had already been purified once.

The reason for the additions of large quantities of base metal was so that upon parting, whether by cementation or by acid dissolution, the debased gold would rapidly become very porous throughout as the attack proceeded along the grain boundaries of the metal, thereby exposing all of the metal to the reagent.[39]

The earliest references to the practice are as a part of the cementation process, well predating acid parting. Even so, it seems to have been only been occasionally practised with cementation, the comments of Ercker showing that it was of uncertain value. However, it was, and remains, a regular feature of acid parting.[40] This was because the refining rate ultimately depends on the diffusion rate of the silver atoms through the gold to a reaction surface, and this rate is proportional to the temperature. Thus the diffusion of the silver is likely to have been much quicker in the hot cementation salt parting process than in the acid processes. Therefore, it would be more important to create an open structure through the metal for the acid process than for the cementation processes. Certainly, graduation, or quartation, as it was known in England in the early nineteenth century,[41] was a regular part of the acid parting process and remains part of the traditional acid parting assay procedure. It is strange that in the earlier descriptions the gold was first to be purified and only then diluted with base metal and purified again. Modern practice adds the extra metal, usually silver rather than copper, at the beginning of the refining process. It seems that the prevalent practice in parting generally was to proceed by stages – thus the gold was only fit for graduation after it had been first purified in the ordinary way.

Later history and descriptions

The salt cementation process continued to be used through the post-medieval period and still seems to have been the main method employed in the sixteenth century, as the descriptions above relate. However, in seventeenth-century Paris, for example, acid parting seems to have been the usual method employed,[42] and in the late eighteenth century Howard could write[43] that the cementation process had in great measure been superseded by acid parting, although his description of the cementation process, quoted in full below, does read like an informed, practical, and above all, first-hand account. The early nineteenth-century technical dictionaries, such as Rees,[44] make no mention of salt cementation as a refining process, only acid parting. However, the great chemist Berzelius, in his classic *Lehrbuch der Chemie*,[45] described acid treatment, cementation with common salt and brick dust, smelting with antimony sulphide or with lead oxide and sulphur, and oxidation with manganese peroxide followed by smelting, as being the standard methods of gold purification. This seems a rather backward view and certainly by the end of the century Percy could state that cementation was obsolete, and his excellent description of cementation parting is written very much as an historical survey.[46]

Howard's account is interesting because he describes the process chemically as a gaseous version of acid parting in terms of the then current phlogiston theory, and the description follows on from that process:

> *Gold* is likewise purified from silver by acids applied in the form of vapour free from phlegm. Thus applied, they extract small proportions of silver as well as large ones; the *gold* being unable to defend the metal against the fumes, as it does against the acid in it's liquid state. The marine vapours also, in this method of application, corrode the silver effectually as the nitrous; the only difference is, that when corroded by the nitrous, such part as may remain in the pores of the *gold* is easily dissolved and washed out by water, whereas with the marine, it forms a concrete not dissoluble in water. The method of performing the process is this: Nitre, or common salt, {not both at once or, for the vapour arising from the mixture would dissolve the *gold* as well as the silver,} is mixed with an equal quantity of calcined vitriol, by which it's acid is extricated in the fire; and twice it's weight of powdered brick, to prevent the matter from melting and adhering to the surface of the metal. The *gold* impregnated with silver, is flatted into plates; cleansed by ignition from any external unctuosity that might impede the action of the acid; then stratified and every where surrounded with the mixture in the crucible or cementing-pot; the vessel closely covered and luted; the fire gradually raised to slight red heat, and kept up for twelve, sixteen, or twenty hours. It is often necessary to repeat the cementation; a part of the silver being apt to escape the action of the fumes, especially when the plates are of considerable thickness: the metal, after one cementation, must be melted and laminated afresh, before it is committed to another. These kinds of operations were formerly often practised; but have now in great measure given place to more advantageous and effec-

tual methods of separation. The mixture with sea-salt is commonly called the regal cement, *gold* being the only metallic body, platina excepted, that is capable of resisting it.

The warning not to mix salt and nitre together because of the dangers of creating a vapour capable of attacking the gold itself is interesting and recalls the similar warning given by Lewis,[47] who was a contemporary of Howard. It does, however, contrast with the instructions given a couple of centuries earlier by Agricola and Ercker (see pp. 59 and 64), where a variety of cements containing both common salt and saltpetre were recommended.

Boussingault gave a good description of the process as it was carried out on freshly mined argentiferous gold at the mint in Bogotá, in South America, in 1833.[48] The account is included here rather than as a separate regional variant because it was basically a survival of the European process rather than a true indigenous South American technology. Boussingault states that:

> The argentiferous gold is granulated, and then subjected to cementation in pots of porous earthenware. The cement is a mixture of 2 parts of brick-dust and 1 of common salt. The vessel is filled with alternate layers of cement-powder, of about an inch in thickness, and granulated gold, beginning and ending with the former. A pot may contain from 10 to 15 lbs. of gold.
>
> The furnace is a hollow cylindrical shaft, 4.5 feet in internal diameter, and 9 feet high; and at the height of 3 feet from the ground there is a grate for the pots to stand upon. At the bottom of the furnace, and level with the ground, there is an opening for charging the fuel, which is wood; but there is neither fire-grate nor chimney, the pots being put in and taken out at the top. The pots are kept at cherry-red heat. Cementation lasts from 24 to 36 hours, according to the quantity of silver to be extracted. After the completion of the process, the contents of the pots are stirred up and washed with water, whereby the cement-powder is separated from the granulated gold; the cement-powder, after settling from the water, is reserved for further treatment, and the gold, which is usually from 21 to 22 carats fine [i.e. it contains from 3 to 2 parts by weight of alloy in 24 parts], is cast into bars suitable for rolling. The cement-powder is ground into a fine paste, which is mixed with one tenth of its weight of common salt, and then amalgamated with about 10 times its own weight of mercury. This operation, which lasts from 4 to 5 days, is effected in large wooden vats, at a temperature ranging from 14 °C to 18 °C. This amalgam is always very 'dry', owing to the

large quantities of mercurous chloride with which it is intermixed. The silver from the amalgam is nearly pure, containing only a few thousandths of gold.

The furnace was very simple, seemingly a large and cylindrical version of the Sardis furnaces, but none the less effective. The main difference was in the recovery of silver where, in South America, the Patio or mercury amalgam process was used in preference to cupellation (see Appendix 3, p. 236).

Boussingault also carried out a number of interesting experiments on the salt parting process, based on what he had observed in Bogotá.[49] He first tried the process in impermeable stoneware 'Cornish' crucibles, but to his surprise and the 'great satisfaction' of the local workmen, no silver was extracted from the gold and he was forced to use the local earthenware crucibles.

Gold refining by sulphur, antimony and mineral acids

During the Middle Ages in Europe, cementation with chlorides, sulphates and nitrates began to be replaced by treatments with sulphur, either elemental or in the form of stibnite, Sb_2S_3, and with mineral acids. All of these materials were, initially at least, more expensive than the more traditional reagents, but they did give a better separation and were much quicker. Thus they were first used for more specialist tasks, such as a final treatment to produce very pure gold, or to recover gold from gilded metalwork, or where speed and convenience were of greater importance than the cost of materials. The processes using these materials will be outlined here in order to complete this survey of the parting of gold.

Sulphur and sulphides

In the sulphur process, the finely divided impure gold and elemental sulphur were reacted together at only moderate heat in a sealed crucible. The gold was not attacked by the sulphur but the other metals readily formed their sulphides. The low heat was essential not only to stop the base gold from melting but also to prevent the sulphur from evaporating away too quickly. An alternative was to use sulphides, where the volatility was less of a problem. A common heating arrangement was to sit the crucible in the centre of a ring of fire (Fig. 3.3). The processes are described in detail in the main Renaissance sources[50] and in Percy.[51]

There is no evidence that the sulphur process was used in antiquity. Copper and silver had been reacted with

sulphur to produce the black inlay material niello since the Roman period at least,[52] and soft amorphous sulphur was regularly used from Roman times on as a backing for gold jewellery,[53] but there are no references to the use of elemental sulphur to refine gold. The only reference to a related process in classical antiquity, gold treatments with *sulphides*, is contained in the late Roman Alexandrian alchemic texts, and even these seem to be more concerned with surface embellishment than true refining. This example is taken from *Sur la très précieuse et célèbre orfèvrerie*:[54]

> Embellissement de l'or. Prenez marcassite, 7 onces, soufre 4 onces, fondez ensemble au creuset et il se forme une efflorescence. Quand vous voulez embellir un grain d'or, mettez l'or au creuset au milieu du feu.

Fig. 3.3 Equipment for parting gold from silver by means of sulphur. All the contemporary illustrations show the parting vessels surrounded by an annulus of fire at a respectable distance in order not to overheat the vessels and volatilise the sulphur. **Key:** *A* view inside a furnace, *B* furnace being readied for use, *C* pits and draught inlets, *D* pot for treating granules with sulphur, *E* crucible, *F* iron tongs for lifting crucible from furnace, *G* crucible holder, *H* iron casting pan. (From *Lazarus Ercker's Treatise on Ores and Assaying*, A.G. Sisco and C.S. Smith, 1951, p. 179)

Ensuite mettez de l'efflorescence autant que vous voulez au milieu du creuset et laissez bouillonner. Quand il refroidit, mettez-le sur une brique grecque dans le feu jusqu'à ce qu'il refroidisse.

Thus even though sulphur springs are found within a short distance of Sardis, it is most unlikely that sulphur was ever used for gold refining in the PN gold refinery.[55]

The earliest unequivocal European reference to sulphur refining is in the treatise of Theophilus,[56] where it is mentioned as a specialist treatment for the refining of gold to be used for making a gold vessel, or in the manufacture of a chalice. These were both small-scale operations, where the gold was needed quickly, rather than a general method for the refining of impure gold in quantity, for which the salt process was preferred. Similarly, Agricola[57] described a cement of sulphur and sal-ammoniac for removing gilding from silver vessels, but turned to the old salt cementation process for the large-scale refining of mined gold. The earliest certain references in the Islamic literature are from the fourteenth century.[58]

By the sixteenth century, the process also seems to have been used in Europe in conjunction with mineral acids especially to recover small quantities of gold from silver or base metals. The sulphur parting was performed as a preliminary to remove the bulk of the silver or base metal, before the final parting using the much more expensive mineral acids.[59] The silver could be recovered by treatment with iron filings to precipitate silver from the sulphides, followed by cupellation.[60] Conversely, the sulphur treatment was also used to remove the final traces of metals from high-grade gold after preliminary refining by salt cementation.

A process of refining gold containing a great deal of silver with sulphur is described in the *Ā-īn-i Akbarī*.[61] The gold–silver alloy was melted with copper in an approximate proportion of 12 parts of the gold alloy to one part copper, and then poured into cold water through a bundle of twigs to granulate it. This finely divided metal was then treated with molten sulphur at moderate temperature, and after three such treatments pure gold was left and a mixture of silver and copper sulphides. Percy[62] was at a loss to understand the function of the copper. The reason was either to aid the granulation of the metal, as described in the *Probierbüchlein* (see pp. 54–5), or more probably to lower the melting point of the mixed sulphides. Molten silver sulphide is unstable, often decomposing back to elemental silver which would have redissolved in the gold, whereas the mixed copper–silver sulphide melts at a much lower temperature and is stable.[63]

At the gold mines on the island of Sado, off the mainland of Japan, silver containing relatively small amounts of gold was treated with sulphur as a preliminary separation, prior to salt cementation, as observed by Gowland in the late nineteenth century.[64]

Antimony process
The process was usually referred to as the antimony process even though the active agent was stibnite. The problem of the volatility of elemental sulphur was overcome by using more stable sulphides, especially antimony sulphide, stibnite (Sb_2S_3),[65] sometimes alone or in conjunction with elemental sulphur. As Agricola states,[66] this enabled higher temperatures to be used than with elemental sulphur alone. As the process was apparently not employed in antiquity, it will not be discussed in detail

Fig. 3.4 *Four types of furnace for distilling parting acid.* **Key:** *A lazy Heinz, B chambers for distilling parting acid, C glass receivers for parting acid, D earthenware receptacle, E furnace operated with a retort. F auxiliary receiver for surplus distilled acid, G long furnace, H side chamber in which parting acid is distilled. (From* Lazarus Ercker's Treatise on Ores and Assaying, *A.G. Sisco and C.S. Smith, 1951, p. 159)*

here.[67] The main advantages of the sulphur or antimony process were that it was much quicker than the salt process, and was believed to give a much purer gold. A disadvantage, aside from the cost, was that it absorbed some of the gold, which, together with the silver, could then only be removed with difficulty. This was done by treating the molten argentiferous stibnite with iron, which caused the precious metals to be rejected whence they could be recovered by liquation followed by cupellation. Ercker stated that the stibnite process was used in the second stages of refining gold after most of the silver had been removed, and Percy noted that the stibnite process had been used for the total removal of silver from high-grade gold.

It must be stressed again that there is no real evidence for this process in antiquity. Davies[68] speculated that the antimony reported in some Early Dynastic gold from Egypt could be evidence of antimony parting. The earliest published descriptions of parting with antimony sulphide are to be found in the *Probierbüchlein*,[69] compiled in south Germany in the early sixteenth century and the process was also supposedly referred to by Basilius Valentinus at about the same time.[70] Parting with stibnite clearly rapidly gained in popularity and was still in use in the early nineteenth century, but by the 1880s Percy could state that the process was 'obsolete unless perhaps with some conservative goldsmith or in some outlandish locality or other'.[71]

Mineral acids

The consensus is that the distillation of mineral acids probably commenced in the eleventh or twelfth century, probably in the Middle East or in Europe, there being clear descriptions of the preparation of acids by distillation both in the Islamic and Christian literature of this period,[72] and shortly afterwards in China.[73] It is also possible that some of the Indian iatrochemical recipes of the Middle Ages refer to and make use of mineral acids.[74]

Nitric acid, the *aqua valens* or *aqua fortis* of the medieval accounts, seems to have been the first to have been widely used, probably during the course of the thirteenth century. It was made by the distillation of saltpetre, KNO_3, with alum or green vitriol, $Fe_2(SO_4)_3$. If common salt was added to the acid already formed and the mixture redistilled, hydrochloric and nitric acids were formed together, the *aqua regia* of the alchemists[75] (Fig. 3.4). This acid alone could dissolve gold, and seems to have been first prepared at much the same time as nitric acid. In fact, given that the sources of saltpetre then available were

usually contaminated with common salt, it is likely that the very first acid produced was a mixture predominantly of nitric acid with a little hydrochloric acid. Sulphuric and hydrochloric acids were prepared somewhat later in Europe, but possibly earlier in the Islamic world.[76]

Nitric acid was the main acid used for gold refining by the dissolution of silver from the finely divided metal, but acid parting was not extensively used prior to the post-medieval period, due mainly to the great expense of the acids. The first descriptions of the parting of gold by acid are contained in the sixteenth-century treatises on metallurgy, and even there the method was not really regarded as cost effective compared with salt cementation.[77] Also, as described above, the sulphur or antimony processes were often performed as a preliminary, especially on low-grade gold alloys, reserving acid dissolution for the final treatment. However, the use of mineral acids made it feasible to remove small quantities of gold from silver which had previously been ignored, and thus European silver by the eighteenth century rarely contains more than 0.1% gold.[78]

An interesting picture of gold refining in the sixteenth and seventeenth centuries in Paris has been compiled from the records preserved in the Archives Nationales.[79] These include many inventories of the contents of the workshops used by the gold refiners, from which it has been possible to deduce the processes normally followed. It seems that metal to be refined was first melted with lead and then cupelled, either in ash-lined cupellation furnaces or in individual cupels, to remove the base metals. The silver was then separated from the gold by dissolution with nitric acid.

The various cementation methods were superseded by acid parting during the course of the post-medieval period as the price of nitric and sulphuric acids dropped. Acid parting was joined by chlorination techniques with the introduction of the Miller process, which was initially developed in Australia in the 1860s, whereby the silver was removed from the gold by bubbling chlorine gas through the molten metal.[80] Electrolytic refining of gold began in the 1870s, in the first instance specifically to deal with the problem of removing platinum from gold, and the work of Wohlwill was especially important in establishing the process.[81] By the end of the century, the process was coming into general use all over the world.[82] Through the twentieth century these three processes – acid treatment, chlorination and electrolysis – have become the main methods by which gold has been refined on an industrial scale.[83]

Notes

1 The particular editions and translations used in this chapter are as follows:

The *Probierbüchlein – Little Book on Assaying* (Sisco and Smith 1949) is a collation of numerous recipes produced for working goldsmiths, dealing mainly with the assay, refining and recovery of precious metals. It appeared in various related editions in German, mainly at centres in southern Germany, through the early sixteenth century. The edition used by Sisco and Smith was published around 1520.

Pirotechnia by Vannoccio Biringuccio (Smith and Gnudi 1942) was published in Italian at Venice in 1540.

De Re Metallica by George Bauer, more commonly known as Agricola (Hoover and Hoover 1912; reprinted 1950), was originally published in Latin at Basel in 1556.

Beschreibung aller fürnemisten mineralischen Erzt- und Berckwerck-sarten (*Treatise on Ores and Assaying*) by Lazarus Ercker (Sisco and Smith 1951) was first published in German at Prague in 1574.

The description of the old parting processes is given by Percy (1880) in *Metallurgy: Silver and Gold*, Part I. As stated already on p. 51, note 5, this work is invaluable.

2 From the preface to the 1741 edition of Cramer's own book on assaying.

3 Sisco and Smith (1949).

4 See Sisco and Smith (1949, pp. 157–63) for a discussion of the likely provenances and dates of the various early editions.

5 For a description of graduation, see the commentary on Ercker (p. 64).

6 Sherd 44401Z. (The numbers with a letter suffix used thoughout this book to identify certain finds were assigned by the British Museum's Department of Scientific Research. See Appendix 6, p. 251, for a concordance.)

7 One *lot* is about 14.5 g (Hoover and Hoover 1912, p. 617).

8 Sisco and Smith (1949, p. 147) note that this 'white gold' must refer to an amalgam of gold with mercury or gold alloyed with lead or tin, rather than to pale silver-bearing gold, which they believed would not be especially dangerous to the refined gold. However, melting the fine gold in crucibles that had previously been used for melting electrum would surely risk contamination, especially if base metals had been added, as suggested by the earlier parts of the *Probierbüchlein* or by the British Museum's investigations of the refractories from the PN refinery (see p. 109). These suggest the parting vessels were not regularly reused to melt the electrum or gold, and so contamination with silver or any base metals present was avoided.

9 See Appendix 3, p. 236.

10 *The Pirotechnia of Vannoccio Biringuccio* (Smith and Gnudi, 1942, pp. 202–5) The original Italian text was used by A. Giumlia Mair (corrections in bold).

11 An equivalent term for an inch, approximately 22.4 mm (Smith and Gnudi 1942, p. 458).

12 One *braccia* was equivalent to about two English feet (Percy 1880, p. 388).

13 See the account of the preparation of *lutum sapientiae* given by Albertus Magnus (Wyckoff 1967, p. 230), and on p. 40. Note that Biringuccio's original text uses the medieval Latin form, omitting the last 'a'.

14 *Georgius Agricola: De Re Metallica* (Hoover and Hoover 1912, pp. 453–90) was based on metallurgical practice in Central Europe,

especially the Erzegebirge. Note: the words in bold are in the original Latin edition but which either are not in the Hoovers' translation, or are alternatives.

15 12 *unciae* = 1 *libra* = 323 g (Hoover and Hoover 1912, p. 616).

16 8 *unciae* = 1 *bes* = 222 g (Hoover and Hoover 1912, p. 616).

17 1 *sicilicus* = 6.7 g (Hoover and Hoover 1912, p. 616).

18 Hall (1896). See also p. 41.

19 Lewis (1763, pp. 154–5) and quoted in Percy (1880, p. 382); Howard (n.d. [1788], p. 1069). See also p. 66.

20 See also the comments on the Indian saltpetre process, p. 43, where the spent cement contained small amounts of gold, due to the use of saltpetre, which almost certainly contained some chlorides as well as the nitrates.

21 Fishlock (1962, pp. 294–5). See also p. 30.

22 Hoover and Hoover (1912, p. 455) omit the descriptive 'white' from granules although it is in the original text. Presumably, Agricola meant to emphasise that they were of high silver content. The Hoover translation describes the lute as being 'artificial', but with no apparent justification in the original text for such a qualification.

23 Notton (1971 and 1974). See also pp. 35 and 178.

24 *Lazarus Ercker's Treatise on Ores and Assaying* (Sisco and Smith, 1951). This translation is taken from the German edition of 1580.

25 Sisco and Smith (1951, pp. 182–90).

26 Hall (1953). See also pp. 179–80.

27 Percy (1880, p. 383). See also pp. 180–1.

28 See Percy (1880, pp. 429–33) or Rose (1915, pp. 300–21) for details of the Miller process of refining gold with chlorine gas.

29 Bayley (1985; 1991a, b; 1992a, b).

30 Quoted by Hoover and Hoover (1912, p. 617).

31 Potts (1987, p. 489).

32 Bayley and Eckstein (1996). See also Ammen (1997, pp. 325–9) for a present-day description of the process for small-scale operators.

33 Sisco and Smith (1951, pp. 37–44).

34 See comments on use of chlorides and nitrates on pp. 43 and 66.

35 See p. 66.

36 Libavius described the process, which he called quartation, of refining gold by adding three parts silver to one part of the gold, prior to treatment with *aqua regia*, although *aqua fortis* is almost certainly meant. *Alchemia*, 1597 edition, Book II, Tract ii, c. 40; 1606 edition, pp. 99–100; Partington (1961, p. 255). A similar ratio of silver to gold is used to this day in acid assay (see Appendix 5, p. 250, note 15).

37 For descriptions of the modern practice, see Appendix 5, p. 249, note 15.

38 See p. 31 for more discussion on the changing concepts of pure materials and on the rise of the concept of elements with intrinsic and unchanging properties.

39 The presence of copper in the metal could also have promoted the production of chlorine from hydrogen chloride, which would have attacked the silver in the gold much more vigorously. Note the comments of Percy (1880, p. 397) and on pp. 180–1, on the in-situ production of chlorine and its role in the process.

40 See Appendix 5, p. 250, note 15, for a description of modern acid parting.

41 Rees (1819–20, Vol. III, p. 107).

42 See Allaire (1996). See also p. 69.

43 Howard (n.d. [1788], p. 1069). See also p. 66.

44 Rees (1819–20, Vol. III, p. 107).

45 Berzelius (1836).

46 Percy (1880, p. 368). Thus, for well over a century, no practical descriptions of salt cementation were published until the detailed instructions for small-scale operators in Ammen's extraordinary book (1997, pp. 351–2). The method given by Ammen is much the same as the traditional process except that the cement includes one part ammonium chloride together with one part common salt and one part brick dust. Earlier recipes also often included urine. The would-be gold refiner is instructed not to melt the gold but 'just cook it', and is advised that the operation should be complete in one or two hours.

47 Lewis (1763, pp. 154–5), quoted in Percy (1880, p. 382).

48 Boussingault (1833), quoted in Percy (1880, p. 390).

49 See pp. 177–8 and Percy (1880, pp. 390–3) for more discussion on these and other early experiments on the chemistry of the process.

50 *Probierbüchlein* recipes 132, 143, 153 and 162 (Sisco and Smith 1949, pp. 139, 144, 149 and 153); Biringuccio, Book 4, Chapter 6 (Smith and Gnudi 1942, pp. 201–2); Agricola, Book 10 (Hoover and Hoover 1912, pp. 448–51); Ercker (Sisco and Smith 1951, pp. 171–8 and 198–200).

51 Percy (1880, pp. 356–67).

52 La Niece (1983). Since then there has been some evidence for the use of niello in the Hellenistic period (Giumlia Mair and La Niece 1998), but its use in Mycenaean times remains problematic. The well-known Mycenaean-style silver cup from Enkomi, Cyprus, has been confirmed as having real niello inlays, but other examples quoted by Laffineur (1974) and Xénaki-Sakellariou and Chatziliou (1989) have proved to be very different. Ogden (1993) and Demakopoulou et al. (1995) have shown that the black inlays on many of the Mycenaean pieces are of a patinated gold–copper alloy similar to the oriental *shakudo* and *wu-tong* metals. See p. 53, note 120, and Craddock and Giumlia Mair (1993).

53 Ogden (1992, p. 44).

54 Halleux (1985, p. 68). See also Berthelot (1888, p. 319) who believed (note 2) the *marcassite* to be stibnite. In fact, it seems more likely to have been an iron sulphide, pyrites or marcasite.

55 Note again that Healy (1978, p. 155) incorrectly refers to the process described by Strabo using sulphates as a sulphur process. Allan (1979, p. 8) refers to *zāj* as sulphides where sulphates were almost certainly meant, and Rackham (1952, p. 67) translates *misy* as a sulphide, which should be iron sulphates.

56 Theophilus, Chapters 69–70 (Hawthorne and Smith 1963, p. 147).

57 Hoover and Hoover (1912, p. 460).

58 ʿAlī ibn-Yúsuf (Levey 1971, p. 23).

59 Ercker (Sisco and Smith 1951, pp. 171–9).

60 Ercker (Sisco and Smith 1951, pp. 176–7) and Percy (1880, pp. 360–1).

61 Phillott (1927, Vol. 1, p. 26).

62 Percy (1880, p. 376).

63 La Niece (1983).

64 Gowland (1915). See also pp. 49–50.

65 The alchemist's name for it was *lupus metallorum*, wolf metal, because it devoured all metals which the gold might contain. The term *antimonium* was apparently first used by Constantine the African in the eleventh century (Halleux 1985, p. 68).

66 Hoover and Hoover (1912, p. 367).

67 See *Probierbüchlein* (Sisco and Smith 1949, pp. 126–30ff.), Biringuccio (Smith and Gnudi 1942, p. 201), Agricola (Hoover and Hoover 1912, pp. 448–53), Ercker (Sisco and Smith 1951, pp. 195–201) for the earliest extant descriptions of the sixteenth century; Schulter (1738), summarised in Percy (1880, p. 368), for the eighteenth century; Percy (1880, pp. 367–8), Schnabel and Louis (1905, pp. 1043–6) or Rose (1915, pp. 432, 436) for descriptions of the process at the very end of its life in the nineteenth century.

68 Davies (1935, p. 56) suggested that the high antimony content encountered in some Early Dynastic gold was evidence that it had been purified by the antimony process, rather than indicating a Translyvanian origin, as had previously been speculated. (See Lucas 1962, pp. 226–7 for various interpretations of these analyses.) In fact, if the analyses are reliable, it is more likely to indicate the gold came from deposits of auriferous pyrites, which are often rich in antimony.

69 Sisco and Smith (1949, pp. 116–24).

70 However, the very existence of Basilius Valentinus is uncertain, and the status of his works is even more unclear (Partington 1961, Chap. V, pp. 183–208). *Triumph Wagen Antimonii …* was not published until 1604 (Thölde 1604), long after his supposed life, and its authenticity as an original late medieval document is very unclear (Stillman 1924, pp. 353–78, esp. p. 375). See also the comments of Kelly (1990) to the republication of the English edition of 1678.

71 Percy (1880, p. 368). Note, however, that Laist (1954, pp. 209–10) states that the process was widely used in Germany throughout the nineteenth century, where it was popularly known as the *Guss und Fluss* method.

72 See Forbes (1948) and Taylor (1957, pp. 90–9), for example; al-Hassan and Hill (1986, pp. 147–8) for the Islamic world; and the brief but wide-ranging summary in Needham (1980, pp. 195–6).

73 Needham (1980, pp. 172–95).

74 Various references collected in Rây (1956, pp. 122, 138, 153, 173–4, 188). See also Needham (1980, p. 198).

75 Ercker (Sisco and Smith 1951, pp. 153–4).

76 Apparently as early as the tenth century according to al-Hassan and Hill (1986, p. 148).

77 Detailed descriptions of parting with nitric acid given in: *Probierbüchlein* (Sisco and Smith 1949, p. 74), which contains various recipes; Biringuccio (Smith and Gnudi 1942, pp. 181–202); Agricola (Hoover and Hoover 1912, pp. 439–48); and Ercker (Sisco and Smith 1951, pp. 124–71), which also has a whole chapter on the preparation of saltpetre from which the nitric (parting) acid was prepared.

78 There are not many published analyses of post-medieval European silver, but see Forbes and Dalladay (1958–9), for example. See Percy (1880, pp. 437–50) and Rose (1915, pp. 422–89) for descriptions of various nineteenth-century and early twentieth-century processes for extracting gold from silver.

79 Allaire (1996).

80 Percy (1880, pp. 402–37) and Rose (1915, pp. 450–64).

81 Wohlwill (1897–8).

82 Borchers (1904, pp. 358–82), which contains a translation of Wohlwill's own work; Rose (1915, pp. 474–89); Allmand and Ellingham (1924).

83 McLaughlin and Wise (1964, pp. 20–4); Laist (1954).

CHAPTER 4

The Excavations and Finds

A. Ramage

The goldworks

Most of the goldworks was found and recognised in 1968.[1] A considerable section of what was eventually identified to have been the industrial area had already been partly excavated by others in 1964, but had been misidentified at that time. The fired mud-brick debris from the furnaces in Unit II (Fig. 4.1) had earlier been taken for destruction remains produced from the burning of Sardis by the Ionians when they revolted against the Persians in 499 BC.

The plan (Fig. 4.2) shows that a Lydian domestic suite of later date was built over a bank of furnaces (A) in the south-western corner of the refinery complex. Thus, most of the walls in this plan did not belong to the refinery, although several wall foundations were in continuous use because they were added to again and again. Larger stones characteristic of the Hellenistic era sit directly on walls of the Archaic and Classical era and even stick out beyond the thickness of the earlier wall line.

A number of hollows in the ground, which were designated 'cupels' (see pp. 81–3), served as hearths rather than crucibles. A few were found during the 1964 season in the area labelled as Unit III in Fig. 4.2. At that time, they were described as rather ill-fired pots of unknown function. Some were, however, taken up and preserved in the laboratory at the site.[2] In addition, a number of miscellaneous pieces ranging from coarse terracotta to lumps of litharge (lead oxide) were also preserved and can now be recognised as tools or raw materials for the goldworks.

The altar

By 1968, a distinctly larger area to the north and to the east of the area supposedly burned by the Ionians had been explored. In particular, a large, solid, rectangular structure was found (Figs 4.3 and 4.4), which was quickly recognised as an altar to the Lydian goddess Cybele.[3] The altar dominated an open area in the midst of domestic structures of no particular distinction to the east and north. They were built of fieldstone and mud mortar for the lower part of the walls and mud brick above. The altar, in contrast to the houses, was built of roughly trimmed rectangular pieces of bedrock from the mountains just to the south of Sardis rather than from the rounded fieldstones derived from the eroding conglomerate of the Acropolis. The altar was 1.75 m high, 3.10 m wide and 2.05 m deep. The long sides ran north–south, but an east–west orientation for the cult practice was determined by a substantial step of schist slabs that was not bonded to the altar at the west side.[4] The step is singular in being on the west side, contrary to the usual Greek practice of putting the step on the east side.

Fig. 4.1 Units I and II, where fired mud-brick debris was found at the end of the 1964 season: looking north.

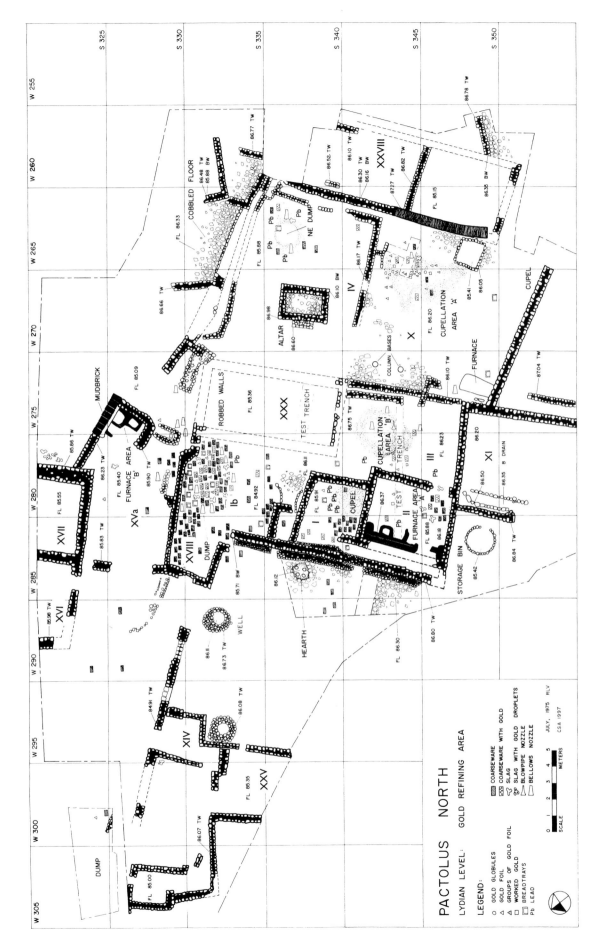

PACTOLUS NORTH

LYDIAN LEVEL: GOLD REFINING AREA

LEGEND:
○ GOLD GLOBULES
△ GOLD FOIL
△ GROUPS OF GOLD FOIL
□ WORKED GOLD
Pb LEAD

▨ COARSEWARE
▧ COARSEWARE WITH GOLD
∀ SLAG
⩔ SLAG WITH GOLD DROPLETS
△ BLOWPIPE NOZZLE
⊔ BELLOWS NOZZLE

JULY, 1975 RLV
CSA 1997

SCALE
0 1 2 3 4 5
METERS

Fig. 4.2 North-west area plan with unit designations.

There was a crust of clay about 80 mm thick on the top of the step, and the clay surface sloped up to it and to the altar itself, indicating that it was in use up to the time of a major flood from later in the sixth century BC that deposited about a metre of gravel over the whole area.

The main structure of the altar shows two stages of building: below, the construction is finer and more regular with small, roughly dressed pieces of gneiss and schist; above, larger, coarsely handled stones were used[5] (Fig. 4.5). The lower stage had a coping of about 0.5 m and a cobbled floor inside[6] (Fig. 4.6). On this floor were many alternating layers of ash and earth (at least 11 major ones). At the bottom, heavily calcined fragments of actual bone and horn were found; higher up were thin layers of whitish ash, which on investigation under a microscope proved to be derived from bones. Later, the whole structure was built up solid with an additional 0.6 m of large fieldstones on the outside, and thin pieces of schist packed with small chips of limestone on the inside. Traces of burning were clear on the top.

The change of design, from small to larger stones, indicates an abrupt change of cult practice. The new Persian rulers, who would have had different religious customs, might have been the impetus for this alteration. They seem to have used the structure as some sort of 'fire altar', in keeping with their religious requirements. It is known that the Persians were extremely sensitive to the difference in religious practice between the Lydians and themselves. For example, an inscription translated into Greek and copied over in Roman Imperial times forbids the priests of Ahura Mazda (who is called Zeus the Law-giver) from consorting with the priests of the local cults because they are unclean, and from participating in rituals requiring the roasting of sacrificial animals.[7]

The stone filling material of the upper section of the altar rested upon the lower layers of ash and earth (Fig. 4.7) except in the south-east and south-west corners, where the ash had been cut away to receive two Archaic sandstone lions that were rather battered but essentially complete.[8] They were packed in carefully with small stones, mostly schist chips (Fig. 4.8). In the north-west corner, the rear half of a similar lion, which had been purposely divided or at least neatly trimmed, was found (Fig. 4.9). Presumably, it had been treated similarly but much of the altar is missing at the north end because of interference from the deep Roman foundations of a mausoleum built nearly 1000 years later.

The lions, as found within the walls, were part of a second, built-up, stage of the altar; but because of their placement in the corners and their Archaic sculptural style (Figs 4.10 and 4.11), it is certain that they had belonged to the altar in its original form, and they are important evidence for a date before the middle of the sixth century.[9] There were probably originally four lions on the top of the altar, each crouching at a corner and facing east towards the rising moon or sun. This supposition is borne out by the fact that one side of each lion is burned. One can imagine that as the altar was used, what would have been the inside faces of the lions were scorched, whereas the outside faces were not.

The use of lions as decoration for the parapet was a strong hint that the altar had been sacred to Cybele, the pre-eminent Lydian goddess. The finding of a potsherd nearby with a graffito reading KUVAV, the Lydian equivalent of the Greek Cybele (Fig. 4.12), made this certain.[10] At first sight, the names Cybele (Kybele) and KUVAV do not appear to have much connection. When one understands, however, that the Lydian form of the name derives from the Hittite *Kubaba*, and that the Greeks used a later form deriving from the Phrygians, the single identity of one goddess under the two names becomes clear.

The altar does not stand on the layer of clay which covered the cupels, but below it on a thin layer of gravel. Further, it does not stand on or in the layer of stones and burnt mud bricks found in a test trench just south of the south face and picked up in another test trench.[11] These facts are the best evidence for putting its first stage in a stratigraphical and chronological context independent of the style of the lions. In 1964, the excavator might have removed a flimsy, relatively unburnt, clay floor just above what turned out to be industrial debris in Units II and III (Fig. 4.2). Thus it seems probable that after the altar was built, a clay floor was laid over what may have become obsolete facilities.

Examination of the altar area

Ultimately, the whole interior of the altar was examined. A test pit that had been dug below the cobbled floor in 1967 (see Fig. 4.4) was extended to remove all the fill and to discover how the altar was built. The inner cavity was 1.8 m long and 1 m wide, and the edges of it were defined by small stones built up like a wall but without any regular coursing. A section cut in the south-east corner showed that there was an inside face to the main wall of the altar. The Lydians seem to have constructed what amounted to a hollow box, and to have begun to fill it with a careful lining of river stones (rather than cut schist). No clay floor or rubble foundation was found at the bottom.

Certainly the 500 or so pieces of pottery found in the test pit in the lower part of the altar were predominantly domestic and consonant with a late seventh- to early

Fig. 4.3 *(above)* Altar of Cybele, east side, with the Necropolis behind.

Fig. 4.4 *(right)* Altar of Cybele: proposed reconstruction and excavation details.

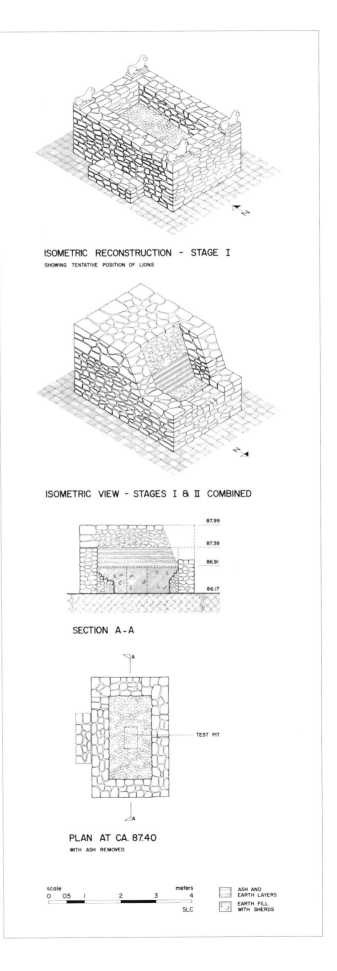

ISOMETRIC RECONSTRUCTION - STAGE I
SHOWING TENTATIVE POSITION OF LIONS

ISOMETRIC VIEW - STAGES I & II COMBINED

87.99
87.39
86.91
86.17

SECTION A-A

A

TEST PIT

A

PLAN AT CA. 87.40
WITH ASH REMOVED

scale   meters
0   0.5   1   2   3   4

SLC

ASH AND
EARTH LAYERS

EARTH FILL
WITH SHERDS

sixth-century date. It is difficult to be much more accurate with pottery that is almost all Lydian patterned ware.[12] There were two pieces with figural decoration amongst this mass of plain coarse and patterned pieces: one obviously Lydian sherd imitating the Wild Goat Style of East Greek pottery and a possible import of the same style (Figs 4.13 and 4.14). Dating these pieces closely is difficult, but about 575 BC seems likely.[13]

The altar stands over a layer of burned debris that is derived from the waste products and floor sweepings of one or more of the industrial processes. This is important for the date of the activity. The sculptural style of the lions,[14] by comparison with Greek styles, is consonant with a date of about 575–550 BC. Even allowing for some lag and religious conservatism, they can hardly be any later. The evidence of the two painted sherds with figural decoration, depending at various removes on the dating of Corinthian pottery, leads to the same estimate: 575–550 BC.

Fig. 4.5 Altar of Cybele: south face showing original construction and additions

Fig. 4.7 Altar of Cybele: looking south with the earth and ash deposit of the interior exposed.

Fig. 4.6 Altar of Cybele: cobbled floor with earth and ash layers at top.

Fig. 4.8 Altar of Cybele: lion in situ in altar.

Fig. 4.9 Altar of Cybele: rear part of half-lion in situ in altar.

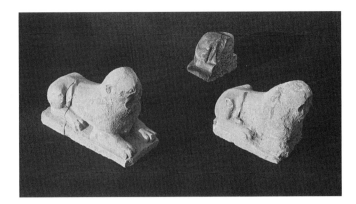

Fig. 4.10 The three lions from the Altar of Cybele.
Half-lion: preserved length 19 cm, width 22 cm. Lion on left: length 49 cm, width 22 cm, height 34 cm. Lion on right: length 54 cm, width 21 cm, height 41 cm.

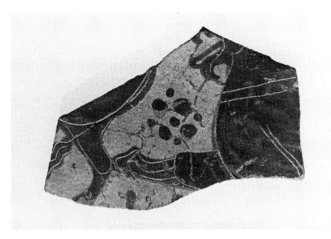

Fig. 4.13 Corinthianising East Greek Wild Goat style sherd: width 52 mm, height 41 mm, thickness 4 mm.

Fig. 4.11 Close-up of one of the three lions.

Fig. 4.14 Lydian Wild Goat style sherd: width 60 mm, height 63 mm, thickness 4 mm.

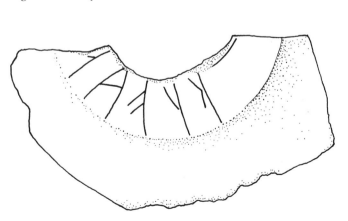

Fig. 4.12 Potsherd with graffito reading KUVAV: height 45 mm, diameter of base 90 mm.

The question of the date of the altar becomes even more important when the contemporary and adjacent process of gold refining is considered. The cementation and cupellation techniques (see Technical Glossary, p. 253) that enable gold and silver to be separated from electrum seem to have been invented at the very era when, according to Herodotus *(History* I.94), the Lydians began the practice of using separate gold and silver coins. This evidence is particularly important for the current controversy about the date of the so-called Croeseid coins of gold and silver, which several scholars would remove from the realm of the Lydian empire altogether in favour of the Persians (see p. 23).[15]

In addition, no one knows why the installations went out of use. The thick layer of gravel indicates the possibility that there was widespread destruction by flooding; part of one house (in the north-east corner) was clearly washed away. Much of the industrial area, however, had already been covered by the clayey surface built up around the altar and was saved from the effects of the flood. One should, therefore, look for some political or economic (rather than catastrophic) explanation for the cessation of activity at the refinery. It is not even known whether the closing down has anything to do with the Persian

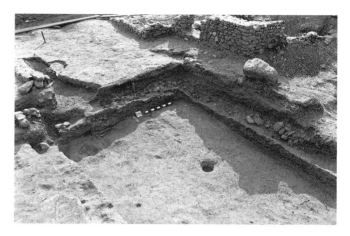

Fig. 4.15 Unit XXX, south end.

Fig. 4.16 *(below)* Two limestone column bases, east of cupellation area B (Unit III).

Fig. 4.17 *(facing page)* Plan of cupels in cupellation area A and the Altar of Cybele.

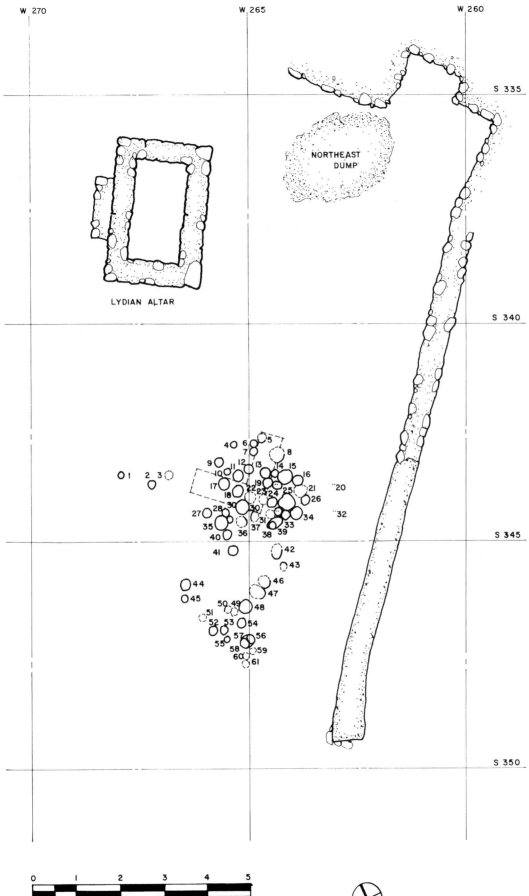

W 270   W 265   W 260

S 335

NORTHEAST
DUMP

LYDIAN ALTAR

S 340

40 6 5
7
8
9
10 11 12 13 14 15
16
17 19
18 22 23 24 25 21 20
26
28 30 30
27 31 32
35 37 33 34
40 36 38 39

S 345

41 42

43

44 46
45 47
50 49 48
51
52 53 54
55 57 56
58 59
60 61

S 350

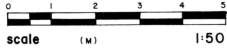

scale   (M)   1:50   MAG. NORTH

occupation, because the chronological markers are not sufficiently precise. One could also imagine that the installation had served its purpose and was becoming a public nuisance right in a built-up area.

Other installations

To the west of the altar is the outline of a large rectangular room (Unit XXX), about 9 × 4.5 m (Fig. 4.15). It was built partly below the ground, and was entirely filled with barren gravel. Not more than a handful of pottery was found in the whole area. The room is associated with the clay surface,[16] – also covered by the same gravel – which is remarkably uniform in this area.[17] The walls stood on a low clay bench 0.5–0.6 m wide and 0.5 m high. Hardly any of the stones remain except at the north end; most of the walls had been removed prior to the major inundation that flooded the area later in the century, although a few stones remained embedded in the clay. No features of the room remain either, except its basic shape and a possible post-

hole which is on neither axis of the room. At present, the relationship of the surrounding structures can give the only clue to the likely location of doors and windows and that indicates something to the west or north. The room is, however, sufficiently close to the altar to suggest some related function, but there is nothing left, apart from its location, that connects it with the goldworks.

Complete buildings closely associated with the cupellation level are absent except for that implied by the robbed out walls at the back of the furnaces which run under Units I and II. A north–south wall, at the east side of Unit III, may have divided the groups of men working the cupels into two units, even though there was no obvious difference between the two dense groupings of cupels. At its north end the wall aligns with the east wall of Unit XXX. The clay floor – that overlying the cupels – runs over it and the cupellation floor runs up to it. It is odd to find this wall, since the hearths are found on both sides of it. Associated with it on the east side were two roughly cut cylindrical column bases of limestone about 25 cm in

Fig. 4.18 *(facing page)* Cupels in the ground.

Fig. 4.19 *(top)* Test trench containing cupels.

Fig. 4.21 *(above)* Detail of overlapping cupels in the ground.

diameter, apparently lying in situ (Fig. 4.16). They are about 1.6 m apart on centre and their distance from the wall is 1.4 m. Since there is no sign of secondary working on the tops or sides, they may have provided the basis for the wooden posts of a formal entrance to a second group of hearths and furnaces at the west. Between the more northerly one and the wall were two cupellation hollows.

It is tempting to associate Unit XXVIII, the long building to the east, with the goldworking, and its alignment certainly fits with that of the complex as a whole. Unfortunately, nothing that was found in it could be directly connected with the industrial activities and there is no sign of access towards the altar space. In fact, there seems to be a north–south enclosure wall just to the west with a corner running east–west that may have been used to screen the industrial area from the altar space.[18]

Finding the altar was in itself a breakthrough, and the examination of the surrounding area to find out more about the circumstances of its building led directly to the uncovering and identification of the goldworks.

Details of the goldworking area

The cupels

To the east and especially to the south of the altar was an uninterrupted clay floor about 10 cm thick. A layer of gravelly earth below was full of bigger stones and burned mud bricks. It also contained a large number of small rings, from about 15 to 25 cm in diameter, that were composed of ash and reddened clay with greenish stains and occasional pieces of litharge (Figs 4.17 and 4.18). This was first taken to be a destruction layer, but the theory was later discarded in favour of industrial debris.

Although it had been concluded that some metallurgical process had taken place in all the small burned rings, there were no serious thoughts (apart from Ramage's suggestion in the field book on 6 July) of a goldworks until Richard Stone, the senior conservator, proposed that the rings were evidence of cupellation.[19]

After considerable extra cleaning over a wider area, a small test trench was laid out (about 2.5 × 0.4 m) with the specific aim to find pieces of gold. This trench contained some of the best-preserved of the ashy rings (Fig. 4.19).[20] Painstaking excavation of this area very soon revealed a tiny piece of gold foil, quickly followed by another. Later that day, a minute globule was found that seemed even more significant, and it remains so, because it meant that the metal was being melted and not just worked at the site. This was important for postulating that a primary

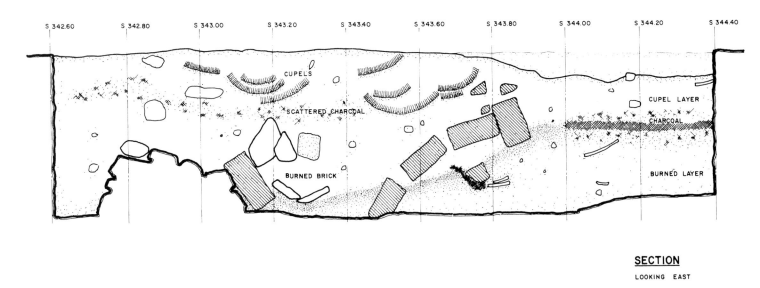

S 342.60   S 342.80   S 343.00   S 343.20   S 343.40   S 343.60   S 343.80   S 344.00   S 344.20   S 344.40

CUPELS

CUPEL LAYER

SCATTERED CHARCOAL

CHARCOAL

BURNED BRICK

BURNED LAYER

SECTION
LOOKING EAST

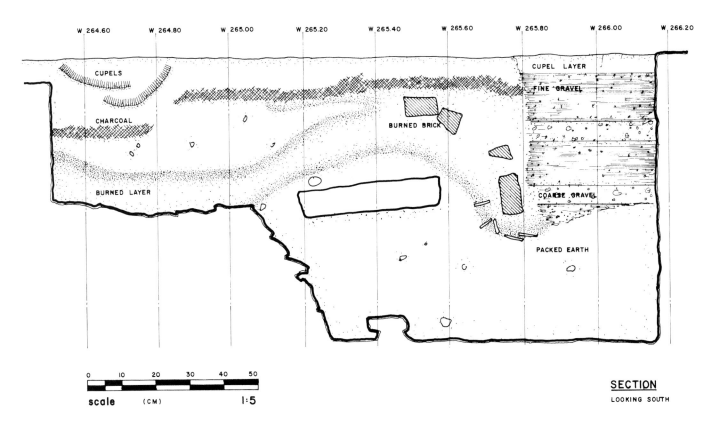

W 264.60   W 264.80   W 265.00   W 265.20   W 265.40   W 265.60   W 265.80   W 266.00   W 266.20

CUPELS

CUPEL LAYER

FINE GRAVEL

CHARCOAL

BURNED BRICK

BURNED LAYER

COARSE GRAVEL

PACKED EARTH

scale   (CM)   1:5

0   10   20   30   40   50

SECTION
LOOKING SOUTH

gold-processing site had been found rather than a jeweller's workshop. This inference depends upon the fact that hardly any Archaic or Classical gold jewellery in Greece and the eastern Mediterranean world was made by casting. The use of cut wire and hammered foil was the rule – operations which should not generate molten metal.[21]

Upon complete excavation, the rings turned into basin-like forms that were designated 'cupels', even though they did not fit the strict definition of the small cup-shaped vessels used for assaying precious metals (Fig. 4.20, p. 130). They are really equivalent to bowl hearths, but do not appear to have had a permanent cover to keep the fire in and conserve the heat (but see pp. 90 and 209).

Fig. 4.22 *(facing page)* Two sections of the test trench.

Fig. 4.23 *(above)* Test trench: view of section looking east.

They had first appeared as multiple concentric rings of fiercely heated earth and clay enclosing a considerable amount of ash. The average diameter was about 25 cm, but there were a few larger ones. It was clear that the cupels had been heated from above and inside because of the gradation in firing colour – from bright red to dark brown – and because of the variation in the hardness of the basic earthen bowl that was apparent when they were fully excavated; they were softer on the outside than the inside.

There were occasional areas of vitrified slag and at least one glassy dribble in the debris. The hearths in some cases were built one upon the other but not in neat layers. Multiple intersections in a rather loose matrix precluded precise understanding of the temporal sequence over the whole area. The intersections were particularly clear in the sides of the original test trench (Figs 4.21 to 4.23) which also showed baked brick debris that must have derived from other processes or installations connected with the goldworking. The most likely explanation for the debris is the dismantling of old furnaces. The cupels were concentrated in two areas particularly (Fig. 4.2); a few others were found just to the west within the area of a later enclosure that originally must have contained a bank of six furnaces. There was little sign of specialised structures, but much of the time these activities could take place in the open or under a light shelter. The walls and the two column bases indicate that some of the area had a more substantial shelter than just a lean-to. Considering the very poisonous nature of the fumes of lead oxide generated by the cupellation operations, an open or at least very well ventilated work area would be essential. The general environmental hazard to the whole area would, however, have been considerable.[22]

The furnaces

Stone's help was further sought to explain the purpose of the squarish, reddened mud-brick furnaces that appeared at the west side of Unit II. Referring to the writings of Biringuccio (see note 19), he concluded that the Lydians had used the furnaces for cementation, which is a method for separating gold and silver from an alloy of the two (see pp. 57–8).

In two areas there were substantial remains of what were believed to be sets of furnaces. Also, there were sufficient traces to locate a third group, although its superstructure apparently had been shaved to ground level.[23]

Furnace area A

Figures 4.24 and 4.25 show furnace area A.

The outline was revealed of a brick structure with a continuous back wall and four bays. The faces of the bays were much harder in texture than the gross structure, which was, in turn, harder than the fill, except for discrete lumps of harder fired brick. No traces of soot were found; the colour of the material behind the faces was a pale orangey-yellow, shading to 'brick' within about 3 mm, whereas the faces themselves were whitish grey with lime incrustation, as was much of the pottery found in the fill. These pieces of pottery were of a distinctive purplish appearance. It was estimated that the maximum temperature reached in the furnace was 700–800 °C.[24] Anything higher would probably have fired the main structure more, making it considerably harder and less liable to decomposition. But, since the colour is so uniform and the whole thickness of the structure (about 0.5 m) reddened, this temperature may have been held for some time (assuming that the furnaces were initially built of raw mud brick, which is supported by the petrographic examination, see p. 158).[25] This structure is therefore interpreted as a series of furnaces, which we numbered from the north.[26]

There were probably originally six furnaces in area A, since four were exposed and there were traces of a similar structure adjacent to the south wall, and the space available between the south wall and furnace 4 (1.8 m) would nicely give room for two more (Fig. 4.2, p. 73). One furnace (no. 3) was cleared of debris and found to have traces of an ashy floor at about *86.00. It seems that plinths stood in the centre of the furnace floors to support a grid or container. In some of the furnaces, these were large pieces of fired clay which stood on end, set in mortar. Their fabric of well-levigated clay, without coarse temper, contrasts with the clay material of the mud bricks forming the body of the furnace, which contained small stones and occasionally small pieces of pottery (see p. 158).

Furnace area B

Remains of at least three furnaces, preserved to a much higher level than those in area A, were found at the north side of the excavation (Figs 4.26, 4.27 and 4.28, p. 130). Many of the features were the same as in the other area (Fig. 4.29) but one of them preserved what seemed to be the entrance mentioned by Agricola in his description of a cementation furnace, as well as the central brick plinth (Fig. 4.30) that could have functioned like Agricola's grid support, only in this case adapted for a single pot.[27] This furnace also gave an excellent idea of the canonical box shape. Of course, the upper parts had not survived, so one cannot say with certainty that the fire chamber and pot layer were precisely comparable to the one in Agricola's description (Fig. 3.1, p. 60).

The small size of the furnace chambers and the fired bricks in their centres, combined with the relatively large proportion of burnt but unvitrified coarse cooking-pot sherds, make it virtually certain that the vessel used for the cementation process was an ordinary cooking pot of the most common jug shape (Figs 4.31 and 5.3, p. 132). No evidence was found for lids (but see p. 203).

These sherds had the distinctive appearance of having been exposed to hot salt vapours for long periods. This impression was scientifically confirmed, suggesting that common salt was the active agent used in cementation. For

Fig. 4.24 *(above)* Furnace area A.

Fig. 4.25 *(facing page)* Furnaces 1 and 2 in area A: plan and oblique view.

this process, a relatively low temperature is required, below the melting point of salt (804 °C). Both the furnace walls and the parting sherds from the parting vessels have a burnt but unvitrified appearance that suggests relatively low temperatures for a pyrometallurgical process. It is known from the admonitions of Biringuccio that the work would be ruined if the cementation furnace was allowed to overheat: '... always taking care, however, not to apply a fire so strong that it melts the gold and materials together (for they would then not act; indeed your labor would be increased) but let it be only enough to keep the vessel always red.'[28]

Chronological indicators were few. In general, all one had were potsherds of the same kind as from the cupellation area to the south-east of the altar and the north-east dump, which is described below. There is, however, no doubt that all the remains (see p. 200) are part of one integrated activity, even if there were changes now and then in the preferred area for any particular process.

Another collapsed furnace, which had a stone surrounding wall, formed the core of a dump of brick

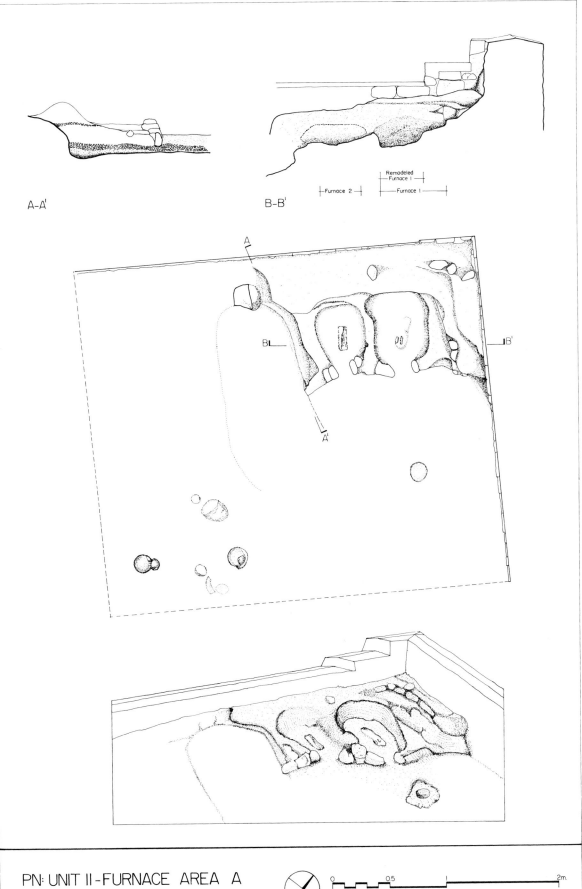

A-A' B-B'

Remodeled
Furnace 1

Furnace 2   Furnace 1

PN: UNIT II-FURNACE AREA A

0   0.5   1   2m.

CJK, PMS   1987

fragments and sherds below the general level of cupellation floor A, right by the original test trench (Fig. 4.32).
There is no reason to think that this was markedly earlier
than the other structures; it is more likely to have been
built partly below ground like some modern pottery kilns
in the neighbourhood. The quantity of debris and
potsherds from this area was quite remarkable.[29]

The two dumps

The finds from two heaps of discarded or stored material
provide us with the most spectacular evidence for reconstructing the process – actually several different but
related activities that were carried on in the hearths and
furnaces. Here, the remains from the various activities
offer a good opportunity to test the different snippets of
ancient testimony dealing with the processing of gold. In
addition, they provide material for analysis of the refractories, which has been so important for understanding the
main installations.

North-east dump

The north-east dump was an irregular circular area, about
2.5 m across, that lay roughly within the area W262.00–
265.00/S336.00–338.50 *86.30–*85.54 (Figs 4.2, 4.33 and
4.34). It consisted of burnt brick, a large quantity of
charcoal, many coarse potsherds, and fragments of an
exceedingly coarse flat pan which became known as 'bread
tray' (see p. 158 for a discussion of the unusual ceramic
fabric of the bread trays). The whole was formlessly mixed
with gravel and ash and it lay in a lens of gravel where no
distinct edge was apparent.[30] In addition, about 12 kg of
litharge in the form of substantial lumps was retrieved
from this area. This is an unusually large amount for a
small area when most of the other examples were scattered
over the workshop area. Most of the lumps preserved part
of the profile of the hearth in which they had been melted,
and many of them showed a step in the centre where the
silver cupellation button sat until it was prised out (Figs
4.35 and 4.36). A complete example was reconstituted
from several joining pieces (see Appendix 2, item 60,
p. 231). Not only was a complete profile of the hearth
bottom recovered but the whole shape of one particular
silver button could be made out. Pieces of bread tray,[31]
many with vitrified edges, were especially common (Fig.
4.37). The edge of a small piece (rarely the finished edge)
was often found heavily vitrified and changed in colour,
whereas the interior retained its original colour and
texture. The direction of the slumped vitrification shows

Fig. 4.26 *(above)* Furnace area B.

Fig. 4.27 *(facing page)* Furnace area B: plan and isometric view.

that the tiles were normally horizontal when heated (for
example, see Appendix 2, item 21, p. 226; Fig. 4.38). The
extreme temperature differential also showed on the brick-
like pieces, where the difference was even more exagger-
ated, for some pieces were found vitrified on the outside
edge but were almost raw clay on the other side. This
indicates extreme local heating from one direction rather
than continuous exposure. This evidence suggests that
fresh bricks were exposed to the intense heat and volatile
chemicals of the melting or cupellation hearths. Many
body sherds from cooking pots were found with gold
globules trapped in a thick purplish glaze of vitrified
ceramic (Fig. 4.39, p. 131).

Some pieces of finer pottery showed intense local heating on the surface, with graded colour bands at the
perimeter, and may have been used, with a blowpipe, as
holders or palettes for some activity distinct from the main
processes. The slagged bricks and bread trays are most
likely to have been built up around the cupellation or
melting hollows. This is discussed on pp. 208–9.

North-west dump

The north-west dump area was much less precisely defined
than that of the north-east dump (*c.* W276.00–280.00/
S330.00–332.00, *85.00) but the random mixture of items
without contexts gave a similar impression. Scattered
pieces of an East Greek oinochoe[32] came from this general
area (Fig. 4.40). Somewhat below it, in what may have been
a pit, a piece of a dinos in the Wild Goat style was found
(Fig. 4.41).[33] There is certainly some connection with the
goldworks, because two pieces of pottery with vitrification

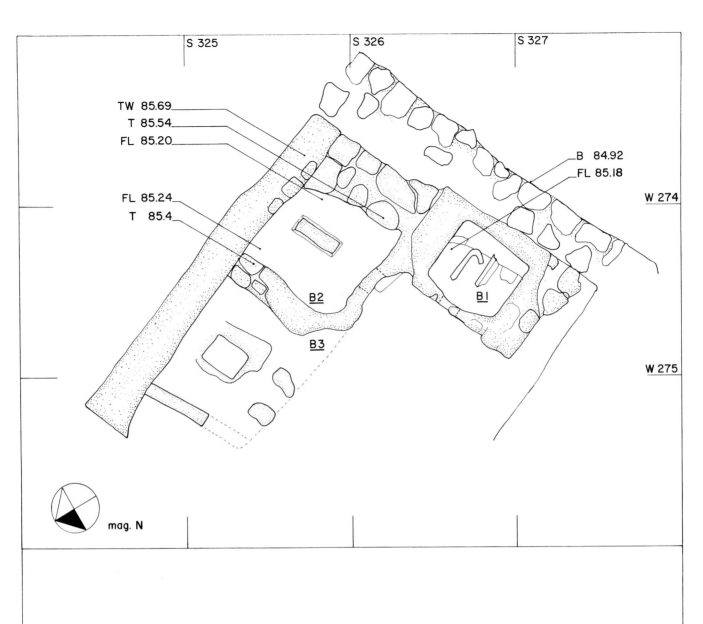

S 325 S 326 S 327

TW 85.69
T 85.54
FL 85.20

B 84.92
FL 85.18

W 274

FL 85.24
T 85.4

B2

B1

W 275

B3

mag. N

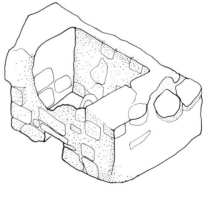

ISOMETRIC   FURNACE B1

scale   meters
0   .2   1

datum 100.00 = 138.38 a.s.l.   NHR

PACTOLUS NORTH

FURNACE AREA B   AUGUST 1969

Fig. 4.29 Furnace area B: furnace 1.

Fig. 4.30 Furnace area B: detail of central plinth of furnace 2.

or lead glazing are recorded from the same place.

A considerable amount of industrial debris also was found in Unit Ib[34] (see Fig. 4.2, p. 73). One area produced a large number of small pieces of gold foil.[35] In addition to the foil, there were many sherds that had either gold globules or a lead glaze adhering to them. The finds were comparable to the pieces found in the north-east dump except that many pieces of gold foil were found but hardly any lumps of litharge. Several pieces of tuyeres (bellows nozzles) were discovered both here and in the north-east dump (see pp. 92–4 and Appendix 2, p. 223).

The finds

The gold pieces

Over the course of two seasons, more than 300 pieces of gold were recovered. They fell into two main categories: hammered foil and round or irregular globules or dribbles (Fig. 4.42).[36] The latter are direct evidence for melting on the premises; the former had been prepared for the cementation process already mentioned on p. 11 and described in detail on pp. 202–8. In general, the foils were closer to the furnaces and the globules and dribbles closer to the cupels, but there is some overlapping (Fig. 4.2, p. 73). The foils outnumber the globules by an enormous margin unless one counts the innumerable spheres trapped in the vitrification on some of the sherds (Fig. 4.39, p. 131). Close inspection of these sherds revealed many 'craters' in the surface of the vitrified surface, which suggested that originally there had been many more gold globules of varying size on some of the pottery pieces.

The evidence for jewellery manufacture at Pactolus North is largely circumstantial, depending on the recovery there of half a mould[37] for preparing parts of earrings, pins and other standard pieces of jewellery; two finished pieces of gold (a circular pendant and an earring in the shape of a lamb, see Fig. 1.3, p. 17);[38] and large quantities of rock crystal. These are rough lumps, which are usually about 50 × 30 mm and 20 mm thick, chips or flakes and the occasional worked or finished piece like the little lion pendant (Fig. 4.43)[39] found in 1964 just to the south of the industrial area.

Notable examples of gold in different states included a natural granule (Fig. 4.44, p. 131, and see p. 148), globules or beads, dribbles and pieces of foil, as well as a relatively large cut lump and another much smaller one. One of the second variety was a spherical bead about 1 mm in diameter (Appendix 1, sample 3, p. 215). Dribbles produced the heaviest and most spectacular examples of the gold found (Appendix 1, sample 39, p. 219). How these escaped from their processors is hard to imagine.

One of the cut lumps is particularly noteworthy because of its regular shape: it appears to have been sliced from a substantial object because it is compact, very heavy (180 mg) and one edge is rounded (see Fig. 9.6a, b, p. 189, and Appendix 1, sample 21, p. 217). This is a unique piece among the gold retrieved. It has been suggested to be part of a coin of Croesus that had been divided for remelting.[40] But neither its supposed resemblance to part of a coin nor the supposed significance of its weight, which is exactly one sixtieth of the standard Lydian coin, are convincing. It would have been of the utmost importance had the piece

Fig. 4.31 Charged cementation pot.

Fig. 4.32 Collapsed furnace.

Fig. 4.33 Main goldworking area, looking south-west, with the Altar of Cybele and the north-east dump at centre foreground. Note burnt patch.

Fig. 4.34 North-east dump.

indeed been cut from a coin of Croesus, but how the Lydians could have achieved such accuracy in dismembering an irregular lump is hard to imagine. Analysis of the metal showed it to contain about 15% by weight of silver (see Table 7.1, p. 170), which is different from either the electrum or pure gold coin series.

There was a considerable variety among the fragments of gold foil in size, thickness and colour, as well as in surface texture (see p. 185; Figs 4.45 and 4.46, p. 131; and Figs 9.1 to 9.5, pp. 139–41). They ranged from small, roughly rectangular pieces crumpled like rice paper, whose typical thickness was 50–60 μm (see p. 185), with a maximum dimension of about 5 mm, to much larger pieces with a palpable thickness like heavy-duty aluminium foil, with maximum dimensions of about 50 × 30 mm. The colour varied from light to deep yellow and some pieces had a pinkish

Fig. 4.35 Reconstituted litharge cake: diameter *c*. 15 cm.

Fig. 4.36 Profile of the same litharge cake.

Fig. 4.37 Three joining pieces of bread tray vitrified on one edge (see Appendix 2, item 25, p. 227).

cast. We associated the colour shifts with varying fineness in the gold, and speculated that they were at different stages in the parting process. Some pieces seemed more porous than others, which would be a natural result of the cementation process. Presumably at the beginning, when there was a higher proportion of silver in the mix, more voids would be produced by the corrosive action of the salt.

Both the horizontal spread of the findspots for each piece of gold and the physical variation noted above indicated that we were dealing with the residue from several batches, and subsequent analysis at the British Museum bore this out convincingly (see pp. 145–51). The shapes of the foils varied considerably, from squarish to rectangular and occasionally strip-like, but it was not possible to find worked edges or traces of designs on the flat surfaces. Thus we did not see any evidence of the recycling of pieces of worked artefacts, such as headbands or garment plaques. Most likely, the irregular pieces are the result of chopping up sheets of foil ready for the first round in the cementation furnace and subsequent breakage of the most delicate examples.

In many ways, finds other than the gold pieces themselves give us more insight into the process. The cemented gold, for example, whether as dust or foils, had to be melted together to bring it into usable form. Once again, there do not seem to be any custom-made crucibles, so one supposes that for melting trays the Lydians utilised curved pieces from the cooking pots used as cementation pots, or melted the foils elsewhere.

However, this is not the whole story. A complication arises from the presence of several rims of broken cooking pots and the edges of the super-coarse ceramic trays that were heavily enough vitrified to have produced drips that showed which way up the trays were set (see p. 161). The conclusion is that many of the ceramic fragments were reused not just once but many times, with the result that what would have been visible traces from the original process have become obscured. To summarise (see also p. 202): the cementation process requires complete cooking pots and does not result in extensive vitrification, either on the inside or the outside. Another process requires the melting of gold in a cooking pot or on a piece of pot, probably prior to assay by touchstone. As evidenced by various ceramic fragments, other processes produced drips of thick lead glazing of the ceramic body but without the presence of visible gold inclusions. The lead-glazed coarseware sherds were probably used for the fire assay of silver by cupellation. The brick and bread-tray fragments seem likely to have been built up like walls around the hearths that were used for cupelling the cementation mix to

Fig. 4.38 Vitrified bread tray fused with a fragment of coarseware (see Appendix 2, item 21, p. 226).

Fig. 4.40 Fragments of an East Greek oinochoe from near the north-west dump: height 67 mm, width 32 mm, thickness 4 mm.

Fig. 4.41 Fragment of a Wild Goat style dinos: height 50 mm, width 49 mm, thickness 9 mm.

retrieve the silver. This would match the procedure described by Biringuccio for increasing the capacity of a bowl hearth.[41]

The first analyses of selected pieces of gold and of the vitrified materials and surrounding earth were carried out by spectrographic methods.[42] Some samples of the gold were also analysed by neutron activation analysis (NAA), including the use of streak samples.[43] Several other items in precious metal from Sardis were also tested using the streak method of NAA, as well as several Lydian coins from the Ashmolean Museum, Oxford (see Table 7.1, p. 170). Other analytical tests, such as those using the scanning electron microscope, were done more recently on equivalent samples of globules and foils (see Table 5.1, pp. 103–7, and pp. 146–51, 185–8). Additional analyses, performed on most of the Lydian gold and silver coins in the British Museum, are described in Tables 7.4 and 7.5, p. 171.

Why there should have been so many pieces (especially foils) of gold apparently just lying around remains a mystery, because there were no concentrations that might suggest a dishonest worker's stash, and the pieces were not all of the same purity, which one might expect if they were from a single batch.

Fig. 4.42 Assorted gold pieces from the refinery, including a sherd with gold globules.

Fig. 4.43 Rock-crystal lion pendant (see note 39).

Fig. 4.45 Gold foil (see Appendix 1, sample 7, p. 215).

The tuyeres

A tuyere is the tip of a bellows nozzle that is specifically intended to withstand the heat of a furnace. The typical piece of tuyere from the Sardis refinery is vitrified, flat-sided, and of tapering square cross-section (Figs 4.47 to 4.50, p. 131). Two types of tuyere are represented in the refinery: a straight configuration and a configuration in which the air passage has a right-angled bend at the end exposed to the fire, thereby directing the air blast vertically downward (Figs 4.51 and 4.52). The angled tuyere would have been specifically useful in the cupellation process, directing the blast onto the surface beneath, in order further to facilitate the oxidation of the metals. Both types have a square external profile. Some joined to make a whole side and one even produced a complete section. Many were scoriated and partly vitrified.

The size of the air hole varies but two main forms can be identified: one has a circular cross-section about 15 mm in diameter, and the other has an oval cross-section about 20 × 30 mm. This would allow for changing the volume and pressure of the air blast into or over the hearth. The bellows end of one unvitrified fragment is 66 mm wide, and the nozzle end of another, which was reconstructed from several smaller fragments, is 50 mm square (Fig. 4.49). One cannot be sure how much of the nozzle was inserted into the furnace, but it is estimated that the nozzles were not much more than 15 cm long. The tuyeres do not seem to have been very long, but rather like tips – quite different from the long snake-like tuyeres envisaged by Zimmer in his recreation of a bronze foundry.[44]

Some of the rear parts of the tuyeres are preserved, showing, in some instances, a flared opening (Fig. 4.53). A flare in the air channel at the bellows end indicates that an intermediate (wooden?) nozzle was inserted and then withdrawn slightly to allow the bag to be refilled with air without drawing hot gases into it. For a bellows bag without a valve, such a loose junction would ensure that flames and hot gases would not be sucked back into the bag on the reverse stroke. Technologically this is of interest because it shows that the Lydians were not yet familiar with the use of a flap valve in the body of the bellows (see Appendix 2, p. 223).

Tylecote suggested that square cross-sections were limited to tuyeres that lay on the ground to serve surface installations such as the cupellation hearths.[45] The straight tuyere would have been employed both in melting the metal and in the cupellation, directing air into the middle of the burning fuel. Two of the straight tuyeres (see Appendix 2, items 8 and 13, pp. 224 and 225) have globules of gold imbedded in their vitrified ends, showing that they were used in the melting of gold.

The tuyeres seem to have been moved around with the bellows, rather than fitted into large-scale permanent furnace installations that are often shown in reconstructions for heavier industrial processes. Certainly, no traces of such installations were found.

Representations in sculpture and vase painting, from the end of the sixth and early fifth centuries BC, suggest the use of a pair of bellows bags that were filled and exhausted alternately.[46] This can be seen on the relief sculpture of the north frieze of the Siphnian Treasury at Delphi, which shows the god Hephaistos working two bellows alternately (Fig. 4.54). The bellows has been installed with two nozzles leading to the forge.[47]

There is no direct evidence for the actual form of the bellows to which these tuyeres were attached. Bellows

Fig. 4.47 Scoriated tuyere (see Appendix 2, item 1, p. 223).

Fig. 4.49 Front view of the same tuyere.

Fig. 4.48 Flat bottom of the same tuyere: hardly any slag or scoriation.

Fig. 4.51 Angled tuyere (see Appendix 2, item 3, p. 223).

seem to have undergone considerable development through the first millennium BC in the Middle and Near East,[48] and pot bellows are attested in the Middle East from at least the third millennium BC.[49] It is likely that simple leather bag bellows were also in use from early times until well into the mid first millennium BC, as exemplified by the representation on the walls of the Siphnian Treasury. The familiar blacksmith's concertina bellows seem to have developed only from the mid first millennium BC,[50] very

probably in the eastern Mediterranean. The Sardis refinery would seem to be rather too early for the concertina's general use and thus, on balance, it seems most likely that the air was supplied from either pot or bag bellows through nozzles which rested in the ends of the surviving tuyeres. The cementation furnaces operated at relatively low temperatures and almost certainly did not require permanent bellows. The air supply through the opening in the front of a furnace would have been sufficient.

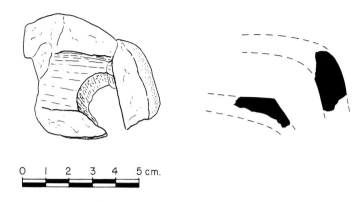

Fig. 4.52 Explanatory diagram and section of the angled tuyere shown in Fig. 4.51.

Fig. 4.54 Hephaistos with his bellows, from the north frieze of the Siphnian Treasury at Delphi.

Fig. 4.53 Flared end of tuyere (see Appendix 2, item 7, p. 224).

Fig. 4.55 Handle of Rhodian oinochoe with incised scale pattern.

The bellows type that best matches the finds is like that seen on a column-krater in the Caltanisetta Museum, Sicily, where a satyr is holding up a goatskin fitted out as bellows.[51] Unfortunately, it does not show what was done with the forelegs, but it gives a good hint of a 'valve' at the neck.

Chronology

Since it is not possible to put a precise date on any particular piece of Lydian pottery, one must rely on fragments of Greek pottery, imported from the various centres of manufacture, to provide the chronological limits for the

Even though it is agreed that Ionic cups are a generic product of Ionia, the best sequences to date have been produced by the finds from the sanctuary site of Taucheira, now Tükrah (Tocra), in Libya, and in Sicily at Megara Hyblaea near Syracuse (Siracusa). Generally speaking, the types are standard for the early part of the sixth century BC, which bolsters the contention that the installation is a product of the reign of Croesus. The most exact chronological indicator comes from the very few imported sherds found in this and associated areas.

Here is a selection of catalogued finds, many from W263.00–265.00/S336.00–343.00 *86.10–*86.00, an area which includes half of the north-east dump (see p. 86) and runs towards the cupellation area.

P68.81:7714   Rim of banded Ionic cup cf. Tocra type IX, 'second and third quarters of sixth century BC'. N-S test trench in cupellation floor.[53]
P68.100:7742   Rim of small Corinthian cup.
P68.101:7743   Rim of banded Ionic cup.
P68.102:7744   Shoulder of Rhodian juglet.
P68.103:7745   Small piece of Corinthian skyphos.
P68.109:7755   Rim of banded Ionic cup. Unit II NE corner *85.90–*85.75.
P68.119:7785   Rim of banded Ionic cup.
P68.120:7786   Shoulder sherd of Corinthian juglet with incised tongues, 'probably EC'.[54]

Some other pieces from the same general level and not far away include:

P69.16:7943   Upper part of handle of Rhodian Corinthianising jug with incised scale pattern (Fig. 4.55).
P69.49:7990   About two thirds of a reserve banded Ionic cup; Tocra type IX (Figs 4.56 and 4.57).[55]
P69.67:8018   Orangey plainware 'coarse bowl' very similar to several from the Lydian kitchen covered by the debris from the fortification wall destroyed by Cyrus the Great of Persia in 547 BC.[56]

If the indications from these sherds are taken with those from the sherds in the fill below the floor of the altar and the lions from the altar itself, a date no later than the middle of the sixth century BC must be inferred for the industrial activity. The thinness of the cupellation layer and the relatively small quantity of waste products indicate a short period of use.

Fig. 4.56 Ionic cup, Tocra type IX.

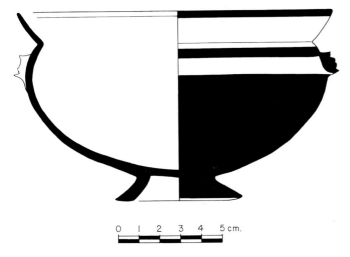

Fig. 4.57 Profile of same Ionic cup.

industrial process. Of course, the nature of the process is not conducive to having large quantities of fine Greek pottery in the area. The best category of fine pots for dating is the series of Ionic cups that have a distinctive but changing profile and are usually decorated with reserved bands and black glaze. Occasionally, they have purple or white bands over the black. These have been put in chronological order by several scholars to provide an estimate of when any particular type was most popular.[52] Given that these sherds are usually small pieces and not part of the actual equipment of the workmen, one has to be cautious but also bold enough to use them appropriately.

A very large quantity of pottery was taken from an area at the north side of cupellation area A that gave the impression of having been another dump with more concentrated furnace material like that shown in the section drawing (see Fig. 4.22, p. 82, and p. 86). For example, on one day eleven sherd boxes (41 × 23 × 14 cm) were filled: nine with sherds and two with brick pieces. Only one whole brick was found, about 40 × 23 × 8 cm. In this small area, which seems to be a partially stone-lined pit or the remains of a semi-underground structure,[57] the painted pottery was quite uniform, consisting almost wholly of Lydian geometric ware. The most prevalent styles for closed shapes were those associated with waveline hydriai, concentric pendent hooks and 'red bichrome' (the last two categories often overlap). For 'plates and dishes', the most frequent style was black on red concentric bands; infrequently there were bichrome dishes with a narrow wavy band on the interior of the rim. Skyphoi were rare; their styles were about evenly divided between 'reserved band' and streaky. This kind of ensemble indicates a rough range of date from the end of the seventh century BC to the middle of the sixth.

When one combines the evidence from the style of the lions, the pottery from inside the altar and the pottery from the cupellation areas, one concludes that the main period of the refining activity should be placed in the second quarter of the sixth century BC. This corresponds closely to the reign of Croesus, who is usually credited with initiating the use of separate gold and silver coins. This requires the capacity to separate one metal from the other, if gold alloy is the primary source. The evidence from the gold-working area at Sardis, then, serves to confirm the literary and historical record.

Technical summary

It is proposed that the facilities discovered were for the purpose of purifying the various grades of gold dust as well as metallic scrap from electrum coins or old jewellery (see p. 200). It is not in fact necessary for the gold dust to have been melted before being submitted to the cementation process. It is perfectly feasible to put the gold dust in a pot with salt and, probably, brick dust as well, and then place the pot in a furnace. It requires a different technique (namely, washing and sieving) to capture the spongy gold grains after treatment instead of pieces of foil. This raises the question: What were all the little hearths for? A probable answer is that they served three purposes:

- The original melting of material for purification
- The remelting of pieces in various stages, principally for assay.
- The retrieving of the silver that was formerly part of the gold alloy. The worker would have heated the used cementation mixture with lead or litharge (lead oxide) and allowed it to take up the silver (now silver chloride), which eventually concentrates as a little 'button' on the surface of the molten litharge. The silver can then be picked out after cooling. Several of the litharge cakes show small circular indentations in their top surfaces, where such a button might have been located (see Fig. 4.35, p. 90), and they have a profile that exactly fits the shape of the bottoms of the bowl hearths. The size of the depressions suggests that the average button weighed about 200 g (see p. 209), which is comparable to the weight of about five Greek dekadrachms. The size, also, is not dissimilar to the irregular lumps of silver used as coins from the Derrones tribe in Macedonia.

The litharge cakes often have a remarkably high silver content – up to 13% by weight (see Table 6.2, p. 160). Apparently, the Lydians were in no hurry to turn that back into precious metal. The large number of pieces of gold foil left lying around serves as a vivid and concrete reminder of their careless attitude towards the metallic treasure.

Notes

Note the abbreviations for the following three frequently cited publications:

Annual of the American Schools of Oriental Research AASOR
Bulletin of the American Schools of Oriental Research BASOR
Sardis from Prehistoric to Roman Times (Hanfmann 1983) SPRT

The prefix * with a decimal number denotes the level in relation to a datum of 100.

1 Although the first gold pieces turned up on 19 July 1968, Andrew Ramage first proposed the idea of a gold refinery in the field book on 6 July: 'Growing probability that these [circles] were for refining gold, as suggested by me yesterday at tea!'

2 The remains, which were inspected in 1968, proved to be identical to the cupels. They had not lasted well, however, and were discarded. Some of the new examples were set in plaster and have been used as exhibits in Ankara and in Manisa, which is the capital of the province.

3 The coordinates were W265.70–268.50/S337.00–339.90 approximately. The level of the top was at *87.99, and of the bottom at *86.24. It had been freed from two thick layers of the surrounding gravel. This gravel indicated that the area had been inundated by two successive floods, apparently not very far apart in time, because the layer of earth that had built up between them was extremely thin.

4 It measured 1.34 × 0.44 × 0.33 m.

5 The height of stage one was 1.15 m; that of stage two 0.6 m.

6 At about *87.00.

7 The inscription, in a wall on the east bank of the River Pactolus, was revealed during the winter of 1973/74 (IN74.1, previously published in *AASOR* 43 (1978), pp. 61–5, Figs 2–5, and *SPRT*, p. 104, Fig. 166). See Robert (1975, pp. 306–30). The inscription begins:

In the thirty-ninth year of the reign of Artaxerxes, Droaphernes, governor of Lydia and son of Barakes, dedicated the statue to Zeus the Law-giver.

It instructs the priestly servants of the god, who have the right of entering the chamber of the mysteries and of crowning the god, not to participate in the mysteries of Sabazios or Angdistis or Ma with those bringing burnt offerings. It also instructs the priest Dorates to keep away from the mysteries.

For a slightly different interpretation, see Sokolowski (1979, pp. 65–9).

8 There were eventually two and a half lions: S67.16:7354, S67.32:7550, S67.33:7559; Hanfmann and Ramage (1978, pp. 66–7, nos 27, 28, 29). S67.32: 7550 was previously published in *BASOR* 191 (1968), pp. 11–12, Figs 10 and 11; *BASOR* 215 (1974), p. 44, no. 25; *SPRT*, p. 37, Fig. 52.

9 Gabelmann (1965).

10 Inventory: P68.140:7831 = IN68.18 (previously published in *BASOR* 199 (1970), p. 22, Fig. 12. See Gusmani (1975, p. 28, no. A II 5; and 1969).

11 The second test trench was *c.* W273.25–277.50/S343.25–344.00 *86.23–*85.29.

12 Most of the finds were fragments of waveline hydriai, with one or two pieces using black on red concentric bands, and 'red bichrome', but only one piece of grey ware. This combination is a general indication of an early sixth-century BC date.

13 Inventory: P68.60:7680 and P67.143:7581. Previously published in *BASOR* 191 (1968), p. 12, Fig. 12.

14 Hanfmann and Ramage (1978, pp. 66–7, Figs 107–17).

15 Breglia (1974), followed now by Price (1984) and Vickers (1985) but rebutted by Root (1988).

16 At *86.30.

17 Its situation was *c.* W269.50–275.00/S331.00–340.00; its floor level was at *85.40; its exact dimensions were 9.3 × 4.8 m ext, and 8.3 × 3.9 m int.

18 The bronze-casting arrangements in the Agora at Athens at a comparable time period required considerable heat to melt the bronze. Their installation seems to have been in the open air without ancillary buildings (see Mattusch 1977). In Hellenistic Greece, for example at Demetrias in Thessaly, bronze casting was carried on within what had been an open peristyle court and adjacent buildings (see Zimmer 1990, pp. 103–15, Figs 41 and 47).

19 When Sidney Goldstein, the assistant conservator, and Andrew Ramage described some of the situation and finds to Stone, his answer was short but telling: 'It sounds like cupellation to me.' Upon being pressed for further details, because this was to them an unfamiliar term, he explained that it had to do with the process of refining precious metals. He explained further that the process had been fully described in the Renaissance by Biringuccio (see pp. 57–8) and offered a summary description whose salient points closely resembled what had been already found. The connection was even more convincing because Stone's remarkable memory of Biringuccio's description offered some circumstantial details of things that had been found but not mentioned.

20 There had been a hint, rather out of context, when Halis Aydıntaşbas, who was in charge of the synagogue restoration, noticed a small piece of gold foil on the surface near the eventual position of the test trench. Approximate coordinates of trench: W263.50–265.50/S343.20–345.00.

21 Williams and Ogden (1994, pp. 17–19); Higgins (1980, p. 11). Although Waldbaum (1983) describes the lamb earring (Fig. 1.3, p. 17) as having been cast, it may well have been formed from gold sheet.

22 Nriagu (1983, esp. pp. 92–8).

23 W271.50–273.00/S347.50–349.50 *86.20.

24 This was the opinion of Richard Stone. These figures were broadly confirmed by the subsequent scientific investigations discussed on pp. 159–61 and 205.

25 This is suggested by the scientific examination and would seem to follow from Agricola's and Biringuccio's instructions, where maintaining red heat for some time is a vital requirement. *De Re Metallica*, Book X (Hoover and Hoover 1912, p. 455); *Pirotechnia*, Book IV, 7 (Smith and Gnudi 1942, p. 204).

26 The dimensions, in metres, are as follows: existing length 3.4 ; preserved height 0.4; preserved depth 0.3; thickness of rear wall 0.4; width of dividing walls *c.* 0.15; internal widths of bays (from N) 0.58, 0.55, 0.70, 0.64.

27 *De Re Metallica*, Book X (Hoover and Hoover 1912, p. 455).

28 *Pirotechnia*, Book IV, 7 (Smith and Gnudi 1942, p. 204). See also pp. 57–8.

29 On 9 August 1968, Sidney Goldstein wrote in the field book: 'We are finding much charcoal mixed with mud brick and incredible amounts of pottery. In 3½ hours, 10 boxes have been removed from

less than one square meter.' (The dimensions of these boxes are a standard 41 × 23 × 14 cm.)

30 Coordinates of lens: W261.50–264.00/S335.50–337.00 *86.10–c. *85.70.

31 This was a very coarse, flat ceramic tray about 0.5 × 0.5 m with raised edges on three sides. Supposedly it was originally used for baking, but once broken served many unrelated purposes. The ceramic composition (see p. 158) is quite specific to this shape. It is highly unusual to find it used for any other purpose and they are without exception thick storage jars or kitchen accessories.

32 Inventory: P67.135:7570.

33 Inventory: P69.30:7962.

34 Coordinates: W276.00–279.00/S330.00–334.00.

35 'As Huseyin Aktaş was cleaning hard stony surface, I [Ramage] noticed a thin piece of foil, swooped upon it, showed it around, took it to çardak [expedition hut], had hardly got there when great shouts of elation: small pieces of gold foil turning up by the dozen, in ground, earth pile and wheelbarrow.'

36 Some 98 individual pieces were collected from 9 July to 15 August 1968. For 1969, the tally is less specific but a figure of at least 200 can be reconstructed from the list.

37 Inventory: S69.8:7982, Waldbaum 949. Previously published in *BASOR* 199 (1970), pp. 25–6, Fig. 15.

38 Inventory for pendant: J67.4:7530, Waldbaum 884. Previously published in *BASOR* 191 (1968), p. 13. Inventory for lamb: J67.3: 7400, Waldbaum 719. Previously published in *BASOR* 191 (1968), pp. 13, 14, Fig. 13; *SPRT*, p. 40, Fig. 62.

39 This little lion (length 22 mm) is drilled transversely through the body as if to be strung as a pendant or in a necklace. The front legs are carved separately but the back legs are done as one piece. The body is stocky with a curling tail. Details of the stomach are indicated with incised lines, as are the eyes.
Inventory: J64.2:6339. Previously published in *BASOR* 177 (1965), p. 7, Fig. 7.
Findspot: W246.00/S354.00 *88.25. Length 22 mm; height 12 mm; depth 6 mm.

40 Goldstein, in Hanfmann (1983, p. 40).

41 *Pirotechnia*, Book VII, 3A (Smith and Gnudi 1942, p. 290). See also p. 211, note 41.

42 The spectrographic analysis was undertaken at Maden Tetkik ve Arama Enstitüsü (Turkish Minerals Research Institute at Ankara), under the direction of Dr Sadrettin Alpan.

43 The first neutron activation analysis (NAA) was carried out at the Istanbul nuclear reactor. In 1969, a number of samples from the gold pieces were tested by NAA by Professor Adon Gordus at the University of Michigan, Ann Arbor, on the basis of streaks taken in the field by Sidney Goldstein. (See Table 7.1, p. 170.)

44 Zimmer (1990, Plate 16: Pheidias' workshop).

45 Tylecote (1981).

46 See Siphnian Treasury, Zimmer (1990, Plate 12). For a fragment in red figure, see Zimmer (1990, Plate 13). For a Douris cup in the Bibliothèque Nationale, Paris, see Arias and Hirmer (1962, Fig. 148) with erroneous explanation. They say that they are fur-lined bellows with the pumping ends on the ground, whereas they should have said they are the blowing ends.

47 Illustrated in Healy (1978, Plate 50), and in Weisgerber and Roden (1986).

48 Craddock (1995, pp. 180–2); Weisgerber and Roden (1985 and 1986).

49 Davey (1988).

50 Weisgerber and Roden (1985).

51 Oddy and Swaddling (1985, Fig. 2).

52 On Ionic cups, see Villard and Vallet (1955, pp. 14–31); Boardman and Hayes (1966, pp. 111–34); Vallet (1978). Waldbaum and Magness (1997, pp. 27–9) clearly show that the type was already being exported at the end of the seventh century.

53 Hayes, in Boardman and Hayes (1966, p. 46).

54 Schaeffer, in Schaeffer et al. (1997, p. 44, Cor 92). Actual findspot: W265.20–266.80/S342.00–343.50 *86.00–*85.90.

55 Actual findspot: W282.80/S322.00 *85.40.

56 Cahill (1990, p. 149).

57 Extent: W263.00–265.50/S342.00–345.00 *85.90–*85.35. Towards the south, the stones are far fewer and form no regular pattern as they do at the north.

CHAPTER 5

Scanning Electron Microscopy of the Refractory Remains and the Gold

N.D. Meeks

Introduction

The scientific study of the refractory remains from Sardis is of fundamental importance in the reconstruction of the processes carried out at the precious metal refinery. A variety of fragments of refractory materials and fragments of gold were sent to the British Museum for chemical and microstructural characterisation. The refractories comprised vitrified coarseware sherds, furnace bricks, tuyeres and flat fragments of the so-called bread trays. The scientific study involved x-ray fluorescence analysis (XRF), thin-section petrography, scanning electron microscopy (SEM) and energy-dispersive x-ray microanalysis (EDX).

The sherds and gold fragments from the excavation at Sardis hold the key to the understanding of the metallurgical processes in operation at the site. The sherds retain the evidence of the chemical and thermal histories of activities associated with the gold-refining process, both on their surfaces and within their microstructures. The gold fragments and grains show the effects and stages of the refining process in their composition and microstructures.

The main subject of this chapter is the detailed SEM observation and analysis of the surfaces and polished sections of the coarseware sherds and furnace bricks, together with some of the associated gold particles and a small selection of foils and granules. Other related chapters are by Geçkinli and her colleagues on the examination of gold foils and the description of parting experiements, which include SEM work done at the British Museum (pp. 184–99), and by Middleton, Hook and Humphrey on the petrographic examination of the refractory remains (pp. 157–68). Some of these areas of examination relate directly to results in this chapter and their integration sets the overall results of the British Museum SEM investiga-

tions within the context of the whole refinery operation.

As part of the overall investigation, a number of the Lydian coins in the British Museum have been analysed by XRF (pp. 169–74).[1] These analytical results can be compared with those of some of the gold foils and the gold on the refractory sherds. The latter have implications for deducing some of the types of alloy processed at Sardis.

EXCAVATED MATERIALS AND METHODS OF EXAMINATION

Excavated material

Refractory sherds and furnace bricks

Excavations at Pactolus North have unearthed parting furnaces, cupellation hearths and spoil deposits of refractory sherds which are associated with the ancient gold-refining and silver-recovery operations at Sardis (pp. 81–96 and Appendix 2, pp. 221–32). The various enigmatic vitrified coarseware sherds that were found were not readily recognisable as the essential refractory parting vessels and crucibles normally used for handling molten purified gold. That purpose-made crucibles seem to be absent was surprising, considering the large amount of gold that is believed to have been processed at the site. Instead, a large number of the receptacles appear to be reused broken coarseware sherds, many having discoloured and vitrified or bloated surfaces. Some sherds had clear evidence of rims, handles and bases that, apart from the bloating and vitrification, do not look like conventional metalworking refractories, but appeared to be standard Lydian coarseware cooking pots. Large numbers of these overheated sherds were recorded, implying the existence of large-scale metallurgical processing.

Thirty-six representative sherds that showed good evidence of either unusual surface discolouration or surface vitrification and bloating were selected for scientific study. Approximately half of the selected sherds had visible evidence of tiny gold particles adhering to their surfaces. Analysis of these gold particles is an important part of the project, as they show the range of alloys handled at Sardis and show evidence of stages in the purification process.

Following initial optical microscopy, these refractories were divided into four main categories. The analytical results are presented below and discussed in the same sequence and cateogories as follows:

Refractory sherds with gold globules

Approximately half of the sherds examined optically have gold globules present within the vitrified surfaces, and in some cases these were clearly visible to the unaided eye (Fig. 5.1, p. 132). The majority of these sherds had bloated, vitrified or bubbly areas only on their concave surfaces, e.g. sherd 45680Q (Fig. 5.1). These areas were often dark in colour. The relationship of bloating to edge fracture indicates that these sherds were parts of larger sherds or possibly of dish-like containers when used. They appear to have been used for melting small quantities of both the impure gold and purified gold, probably prior to assay. The broken sherds were presumably assigned to the spoil heap because they were no longer of sufficient size to be able to melt even a small amount of gold.

Refractory sherds with lead-rich surfaces and no gold globules

Most of the remaining sherds examined optically also have surfaces that are vitrified and bloated to various degrees and of a range of colours, but under the binocular microscope no gold globules were found. It was not until surface analyses were carried out that it was found that these sherds appear to have been used for a different purpose. The surfaces were rich in lead, and contained silver to varying degrees. The sherds thus seem to be associated with cupellation, once again on very small quantities of metal, probably for assay purposes rather than the recovery of silver.

Parting-vessel sherds

A few sherds have a purple or pink coloration but without any bloating and are thought to be from parting vessels. For example, sherd 44401Z (Fig. 5.2, p. 132) is an important sherd that provides the strongest evidence of the parting process and shows that alluvial gold grains were used

directly in the process. Only three such sherds were identified as being from parting vessels although many more similar sherds were recovered from the north-east dump in the excavations but were not retained for study. It is likely that they were awaiting processing to recover the silver they contained. The distinctive purple coloration is mainly due to the presence of silver chloride. The pink colour, which is also found on the furnace bricks, is due to the leaching of the iron salts in the clays, made mobile by the action of the permeating salt vapours, forming the volatile ferric chloride, leaving the clay a pale pink colour.[2]

Furnace bricks

Excavation has exposed a number of incomplete rectangular furnaces varying between 0.6 and 1.2 m across and each having a large, low air inlet and a raised central floor block on which the parting vessel stood (Fig. 5.3, p. 132). The furnaces were made of mud bricks having a friable texture (Fig. 5.4, p. 133). The furnace structures were a pink colour, due to the mobility of the iron chlorides.[3]

Gold foils and gold dust

Gold foils and other fragments associated with the purification process were found at the refinery site. The following samples of gold were examined at the British Museum in parallel with other samples examined at the Istanbul Technical University (see pp. 184–7).

Gold fragments from the Archaeological Museum at Manisa

The majority of excavated gold foils are thin (approximately 60–100 μm) and the pieces are quite small. They appear to have been made specifically to maximise the surface area of the gold before being put into the parting vessel to allow the chemistry of parting to occur efficiently. Foils are mentioned in similar processes in the *Arthaśāstra* (see p. 33) and in the more recent past (pp. 57–65). Five of these gold fragments from the excavations at the PN refinery, now in the Archaeological Museum at Manisa, were mounted in resin and polished in section for examination at the British Museum, and these results show various stages of the refining process for the gold.

Gold grains

In 1968, at the Cekmece Nuclear Research Center, Istanbul, natural grains of alluvial gold panned from the Pactolus, near Sardis, were analysed by neutron activation. They were found to have a composition in the range of 74–86% gold and 17–24% silver. The grains examined at the

British Museum purport to be from the same batch, according to their label, but must in fact be from a different batch. The grains are of pure gold showing they had been treated directly without having been made into foils, the grains having a naturally large surface area to volume which allows efficient purification. Similar direct processing of gold grains was noted by Agatharchides at the gold mines in the Eastern Desert of Egypt (see p. 34).

Methods used to examine the materials

The following microscopy techniques have been used in the study of the refractory remains and the gold particles.

Low-power binocular microscopy This was used as the first stage of examination for the visual observation of the colour, texture, vitrification and gold droplet distribution of the sherds. Colour photographs of the majority of these sherds can be found on pp. 132–8.

Petrographic microscopy Polished thin sections of the key ceramic materials were examined using petrographic microscopy. The full details of the examination of these samples are given on pp. 157–68. The same thin-section samples were also examined in the SEM for fabric and vitrification studies, microanalysis and observation of the impregnation of silver chloride reaction products from the parting process within sherds. (Some particular SEM results of thin-section examination are given here.)

Scanning electron microscopy The major part of this chapter concerns the investigations carried out using SEM and EDX on the refractory sherds and gold fragments. The SEM used was a JEOL JSM 840 and the EDX analyser was a Link 860/500 having a SiLi detector with a beryllium window. Later in the project, an Oxford Instruments ISIS 200 EDX system with a germanium detector and light element (SATW) window was used. This latter system was also used for digital imaging. The combination of SEM/EDX allowed the full characterisation of the surfaces and polished sections of sherds to be made both morphologically and analytically for the evidence of heating and vitrification of the ceramic fabrics, and for the presence and distribution of chemical reaction products of the parting processes. Location and analysis of gold particles of a wide range of sizes and distributions on the sherds were made in situ. Polished sections were made through gold globules removed from two sherds for quantitative analysis. Microanalysis and photomicrography were carried out on gold fragments from excavation.

Procedures used to examine the samples by SEM

Imaging
In order to observe the surfaces of the non-conducting ceramic sherds in the SEM, without the need for the normal vacuum coating with carbon usually applied to such materials, a scintillator backscattered electron detector was used.[4] This detector has two principal advantages over conventional SEM imaging:

- Stable topographic images are obtained without 'charging' in the electron beam, which allowed examination of surface structures and clear photographic records to be made.

- The detector has the characteristic of increased image brightness where there are areas of high mean atomic number. That is, it enhances contrast due to variations of composition. This makes the gold particles appear bright on the surface of the sample and hence enables even tiny gold particles down to a few micrometres to be located on the surfaces of the sherds. In this way, the presence or absence of gold particles on each sherd was clearly determined.

Analytical techniques
The technique of analysis used in the SEM was EDX. The system was run with the SEM at 25 kV accelerating voltage and the analyser was standardised using both pure element and mineral standards. For each analysis, the following 22 elements were sought: Na, Mg, Al, Si, P, S, Cl, K, Ca, Ti, Mn, Ni, Fe, Zn, Cu, As, Sn, Sb, Ag, Au, Hg, Pb. In the descriptions of the sherds, only the elements that were found are listed, and in those cases where certain elements occurred only on some analyses on a particular sherd these are marked with an asterisk (*). The analyser automatically gave statistical data relating to the presence or absence of elements, within the detection limits. For quantitative analysis, precise geometry and polished surfaces are normally needed for EDX. Because of the variety of rough, textured surfaces on the sherds, these strict requirements were not met. Thus, all the analyses of the sherd surfaces are regarded as semiquantitative. However, the presence and absence of elements within the detection limits of the analyser are valid. Also, the relative concentration values between analyses on different areas of the samples and between different samples are generally valid, even if the quantitative accuracy is not high. The samples that were mounted in resin and polished to flat surfaces did fulfil the strict geometrical requirements for quantitative analysis.

At the time of examination it was not possible to analyse for oxygen, which is present in all ceramics, and hence is not included in Table 5.1

Bulk detection limits vary among elements but range approximately from 0.15% for Cu to 0.5% for Pb. However, area analyses frequently did not show the presence of gold globules because of their small size compared with the large areas analysed and hence their low 'bulk' concentration. However, as mentioned above, even the tiniest gold globules and particles (down to a few micrometres) were easily found using the backscattered-electron detector and the globules were then analysed in situ.

Surface analyses of sherds
SEM analyses of the sherd surfaces were made on areas considered to be typical in order to get an average general composition of the surface. Some analyses were also made on particularly vitreous areas which appeared significant for the understanding of the ancient thermal processes.

Surface analyses of gold globules
The gold globules were analysed in situ on their as-received surfaces. No attempt was made to clean them as they were too small.

Surface enrichment of gold
A few larger gold globules from two sherds were removed and mounted in resin then polished to provide cross-sections for analysis (see Table 5.2). They showed that only a little surface enrichment had occurred due to burial. Typically, within 2 μm of the surface the loss of silver was less than 1% from an initial composition of about 18% silver in the gold. The loss of copper was less easy to determine as the object had only 0.8% in the body metal and the results were variable. Within 2 μm of the surface some analyses showed loss of all the (0.8%) copper and in others very little was lost. The results indicated that generally surface analyses of the gold particles on sherds would probably be within 1–2% of core values. Hence the composition of the surfaces of the gold from Sardis appears not to have been significantly affected by burial. This is particularly important when examining gold foils and fragments that have been through the parting process, in that the surface composition of these gold pieces represents the effects of the parting process and not the effects of burial.

Resin-mounted samples
Some key ceramic samples and some of the larger gold globules were mounted in resin and polished to flat surfaces by standard procedures for the thin-section and metallurgical sample preparation. This allowed the full quantitative microstructural and microanalytical characterisation of these samples to be made.

Summary
By the above methods of microscopy and microanalysis, the sherds and associated gold were fully characterised both physically and chemically, and their thermal histories determined. Interpretation of the data allows reconstruction of aspects of the gold refining process.

EXAMINATION OF REFRACTORY SHERDS AND FURNACE BRICKS

Descriptions of sherds

The following descriptions of the sherds can be compared with the images (Figs 5.5 to 5.49). Many of the sherds clearly have a red core, exposed by fractures after their final use in antiquity. Some fractures may be relatively recent. With other sherds the core colour is indistinct and pale which may be due to soil staining during burial. The sherds have been washed since excavation.

Table 5.1 presents the analyses of the surfaces of the sherds listed below and, where present, surface analyses of the gold globules. Where there were many globules, only a representative number were analysed. The SEM examination and analyses were non-destructive and hence all of the analysed gold globules and particles remain on the sherds (except for the few removed for resin mounting).

The next four subsections deal respectively with sherds with gold globules, sherds with lead-rich surfaces but no globules, parting-vessel sherds, and furnace bricks. The numbers with a letter suffix used here to identify the sherds and other specimens were assigned by the British Museum's Department of Scientific Research.

Characterisation of refractory sherds with gold globules

Sherd 44396P (Fig. 5.5, p. 133)
This is an irregular oblong-shaped sherd. The fabric is generally red in colour throughout the thickness of the sherd, and it appears to be low-fired. The concave surface of the sherd has an undulating, generally blackened, vitreous, bloated appearance with fine gas bubbles in the vitrified surface. Sandy-coloured bloating is also present. White, probably calcareous, grains and deposits are fused in the surface. These may have originally been associated with the fabric of the sherd, as an area of exposed red core material has similar white grains within it, or they may be

Table 5.1 Surface SEM–EDX analysis of gold globules and sherds in elemental weight percentage*

Sample	Au%	Ag%	Cu%	Fe%	Pb%	Na%	Mg%	Al%	Si%	Cl%	K%	Ca%	Ti%	Mn%	P%
44396P															
gold globule An1*†	100.0														
gold globule An3	100.0														
gold globule An4	100.0														
vitrified surface An2				11.6		4.7	3.9	20.5	43.0		4.3	8.5	0.6		1.5
44397Y															
silver chloride An4		74.5		0.4			1.0	7.4	0.6	14.8					
vitrified surface An1		12.1	1.0	10.4	34.6		1.3	7.6	20.5	1.7	2.5	7.6			
vitrified surface An3		1.6	5.2	6.7	34.9		0.9	6.7	24.1		2.5	9.8	0.7	5.3	
44398W															
gold globule An1	100.0														
gold globule An3	99.3		0.7												
gold globule An4	99.0		1.0												
gold globule An5	100.0														
gold globule An6	99.5		0.5												
gold globule An7	100.0														
gold globule An8	100.0														
gold globule An9	100.0														
gold globule An10	100.0														
gold globule An11	100.0														
gold globule An12	100.0														
gold globule An13	100.0														
vitrified surface An14				13.6		1.5	2.5	12.5	34.0		6.2	24.8	0.8	1.5	
vitrified surface An15				18.6			2.9	12.4	28.2		4.8	28.1	1.0	0.6	
44399U															
gold globule An1	98.1	1.9													
gold globule An3	97.6	2.4													
gold globule An4	98.1	1.9													
gold globule An5	99.1	0.9													
gold globule An6	98.8	1.2													
gold globule An7	100.0														
gold globule An8	99.2	0.8													
gold globule An9	99.1	0.9													
vitrified surface An2		1.0	0.6	16.7			3.7	8.1	32.8	0.4	3.3	24.3	1.6	1.6	2.1
sherd section: centre An1		1.3		9.0	**S**	6.1	1.0	12.5	61.8	1.0	3.5	2.2	1.0	0.3	
sherd section: new area An2		1.7		8.7	0.4	6.0	0.8	11.3	66.7	0.7	2.6	0.6	0.7		
44400Q															
sherd A															
non-metallic inclusion An2	5.6	54.8			19.3		5.8	0.8	7.2			4.2			2.1
non-metallic inclusion An3	4.2	1.7			66.8			1.0	2.5			12.7			7.5

continued on p. 104

* For the gold globules, only significant metallic elements are listed.
  Surface soil elements on the gold are ignored as insignificant contamination.
† An = Analysis. The suffixed number is its order in a sequence of analyses.

Sample	Au%	Ag%	Cu%	Fe%	Pb%	Na%	Mg%	Al%	Si%	Cl%	K%	Ca%	Ti%	Mn%	P%
44400Q															
non-metllic inclusion An4	4.4	2.0			67.5				0.6	3.0		13.2			7.5
non-metallic inclusion An5	4.8	16.9			60.9			2.2	1.2	3.0		10.7			
vitrified surface An1				3.4	13.2	3.7	3.5	12.3	38.0	0.5	6.9	13.2	0.4		3.7
44401Z															
parting vessel															
gold particle An1	74.1	25.9													
gold particle An2	77.5	22.5													
gold particle An3	79.4	20.6													
gold particle An4	98.5	1.5													
gold particle An5	93.5	5.6	0.9												
gold particle An6	58.2	41.8													
gold particle An7	73.1	26.9													
sherd surface An8		11.9		13.3	4.3	3.2	2.4	9.4	37.4	1.7	3.1	7.6	0.8		1.3
cross-section: surface layer An1		55.3		2.3	1.5 S	4.9		11.0	16.2	5.3	0.7	1.3	0.3		0.3
cross-section: below surface An2		13.4		2.1	0.3	12.4		16.8	47.2	2.5	1.5	1.7	1.0		0.7
cross-section: centre An3		2.7		8.0		7.9		12.5	54.6	4.1	2.5	2.5	1.0		
cross-section: inner edge An4		2.4		8.0		7.4		13.8	57.9	3.7	3.7	2.0	0.9		
44402X															
gold globule sherd B An1	93.7	6.3													
gold globule sherd B An2	94.9	5.1													
gold globule sherd B An3	95.5	4.5													
gold globule sherd B An4	94.4	5.6													
gold globule sherd C An6	93.2	6.8													
vitrified surface sherd B An5				3.5		3.4	1.4	27.3	50.1		10.5	2.2	0.9		
vitrified surface sherd C An7				7.6		3.3	3.7	21.9	45.2		4.5	7.6	0.8		1.5
45670U															
gold globule An1	74.3	24.7	0.9												
gold globule An3	100.0														
gold globule An4	100.0														
gold globule An5	100.0														
gold globule An7	100.0														
gold globule An8	100.0														
gold globule An9	100.0														
gold globule An10	100.0														
gold globule An12	100.0														
gold globule An13	100.0														
gold globule An14	100.0														
gold globule An15	100.0														
vitrified surface An2				10.3		2.3	5.3	14.8	41.3	0.5	4.9	13.9	0.8		3.4
45671S															
parting vessel															
sherd surface An1		0.9	0.6	15.8	0.5 S	2.2	4.5	9.9	38.7	0.2	5.2	18.3	1.3	0.5	1.3
sherd surface An2		0.9	0.8	16.4	0.6 S	1.9	4.8	9.4	36.0	0.5	5.2	18.0	1.2	0.7	1.1

Sample	Au%	Ag%	Cu%	Fe%	Pb%	Na%	Mg%	Al%	Si%	Cl%	K%	Ca%	Ti%	Mn%	P%
45672Q															
vitrified surface An1				3.2	58.8	1.8	1.3	6.0	16.9	0.9	2.1	3.2	0.6		1.1
vitrified surface An2				5.6	29.2 S 0.1		1.7	6.0	17.3	1.6	3.4	27.0	0.6		2.9
45673Z															
gold globule An1	84.2	3.6			Sn 8.6										
gold globule An2	93.7	3.1			Sn 3.2										
gold globule An3	77.6	3.1			Sn 19.3										
gold globule surface An5	84.8	3.1		0.8	3.3 Sn 3.4		1.7	1.8			0.7	0.4			
gold outer surface layer An4	2.0		0.7	24.1	Sn 58.1		2.7	1.9	3.4		0.6	0.8		0.9	
gold outer surface layer An6	6.1		1.1	24.1	1.8 Sn 36.2		5.8 Zn 0.5	4.5 Ni 0.5	8.7	0.4	1.2	2.9	0.4	2.0	0.6
vitrified sherd surface An7				7.3	Sn 3.1	3.1 S 0.6	3.1	12.7	34.4	0.7	5.0	24.4	0.6	0.4	2.1
45674X															
sherd A															
vitrified surface An1			3.2	2.7	59.2		1.2	6.2	15.3	0.5	2.2	4.9	0.4		
vitrified surface An2			7.8	5.8	52.1		1.4	6.5	16.2		3.0	3.9	0.6		0.8
45675V															
gold globule An1	96.4	3.6													
gold globule An3	98.6	1.4													
gold globule An4	97.4	2.6													
gold globule An5	96.8	3.2													
gold globule An6	98.6	1.4													
gold globule An7	97.8	2.2													
gold globule An8	96.0	4.0													
gold globule An9	95.5	4.5													
gold globule An10	98.7	1.3													
gold globule An11	98.7	1.3													
gold globule An12	95.8	2.3			Sn 1.9										
gold globule An13	95.4	2.6	0.6		Sn 1.4										
vitrified surface An14				5.6	S 0.5	4.8	2.4	12.7	38.2	0.3	4.4	25.8	0.5	0.2	2.6
45676T															
vitrified surface An1				22.7	S 0.4	8.0	2.0	16.8	36.2	0.7	2.1	7.4	0.8		0.5
cross-section: gold flake 1	99.4	0.6													
gold flake 2	100.0														
gold flake 3	99.1	0.9													
gold flake on surface An5	99.2	0.8													
gold flake inside An7	100.0														
gold deep inside An8	100.0														
cross-section: surface An4				71.5		2.0	17.4	0.8	0.6			0.5		7.1	
cross-section: body 1 mm in, An6				12.0		6.9	1.7	19.6	46.3	0.9	2.8	8.9	0.9		

continued on p. 106

Sample	Au%	Ag%	Cu%	Fe%	Pb%	Na%	Mg%	Al%	Si%	Cl%	K%	Ca%	Ti%	Mn%	P%
45677R															
gold globule An2	95.7	3.6	0.7												
gold globule An3	95.5	3.6	0.9												
gold globule An4	92.9	5.0	2.1												
gold globule An5	95.3	3.6	1.1												
vitrified surface An1				7.4	**S** 0.5	1.5	2.5	10.1	32.1	0.3	3.1	36.7	0.6	0.3	2.5
45678P															
vitrified cream surface An1		1.1	0.8	13.7	5.0 **S** 0.9	3.5	3.4	8.4	25.6	1.1	2.9	26.7	1.3	0.6	2.9
Pb-rich glaze An2			0.5	1.8	65.0		0.6	1.0	1.1	2.2		15.6	0.2		9.0
Pb oxide? crystal An3				2.0	75.4			1.2	3.3	2.4	0.6	5.9			5.9
45679Y															
vitrified surface An1				5.8	40.4		2.0	7.4	15.6	1.3	2.2	15.6	0.5		5.7
45680Q															
gold globule An3	58.9	38.9	0.6	**Sn** 1.6											
gold globule An5	58.4	38.9	1.1	**Sn** 1.6											
gold globule An6	58.0	41.0	0.8												
vitrified surface An7				15.5		2.9	2.5	16.5	47.7		6.6	4.2	1.0		1.2
cross-section: globule 3	57.1	41.1	1.8		**Pb+Bi** trace										
cross-section: globule 4	57.3	40.9	1.8		**Pb+Bi** trace										
cross-section: globule 5	57.9	40.2	1.9		**Pb+Bi** trace										
45681Z															
vitrified sherd I surface inc. gold globule An1	2.8	0.7		15.0	**S** 0.4	2.1	2.6	12.9	32.6	0.5	3.9	20.3	1.1		3.6
gold globule sherd I An2	94.6	4.8	0.6												
gold globule sherd I An3	95.5	4.5													
gold globule sherd II An5	100.0														
gold globule sherd II An6	100.0														
gold globule sherd III An7	100.0														
vitrified sherd III surface An8				9.0		5.7	2.9	13.9	42.9		6.5	11.4	0.7		2.0
gold globule sherd IV An9	69.5	30.5													
gold globule sherd IV An10	85.7	14.3													
vitrified sherd IV surface An11				14.1		2.0	3.4	16.9	37.6		7.6	13.3	1.0	0.4	1.3
gold globule sherd IV An12	68.5	30.9	0.6												
gold globule sherd IV An13	92.3	7.7													
gold globule sherd V An14	97.1	2.9													
vitrified sherd V surface An15				17.2		2.5	5.0	13.4	32.5		4.7	16.1	0.9	1.0	2.0
gold globule sherd VI An16	98.5	1.5													
gold globule sherd VI An17	98.3	1.7													
gold globule sherd VI An18	96.4	3.6													
vitrified sherd VI surface An19				16.0		3.7	4.8	10.5	29.7	0.4	2.1	24.7	1.1	1.2	3.4
45682X															
vitrified black surface An1		3.2		2.3	31.0	6.9	1.2	9.5	34.2		4.1	5.0	0.4		
vitrified surface An3			0.4	12.1	4.1 **S** 0.7	1.5	2.8	17.8	39.8	0.3	6.7	9.3	0.9	1.7	1.0

Sample	Au%	Ag%	Cu%	Fe%	Pb%	Na%	Mg%	Al%	Si%	Cl%	K%	Ca%	Ti%	Mn%	P%
45683V															
vitrified surface An1				14.1	**S**	3.8	2.7	12.7	41.7	0.3	7.2	14.0	1.2	0.9	0.7
vitrified surface An2				24.1	0.5	2.3	2.3	11.6	36.7		6.4	11.7	1.3	2.4	0.5
45684T															
gold globule An1	98.9	1.1													
gold globule An2	82.6	17.4													
cross-section: globule 1	80.8	18.4	0.7												
cross-section: globule 2	81.1	18.0	0.9												
45688W															
XRF centre An1		**Ag** **trace**													
XRF edge An2		**Ag** **trace**			**Pb** **high**										
XRF near edge An3		**Ag** **trace**			**Pb** **high**										
46849Y															
parting vessel					**Zn**										
vitrified black surface An2		0.6	0.3	24.9	0.4	4.2	5.5	10.6	24.3	0.6	1.7	16.4	0.5	7.5	1.6
core pink area An3		1.0		13.4		8.5	2.0	19.2	41.1	0.7	2.3	9.3	1.0		
core Fe-rich pink An5				59.0		5.6	2.2	9.1	14.3	0.2	0.9	4.2	3.2	0.2	0.3
vitrified surface An1		0.6	0.4	27.8		2.4	4.8	9.8	26.3	0.3	3.3	10.8	0.6	11.2	1.2
vitrified surface An2		1.2		13.1		3.0	3.2	10.9	30.0	0.8	3.7	25.0	0.8	3.9	3.0
vitrified surface An3		0.7		17.4		2.7	4.0	11.4	31.0	0.6	3.4	17.7	0.8	7.3	2.5
Pb particle An6				2.5	90.7			0.6	1.7			4.0		0.6	
AgCl particle 1 An7		75.7		6.6			1.4	2.5	1.8	7.5		1.6		2.6	0.3
AgCl particle 2 An8		71.1		4.6			1.5	3.2	6.4	7.7	0.8	2.8		1.8	
Pb-rich grain An9				10.2	40.0	3.1	2.0	7.5	17.8	0.9	2.1	10.3	0.7	3.2	0.6
Ag-rich grain An10		89.6	0.4	2.6			1.2	1.3	2.4			1.4		1.1	
gold flake An12	98.6	1.4													
Au/Ag particle An5	76.7	23.0	0.3												
46851Z															
vitrified surface An1				7.6	16.0	1.2	2.5	14.7	40.8	0.8	6.2	5.5	0.8		3.4
vitrified surface An2				7.8	8.7 **S** 0.5	1.5	2.4	16.8	47.6	0.5	7.2	4.0	0.9		1.9
vitrified surface 1 An3		0.4	2.3	4.9	47.6 **S** 0.8	1.4	1.0	5.3	13.9	1.5	1.7	12.1	1.0	0.2	5.6
vitrified surface 2 An4		0.9	0.4	1.1	60.5 **S** 0.5	**As** 1.7	0.6	2.3	3.9	2.3	0.6	16.0	0.2		7.9
47668S															
furnace brick 20: section		3.6		8.8		8.4	0.9	19.5	44.2	1.0	2.0	7.9	0.6		1.1
Pactolus gold grains															
KZP grains 1–10 at the same purity	100.0														

separate additions such as occur on other sherds. The vitrification is fractured at the edges of the sherd, showing that this was part of a larger piece when in use. A few gold globules are seen trapped in the bloated surface. The sherd has been heated strongly only from the concave side.

ANALYSES

Three gold globules were analysed, one large and two small (*c.* 15 μm diameter). They were spherical, having solidified from the molten state. These three globules were 100% gold, and therefore represent purified gold at the end of the refining process. The vitrified surface of the sherd shows the presence of Fe, Na, Mg, Al, Si, K, Ca, Ti, P.

Sherd 44398W (Figs 5.6, p. 133, and 5.7)

This is an irregular oblong-shaped sherd. The fabric colour is sandy rather than red. The concave surface is about half covered with thick, black, bloated vitrification. Other areas have a sandy-coloured fused surface overlying black, which may be a reduction stain rather than vitrification. The two different areas seem to represent different temperature zones, the bloating being the limit of extreme heat. The vitrification is broken at the edges, showing that the sherd was part of a larger piece when in use. Heating was only from the concave side. There are several gold globules in the thickly vitrified surface.

The globules have a wide size range (20–300 μm) within close proximity to each other. The larger ones are dendritic and spherical, having solidified from the molten state within the once-soft vitreous sherd surface (Fig. 5.7).

ANALYSES

Twelve globules were analysed. All but three are pure gold, the others having between 0.5% and 1.0% copper, but no silver. Clearly, the gold is at the end of the purification process. The vitrified surface shows the presence of Fe, Na*, Mg, Al, Si, K, Ca, Ti, Mn. Calcium is high and this is something that seems to occur on many of the gold-melting sherds associated with the white grains of calcareous material found on some of the other sherds such as 44396P above.

Sherd 44399U (Figs 5.8, p. 134, and 5.9 to 5.11)

A thin section was made from this sherd to study the fabric. This is an irregular oblong-shaped sherd. The internal fabric is exposed by a relatively new fracture and is red coloured throughout. The slightly concave surface has an unusual white/cream coloured finely bloated/solid structure that has cracked through thermal shock. White calcareous grains are seen in the bloated surface, one grain is particularly large. The vitrification is broken at the edges, showing that the sherd was part of a larger piece when used. Heating was only from the concave side. A number of gold globules are seen optically to be distributed across the surface and in the SEM are seen to have a large range of sizes (20–200 μm). On the larger gold globules, porosity and dendrites are clearly visible (Fig. 5.9).

ANALYSES

Eight globules were analysed, one was pure gold and the others contained between 0.8% and 2.4% silver. The

Fig. 5.7 SEM detail of gold globules in the surface of sherd 44398W.

range of silver content in the gold from the Pactolus is believed to lie in the range of 10% to 35% and thus the globules are clearly of gold that has been through the purification process. The white vitrified surface of the sherd contains Ag, Cu, Fe, Mg, Al, Si, Cl, K, Ca, Ti, Mn, P. The calcium content is again particularly high.

THIN SECTION

A polished thin section was made from this sherd, which is approximately 8–9 mm thick. The section shows a moderately fired ceramic matrix with a lot of granular quartz inclusions (Fig. 5.10). The fabrics comprise micaceous clay matrixes with grains of angular silt and fine sand (mainly quartz). Inclusions of quartz and rock up to several millimetres in diameter were also noted. The composition does not significantly change between the medium-fired body and the bloated upper regions.

Analysis by EDX showed the sherd to be relatively low in refractory properties and therefore the highly vitrified and bloated structure indicates the convex surface was subject to intense heat of about 1150 °C. The upper concave surface of the sherd is severely bloated with round glassy pores caused by gas bubbles in the once very hot and semi-molten vitreous surface. This visible thermal gradient penetrates only to a maximum of about 2 mm into the sherd from the heated surface in one area of the section and to only 1 mm in other places (Fig. 5.11). At this point there is a rapid change of vitrification to the normal fired state of the pottery (Fig. 5.10), which suggests that the period of heating was very short, probably about 30 minutes.[5] Quartz grains were cracked by the high temperatures reached in this bloated zone and beyond into the normal structure regions of the sherd. At high magnification, the bloated areas show solid glass between the bloated pores and trapped mineral grains.

The ceramic texture of the sherd in the body away from the heated surface is very similar to that of the parting-vessel sherd 44401Z (discussed on pp. 122–4), but the bloated surface is quite unlike the surface of the parting sherd. The sherd texture is also similar to the common coarseware cooking vessels.

Analysis The body composition is quite different from that of the vitrified surface, being of more usual ceramic composition. In particular, the calcium is quite low while the silicon is high, reflecting the presence of the quartz grains in the basic fabric and possibly the addition of calcium (limestone) to the receptacle surface during melting. The body of the sherd contains Ag, Fe, S*, Na, Mg, Al, Si, Cl, K, Ca, Ti, Mn*. There is a significant silver content and also some chlorine and sodium. This would suggest that the sherd came from a vessel that had previously been used for parting and was thus impregnated with silver chloride. This reuse is unusual (but see sherds 45681Z, pp. 113 and 116–17), and not good practice for melting pure gold because of the danger of contamination. The variable silver content of the gold globules could result from picking up silver from the sherd.

Sherds 44402X

This sample comprises three small sherds. All have vitrified, bloated surfaces. Sherd B is particularly highly vitrified with a clear thermal gradient in the fractured section. This sherd also has the most gold globules, which appear semi-buried in the vitrified surface. The globules have clearly solidified from the molten state; the larger gold globules are dendritic. Sherd C has the largest globule, 400 μm diameter. Sherd A has no globules.

ANALYSES

The sherd surfaces show no unusual compositions and the following elements are present: Fe, Na, Mg, Al, Si, K, Ca, Ti, P*.

The gold globules are all of similar composition of partly purified gold containing between 4.5% and 6.8% silver.

Sherd 45670U (Fig. 5.12)

This is an irregular oblong-shaped sherd. The fabric is red coloured throughout and of the usual low–medium fired texture. The concave surface has only a little evidence of surface vitrification in two small bloated areas on opposite edges. One of these has a blackened glassy appearance. Many gold globules are associated with these vitrified areas. The sherd has only been heated from this side. The sherd surface contains a number of cream-coloured mineral inclusions (calcareous) and the fractured edges of the sherd show it was part of a larger piece when used. The SEM micrographs show the gold globules are somewhat sunk within the surface although this does not look particularly glassy. The globules have a range of sizes from under 10 to 120 μm (Fig. 5.12). The larger globules are dendritic and have solidified from the melt.

ANALYSES

The surface of the sherd is calcareous and the phosphorus content is relatively high. The surface has the following elements present: Fe, Na, Mg, Al, Si, Cl, K, Ca, Ti, P. Of the 12 globules analysed, all but one are of pure gold. The last globule is essentially impure gold with 24.7% silver and 0.9% copper. This can only mean that the sherd had been used more than once to melt gold.

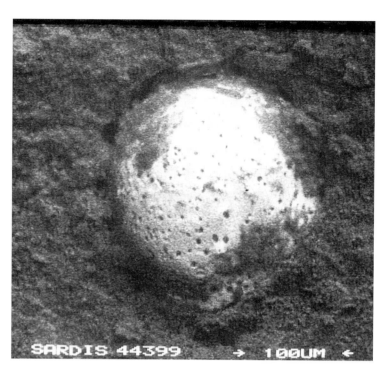

Fig. 5.9 Gold globule in surface of sherd 44399U. Note the dendritic porosity in the gold surface.

Fig. 5.10 SEM micrograph of an area within a thin section through sherd 44399U, showing the as-fired structure.

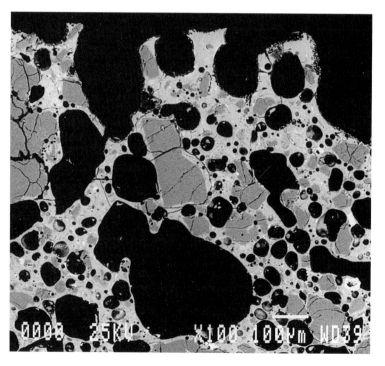

Fig. 5.11 SEM micrograph of an area of the thin section through sherd 44399U, showing the highly fired, bloated nature of the sherd surface. The relatively shallow depth of bloating shows that heating was intense but for a short period of time, and that the sherd was only heated from the top surface.

Fig. 5.12 SEM detail micrograph of gold globules in the vitrified surface of sherd 45670U.

Fig. 5.14 SEM micrograph showing the distribution of gold globules having a range of sizes in the vitrified surface of sherd 45673Z.

Fig. 5.15 SEM detail micrograph of a single gold globule in the vitrified surface of sherd 45673Z, revealing a surface layer of fused mineralised tin-rich material on the globule.

Fig. 5.16 SEM detail micrograph of gold globules of a wide range of sizes in the vitrified surface of sherd 45675V.

Fig. 5.18 SEM micrograph of the iron-rich surface of sherd 45676T.

Sherd 45673Z (Figs 5.13, p. 134, to 5.15)

This is a relatively thick, butterfly-shaped sherd. The body is red but it is very deeply vitrified, bloated and black on the concave surface. A sandy-coloured vitrified surface covers part of the underlying blackened sherd surface. There are white (calcareous) nodules in the surface. The thickness of the vitrification suggests either longer heating from this upper side than was usual or perhaps more than one use of the sherd. There are many gold globules in the vitrified surface, in both the black and sandy-coloured areas. The SEM micrographs show a range of sizes of gold globules in the vitrified surface (1–150 μm) (Fig. 5.14). The larger ones are dendritic, having cooled from the melt. What is different on these globules is a surface covering of an apparently vitrified material which has broken away in areas to expose the gold below (Fig. 5.15).

ANALYSES

All analyses show substantial quantities of tin. The surface of the sherd has the following elements present: Fe, Sn (3.1%), Na, S, Mg, Al, Si, Cl, K, Ca (24.4%), Ti, Mn, P. Once again, there is a very high calcium content at the surface. Three of the gold globules were analysed on apparently uncontaminated surfaces: that is, no clay mineral elements were detected in the analyses, and all contain high levels of tin – 3.2%, 8.6% and 19.3%. Presumably, the tin is in solution in the gold. Some gold globules are covered with a fused mineralised layer (Fig. 5.15) that contains very high tin and iron plus some lead, zinc, nickel and other mineral elements: for example, Au, Cu, Mg, Al, Si, Cl, K, Ca, Ti, Mn, P. This sherd clearly has an unusual composition with such a high tin content, compared with the other sherds, but the gold globules on two other sherds also show traces of tin (45675V and 45680Q), and tin was also found on the surface of foil sample 29A. The origin of the tin is not at all certain, but does recall Agatharchides' description of the cement used by the Egyptians, which included both lead and tin.[6]

Sherd 45675V (Fig. 5.16)

This is an irregular-shaped sherd, red in colour throughout the core. It has a matt black surface glaze covered with sandy- and cream-coloured granular deposits. Gold globules were distributed around the sherd and fused to its surface. It was part of a larger sherd when used and was heated from the concave side. The SEM micrographs show a consolidated, textured surface with no bloating, and with many spherical gold globules sunk within the vitrified surface. The gold globules have a wide range of sizes (1–300 μm) and are found in mixed size groups (Fig. 5.16).

ANALYSES

Twelve globules were analysed on their surfaces and found to be quite pure, but all had some silver, between 1.3 and 4.5%. Two globules also contained tin (1.4% and 1.9% respectively) and one contained 0.6% copper. These small globules were found between much larger globules of gold and within 100 μm of them. It is unclear why globules of different composition lie so close together, and there is no tin in the surface of the sherd that could have been the source of the tin in the globule. The surface is highly calcareous and has some sulphur, sodium and chlorine but no silver, and is similar to sherd 45673Z (but without tin). The surface contains the following elements: Fe, S, Na, Mg, Al, Si, Cl, K, Ca, Ti, Mn, P.

Sherd 45676T (Figs 5.17, p. 134, 5.18 and 5.19)

A thin section was made from this sherd, whose function is puzzling. It has an irregular shape and a red fabric towards the outer side. The convex surface is mainly a pink/purple colour with some grey and light brown areas. The colour is that of a parting sherd rather than a gold-melting sherd (although this is not supported by analysis). The sherd has turning lines in the original clay. It looks more like a fineware sherd than the other coarseware sherds. The surface appears consolidated rather than vitrified and the colour suggests some chemical process has occurred. Tiny reflective crystals are seen on the surface and are either of iron (oxide)-rich crystals or mica (Fig. 5.18), but no gold globules were visible optically. In the SEM, the surface appears compact with tiny iron-rich crystals. The sherd was part of a larger piece when used.

THIN SECTION

The ceramic body has some large grog inclusions, indicating a coarseware body rather than the apparent fineware seen on the surface. A group of tiny flakes of purified gold (2–20 μm) were found *in* the body near the surface and others were found deeper within (Fig. 5.19). They are clearly associated with the fabric of the sherd and align along the direction of pot manufacture. The flakes are not globular but may well be alluvial, or possibly filings. This implies manufacture of the pottery probably within the refinery site itself, where flakes of pure gold would be relatively easy to pick up and could become incorporated in the clay. The sherd has clearly not reached the melting temperature of gold either during its original firing or during subsequent use. There is no bloating of the sherd surface so it was not used as a gold-melting sherd.

ANALYSES

Surface analysis of the sherd shows very high iron

(71.5%), because of the fine iron (oxide)-rich crystals that are present. Cross-section analysis shows the presence of Fe, Na, Mg, Al, Si, Cl, K, Ca, Ti. The body of the sherd, like all of the sherds, is iron-rich (red body), which could be the source of the iron in the surface crystals if the sherd had been exposed to salt vapour. There is no silver in the sherd, so despite the purple colour that might have suggested that it was a used parting-vessel sherd, the chemical evidence shows it was not, unless paradoxically the gold treated was already pure. The gold flakes within the surface were either pure gold or contained up to 0.9% silver. These flakes therefore probably represent already purified alluvial gold trapped within the clay before firing. The absence of silver chloride in the sherd precludes the possibility of their in-situ purification.

Sherd 45677R (Fig. 5.20)
This is an irregular-shaped trapezoid sherd with a red matrix. The concave surface has a brown and grey surface or crust. This has an underlying vitrified and partly bloated texture and is black, particularly at one corner, where it is thickest. There are a number of relatively large, white (calcareous) grains fused in the surface. There are also some sinuous features that look like mineralised organic material, which may have developed during burial. The sherd has only been heated from this side and was part of a larger object when used. A large crack runs into the sherd. There are a large number of gold globules on this sherd. The SEM micrograph shows a larger number of spherical gold globules between 1 and 120 μm diameter. These are fused within the roughly textured surface.

ANALYSES
Surface analysis shows high calcium. The elements present are Fe, S, Na, Mg, Al, Si, Cl, K, Ca, Ti, Mn, P. The gold globules are all relatively pure and contain between 3.6 and 5% silver and 0.7 to 2.1% copper.

Sherd 45680Q (Figs 5.1, p. 132, and 5.21 to 5.23)
A polished cross-section was made of three of the gold globules.

Both the black, bloated vitrification of the sherd surface and the concentration and distribution of gold globules are very visible on this sherd. The trapezoid-shaped sherd, whose body is not particularly red in colour, shows white (calcareous) grains in the broken edges. About two thirds of the concave surface is covered in black (charcoal grey), bloated vitrification with open bubbles. The remaining surface has the same colour but the matt vitrification is either less bubbly or there is a continuous surface covering

underlying bubbles. It looks as if part of the continuous surface layer has broken away and exposed the inner bubbles. There are angular white (calcareous) grains in the vitrification of both areas. Overlying some of the continuous grey surface is a thin, fused, ginger-coloured layer which contains many gold globules (Fig. 5.1, p. 132). Other gold globules are seen in the open bloating pores. The broken edges of the sherd show that it was part of a larger piece when used.

The SEM micrographs show the numerous gold globules and their distribution, which is more grouped than uniformly spread (Fig. 5.21). The globules are trapped within the vitrified surface and are all spherodised from solidification from the melt. Bloating is seen adjacent to the globules showing the intensity of heating on this concave side. The size range of globules is large, from 300 μm down to a few micrometres (Fig. 5.22). Separation between large globules is often less than 200 μm and much less for smaller ones.

The question arises how the gold became trapped during melting. Perhaps the molten gold covered the overheated sherd surface and the bloating allowed gold to fill the burst bubbles, while surface tension prevented the molten gold from escaping during pouring.

SECTIONED GOLD GLOBULES
Three of the larger globules were taken from the sherd surface, mounted in resin and polished. None is perfectly spherical, with various degrees of irregularity (Fig. 5.23). Evidence of dendritic porosity is seen and one has some porosity in the core. There is no dendritic coring and the globules appear homogeneous despite the base composition of around 41% Ag and 1.8% Cu (Table 5.2). White metallic precipitates, which are rich in lead and contain some bismuth, are seen in all three globules. The lead and bismuth may indicate a source of silver other than from natural alluvial gold, and the silver content might indicate the gold came from the melting of Lydian electrum coins (see Table 7.1, p. 170). This is one of the few instances where base gold was being melted on the sherds. For some reason, the gold was being assayed prior to treatment.

Sherds 45681Z (Figs 5.24 to 5.26)
This sample comprises a group of six sherds, all of which have gold globules on their surfaces.

SHERD I
This is an irregular-shaped sherd with a pale red core. One third is covered with porous, bloated, black vitrification, which is thickest at one end. Elsewhere the surface is covered with a light brown layer in which gold globules are

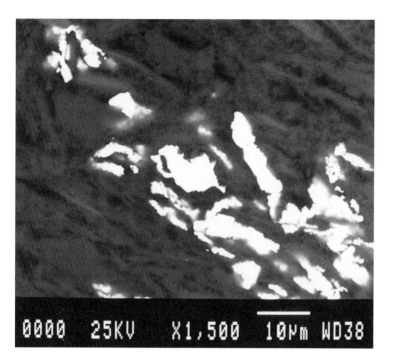

Fig. 5.19 Gold flakes found within the body of sherd 45676T near the surface, shown in thin section.

Fig. 5.20 SEM micrograph of gold globules in the vitrified surface of sherd 45677R.

Fig. 5.21 SEM low-magnification image of the many gold globules in the vitrified surface of sherd 45680Q, shown in Fig. 5.1, p. 132.

Fig. 5.22 SEM detail micrograph of a group of large gold globules in the vitrified surface of sherd 45680Q.

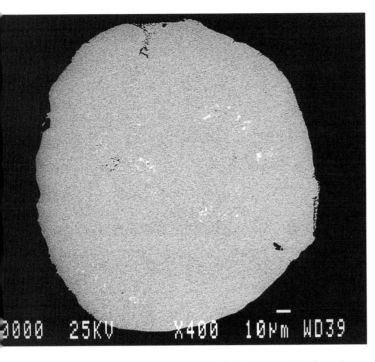

Fig. 5.23 Sherd 45680Q: SEM micrograph of a cross-section through one of the gold globules removed from the surface of the sherd, showing a little interdendritic porosity and white regions of lead and bismuth enrichment.

Fig. 5.24 SEM low-magnification micrograph of the gold globules in the vitrified surface of sherd I of 45681Z.

Fig. 5.25 SEM micrograph of the highly vitrified and bloated surface of sherd II of 45681Z. Gold globules are trapped in the surface.

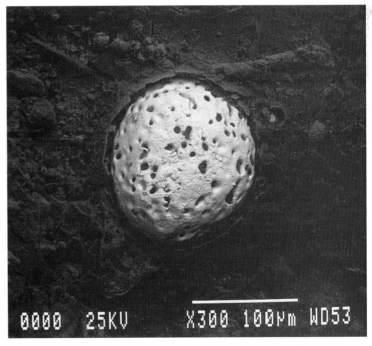

Fig. 5.26 SEM detail micrograph of one of the large (120 μm diameter) gold globules in the vitrified surface of sherd VI of 45681Z. Note the interdendritic surface porosity in the globule.

Table 5.2 Composition of sectioned gold globules from gold-melting sherds (EDX).

Sherd 45684T	Au%	Ag%	Cu%	Sherd 45680Q	Au%	Ag%	Cu%	Inclusions
globule no. 1	80.8	18.4	0.7	globule no. 3	57.1	41.1	1.8	Pb + Bi
globule no. 2	81.1	18.0	0.9	globule no. 4	57.3	40.9	1.8	Pb + Bi
near surface of no. 1	81.5	17.8	0.7	globule no. 5	57.9	40.2	1.9	Pb + Bi

Detection limit for Cu = 0.15%.

Precision: Au = ±0.5%, Ag and Cu = ±0.3%.

seen. In some areas, it appears that the brown covering layer has broken off, revealing the underlying black, bloated surface (as with other sherds). The sherd was heated only from the concave side and was part of a larger sherd when used. The SEM micrograph shows many gold globules in the consolidated surface which also has a few bloating pores (Fig. 5.24).

Analyses The sherd surface showed, excluding the gold globules, the presence of the following elements: Ag, Fe, S, Na, Mg, Al, Si, Cl, K, Ca, Ti, P. The calcium content is relatively high on the surface compared with the body of the sherd, as is the case with a number of the other sherds. The presence of silver in the surface of a gold melting sherd is unusual (perhaps this is a reused parting sherd, cf. sherd 44399U above). The globules contain around 95% gold, the remainder being silver, and one had 0.6% Cu.

SHERD II

This is an irregular-shaped oblong sherd, probably with a pale red core, covered with black (charcoal grey) bloated, porous vitrification. The light brown surface deposit visible in some areas is also bloated and contains gold globules. White (calcareous) grains are also present in the surface. The sherd was heated only from this side and was part of a larger sherd when used. The SEM micrograph shows a highly vitrified and bloated area with a number of gold globules fused within the surface (Fig. 5.25).

Analyses The globules are of pure gold.

SHERD III

This is an irregular-shaped sherd with a red core. Glassy black vitrification covers most of the sherd and is not bloated except at the edges. It again appears as if the bloating is below the surface and the outer glassy layer is unbroken. Fused brown deposits lie at places on the surface, as do white (calcareous) grains. Gold globules are

visible in the surface. The sherd was heated only from the concave side and was part of a larger piece when used. The SEM micrograph shows a solid, glassy, thick, fractured surface with a single gold globule firmly embedded in the solid glass.

Analyses The sherd surface contains the following elements: Fe, Na, Mg, Al, Si, K, Ca, Ti, P. The calcium content is relatively high. The globule is of pure gold.

SHERD IV

This is a trapezoidal sherd probably with a red core, two thirds covered on the concave surface with a black vitrification which is not itself bloated, but which has broken away in one area, exposing the thick, black, bloated underlying material. Some white (calcareous) grains are seen in the surface. Areas of fused surface away from the black zones are grey and brown in which there are a few gold globules. The sherd was heated only from this surface and was part of a larger piece when used. The SEM micrograph shows a consolidated surface similar to sherd I, and the angular surface features are probably crystalline inclusions. A few gold globules are fused to the vitrified surface.

Analyses The sherd surface contains the following elements: Fe, Na, Mg, Al, Si, K, Ca, Ti, Mn, P. The gold globules are of variable composition: one is relatively pure, one has 14.3% Ag, while the other two are essentially of unrefined gold composition (30.5% Ag): see Table 5.1. This probably represents more than one use of the sherd or receptacle for melting gold.

SHERD V

This is a shield-shaped sherd with angular fractures around the edges, leaving a central area of black, bloated vitrification. There are white (calcareous) grains in the surface. The sherd core is pale and like several others is not particularly red. Very few gold globules were found on this sherd. It was heated only from the concave side and was part of a larger piece when in use. The SEM micrograph shows a highly vitrified, crazed but not bloated surface with a gold globule embedded.

Analyses The sherd surface contains the following elements: Fe, Na, Mg, Al, Si, K, Ca, Ti, Mn, P. The calcium content is again high due to the presence of the white calcareous grains in the surface. The analysis of a globule shows it to be of purified gold with only 2.9% silver.

SHERD VI

This is the largest of the sherds examined. It is of trapezoidal shape and with a red core where it is exposed by a recent fracture. This angled fracture shows a darkening of the core towards a black colour from about halfway through the sherd towards the surface. The surface is covered with a relatively smooth, pale fawn-coloured, consolidated layer that looks glassy at one corner and has a brittle fracture running through it. There are possibly a few small, white, calcareous grains in the surface. A few gold globules are visible in the surface. It was heated only from the concave side and was part of a larger object when used. The SEM micrograph shows a 120 µm diameter gold globule in the vitrifed surface. The globule has a surface texture similar to that of a golf ball due to the interdendritic porosity (Fig. 5.26).

Analyses Surface analysis shows the presence of the following elements: Fe, Na, Mg, Al, Si, Cl, K, Ca, Ti, Mn, P. The calcium content is again very high. The gold globules are almost pure, containing only 1.5%, 1.7% and 3.6% silver.

In summary, all of the six sherds have similar surface compositions, being highly calcareous and ferruginous (and sherd I has low silver). All have gold globules on their surfaces containing varying amounts of silver.

Sherd 45683V

This trapezoid-shaped sherd has a thin whitish layer on top of black-coloured material.

ANALYSES

The elements present on the surface of this sherd are: Fe, S*, Cl*, Na, Mg, Al, Si, K, Ca, Ti, Mn, P. XRF analysis showed the presence of a trace of silver (as with sherd 45688W). No gold globules were found on sherd 45683V. No lead was detected either, so it was decided to include this sherd in the present category.

Sherd 45684T (Figs 5.27 and 5.28)

The SEM micrographs reveal a number of gold globules on the sherd surface of a wide size range and in close proximity to each other (Fig. 5.27). The globules are fused to the surface and their spherical shape shows that they were molten before being trapped.

ANALYSES

No surface analysis of this sherd was made.

Surface analysis of two gold globules shows one which is quite pure while the other contained 17% silver. This suggests that the sherd was used more than once.

SECTIONED GOLD GLOBULES

Two of the larger gold globules were removed from the surface of the sherd and mounted in resin and polished in a manner similar to those on sherd 45680Q. The globules are not perfectly spherical and are 250–300 µm in diameter (Fig. 5.28). In cross-section, the edges (globule surfaces) are indented with interdendritic casting textures, and some fine interdendritic porosity is present in the core of one of the globules.

Analyses In cross-section, the two globules appeared similar to each other and both contained around 18% silver and 0.8% copper, which is about half the amounts found in the three sectioned globules of sherd 45680Q (Table 5.2). Thus it is possible that these two globules may represent the assay composition of the gold alloy which has reached half purity. Neither globule shows any coring nor do they show any heavy metal precipitates, as did those of sample 45680Q.

Characterisation of refractory sherds with lead-rich surfaces and no gold globules

Sherd 44397Y (Fig. 5.29)

This is an irregular square sherd. The surface of this sherd is rather complex. It has a highly glazed appearance, which is very bubbly in places. Elsewhere the surface is more like glass than fused ceramic and is cracked; the colour is red-brown. There are some golden-coloured streaks on the glazed surface but no gold particles. Some unusual, relatively large (millimetre size) nodular brown lumps on the surface show up clearly in the SEM at high magnification. The nodules are quite soft and easily scraped to provide a clean surface for analysis. The sherd has only been heated from the concave side.

ANALYSES

The brown nodules appear to be of silver chloride. There

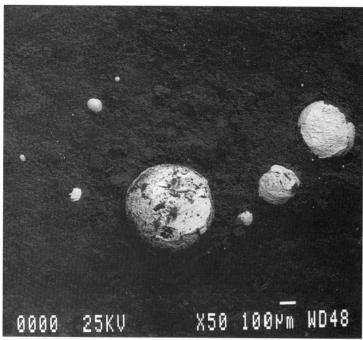

Fig. 5.27 Sherd 45684T: SEM micrograph of a group of gold globules in the vitrified surface, having a large range of diameters.

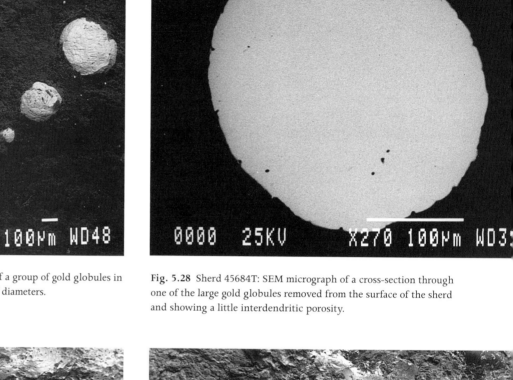

Fig. 5.28 Sherd 45684T: SEM micrograph of a cross-section through one of the large gold globules removed from the surface of the sherd and showing a little interdendritic porosity.

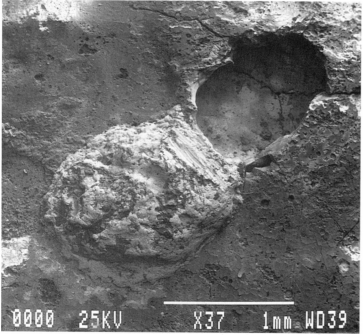

Fig. 5.29 Sherd 44397Y: sherd with a vitrified surface rich in lead and silver. The SEM micrograph shows a large globule of silver chloride.

Fig. 5.32 SEM micrograph of the lead-rich glaze on the surface of sherd 45672Q.

were no gold globules on the sherd surface. The glazed surface is a lead-rich glass. There is also copper in this glaze. The glazed surface shows the presence of Ag, Cu, Fe, Pb, Mg, Al, Si, Cl*, K, Ca, Ti*, Mn*. Silver chloride nodules were found only on this sherd.

Sherds 44400Q (Fig. 5.30, p. 135)
There are three sherds in this sample, designated A, B and C. They do not appear to be from the same original piece, as they each have different surface characteristics.

SHERD A
This is an irregular oblong with a partially red body although the front concave surface is largely black and vitrified with one area being thick, bloated and glassy. Chips of white (calcareous) minerals are stuck in the black glaze. Several round non-metallic inclusions were found in the surface. These are lead-rich. There is a deep crack running across the sherd. The glaze is fractured at the edges showing that it was part of a larger sherd when used. No gold globules were seen on this sherd.

Analyses Surface analysis of the black glaze shows a lead-rich calcareous composition. The following elements were present: Fe, Pb, Na, Mg, Al, Si, Cl, K, Ca, Ti, P. The non-metallic inclusions had broadly similar compositions to each other; they were high in lead and very low in silica, and are likely to be composed principally of litharge (PbO). They also contain significant quantities of gold, up to 5.3%, and silver, up to 15.3%. One is mainly silver chloride. The inclusions contain the following elements: Au, Ag, Pb, Al, Si, Cl, Ca, P.

SHERD B
This irregular-shaped sherd was not examined in the SEM. It has a bloated, concave vitrified surface, which is both brown and black in colour. Some areas of thicker black glaze are visible. Some white grains of calcareous mineral are located in the vitrified surface. Most edges have broken vitrification but one corner appears to have a rounding of the black glaze over the edge. The broken corner appears partially vitrified showing it was used as a sherd rather than as a complete vessel. No gold globules are visible on the surface.

SHERD C
This smaller irregular-shaped sherd, with a prominent pale, cream and grey vitrified surface, was also not examined in the SEM. There is some fine bloating but no gold globules. The sherd body is red and contains white, calcareous grains which are visible at the fractured edge.

Sherd 45672Q (Figs 5.31, p. 135, and 5.32)
This is an elongated, irregular, trapezoid-shaped sherd. It has a red body-colour throughout, while the concave surface has a range of colours from the exposed red core through sandy brown with a bright yellow crust, to light and dark blue-grey and some touches of black. Some vitrification and bloating are visible towards the edge in the grey area. There are nodular grains of off-white mineral in the surface. Visually, the surface appears rather complex, indicating more complex surface chemistry. It was clearly heated only from the concave side, and one or two straight edges may have been the outer edges of the sherd during use because the bloated, vitrified surface runs over these edges, showing it was used as a sherd. No gold globules are seen on this sherd. The SEM micrographs show solid glassy areas on the textured surface (Fig. 5.32).

ANALYSIS
The solid glazed area of the sherd has a very high lead content (58.8%). Clearly, a lead glass has formed on the surface. The textured surface gives lower lead (29.2%) and also shows some sulphur. The calcium content is also high in this area, thus showing that the off-white coloured nodules are calcareous.

Sherds 45674X (Fig. 5.33)
These are two sherds (designated A and B) of rather different appearance.

SHERD A
Sherd A is an irregular oblong shape with a red core. The concave side has a thick and lumpy vitrified surface which is mainly black in colour, but some areas of the surface have cracked off (recently?) exposing a bloated, rust-red, vitrified layer below (Fig. 5.33). The vitrification is broken at the edges showing that the sherd was larger when used. No gold globules were found. The SEM micrograph of the freshly exposed surface vitrification shows a solid glass with some gas bloating.

Analyses Area analyses of the surface (black) and of the freshly exposed glaze (red) show similar compositions with very high lead (59.2%). The following elements are present: Cu, Fe, Pb, Mg, Al, Cl*, Si, K, Ca, Ti, P*. Clearly, a lead glaze has formed, but the presence of significant copper is different than on other sherd surfaces (but see sherd B). The surface chemistry of this sherd indicates that it was not for melting gold.

SHERD B
Sherd B is of semi-triangular shape with a red core. It has

Fig. 5.33 SEM micrograph of the lead-rich glazed surface of sherd A of 45674X.

Fig. 5.35 SEM micrograph of the crazed, glassy surface of sherd 45678P.

Fig. 5.36 SEM micrograph of the heavily vitrified surface of sherd A of 45679Y.

Fig. 5.37 Sherd 44401Z: SEM micrograph of the gold-alloy flakes on the surface of this parting-vessel sherd.

been heated from the concave side. There are two distinct areas of the surface. Two thirds of the sherd appears to have black vitrification that is relatively thick towards one side but appears to thin towards the top and left sides on Fig. 5.33. The colour also changes from black to a more sandy colour. The top edge of the sherd shows little effect of heat. The lower third of the sherd has a cream-coloured knobbly texture as if a subsequent reaction has occurred on top of the vitrified surface. There are also some relatively large green-coloured nodules in this area, presumably copper-rich although this sherd was not analysed. (Sherd A has significant copper.)

Sherd 45678P (Figs 5.34, p. 136, and 5.35)

This is an irregular triangular sherd. The original surface and section is exposed at one corner and is red in colour, extending throughout the sherd. The concave surface of the sherd is covered in a thick layer of cream to light-brown-coloured, fused material with cracks (Fig. 5.34). Some porosity is seen in the material which appears to be a product of thermal activity. The edges of the sherd and the thick material are fractured through, which shows that the sherd was part of a larger piece when used. The SEM micrographs show the fused surface and also more glassy, crazed areas with burst gas bubbles (Fig. 5.35). No gold globules were found on the sherd.

ANALYSES

The cream-coloured surface contains silver and copper with some lead and high calcium. The elements present are Ag, Cu, Fe, Pb, S, Na, Mg, Al, Si, Cl, K, Ca, Ti, Mn, P. The glazed area is a lead-rich glass with the following elements: Cu, Fe, Pb (65%), Mg, Al, Si, Cl, Ca, Ti, P (9%). Some crystals on the surface are even higher in lead; presumably they are predominantly of litharge. It is not clear whether the bulk of the surface deposit is litharge. The high P content is notable, but seems to be unique to this sherd and thus does not suggest the regular deliberate addition of bone ash to the material to be assayed or cupelled.

Sherds 45679Y (Fig. 5.36)

There are three sherds in this sample, designated A, B and C.

SHERD A

This is an irregular oblong sherd with a red matrix. There is a primary brown to black porous surface vitrification extending over one quarter of the sherd, the remaining areas being covered by a thick layer of cream-coloured material. The cracked edges show that this is brittle and somewhat porous; it looks rather like the material on sherd 45678P. A number of white (calcareous) grains are visible

in the brown vitrified area. The curved sherd was part of a larger piece when used, and was heated only from the concave side. No gold globules were seen on this sherd. The SEM micrograph shows solid vitrification which is cracked with some porosity (Fig. 5.36).

Analyses Only this sherd of the three was analysed. It has a lead-rich glaze with high calcium. The elements present are: Fe, Pb (40.4%), Mg, Al, Si, Cl, K, Ca, Ti, P.

SHERD B

This is an irregular oblong sherd with a vitreous, bloated, black, concave surface but which is largely covered with a fused sandy-brown-coloured surface layer. The edges of the sherd appear to have white (calcareous) grains within the matrix. This sherd was once part of a larger piece. There are no gold grains on it.

SHERD C

This is an irregular-shaped sherd, possibly with a red core. One concave area has a vitreous, black, bloated surface. Elsewhere the sherd has a black knobbly appearance with brown between the high points. Perhaps it is calcareous grog which has been blackened through heating effects.

Sherd 45682X

An area of black glaze passes over one edge while the other edges have fresh breaks, indicating that this was part of a larger sherd when used. No gold particles were seen optically.

ANALYSES

The black glaze contains the following elements: Ag, Fe, Pb, Na, Mg, Al, Si, K, Ca, Ti. Lead is very high while calcium is low on this area of the sherd. Other areas of the surface are much lower in lead but contain small amounts of copper and sulphur. The elements present in these areas include Cu, Fe, Pb, S, Na, Mg, Al, Si, Cl, K, Ca, Ti, Mn, P.

Sherd 45688W

Only XRF analyses were carried out. Traces of silver were detected all over the sherd surface. High lead concentrations were detected at the edges of the sherd, but not near the centre.

Sherd 46851Z

This is a thick, pointed trapezoid-shaped sherd. The internal fabric is light brown. The concave surface has a substantial coverage of black (matt) vitrification with some bloating seen within a broken area, the unbroken surface being essentially not bloated and, like other sherds, covering the underlying bloating pores. No gold globules were visible on the surface. White (calcareous) grains are present in the vitrified surface. The sherd had been heated only on the concave side and the fractured edges, showing that it was part of a larger object when used. There is an area of brown soil on the surface.

ANALYSES

Analyses were made on various areas of the surface. In all cases, lead was detected, of which two had high lead contents as well as some silver and copper, and in one case arsenic. The other elements found on the surface were S (in three areas), Fe, Na (in three areas), Mg, Al, Si, Cl, K, Ca, Ti, Mn (in one area), P. The calcium content is high due to the presence of the calcareous grains.

Characterisation of the parting-vessel sherds

Sherd 44401Z (Figs 5.2, p. 132, and 5.37 to 5.42)

A thin section was also prepared from this sherd.

The sherd is oblong shaped. The cross-section is generally red coloured towards the outer (convex) side and is moderately fired, as seen at the fractured edge. The surface appearance is different from the sherds examined previously, being consolidated but not showing vitrification or bloating. The concave surface is purple but with areas of grey, black and brown. The purple colour penetrates into the sherd. The coloration of the sherd suggests that it has been in contact with reactions taking place at high temperature, and this is supported by the presence of thermal cracks. Some tiny bright flecks could be seen optically, but no round globules of gold. The heated surface is broken at the edges, showing it was part of a larger piece when in use.

The SEM micrograph shows a textured, consolidated surface on which are distributed a number of tiny irregular-shaped gold-alloy flakes (Fig. 5.37), which are not fused to the surface of the sherd. The gold flakes are very different from the gold globules described above; they are not spherical (from melting) but are of irregular and somewhat flattened shape. The flakes are generally quite small, 3–20 μm across (Fig. 5.38), and have the physical appearance of natural alluvial gold grains.

ANALYSES

Surface analysis of the sherd shows a high but variable silver content (up to 11.9%, although only one was quantified – see Table 5.1). X-ray distribution maps on the surface of the sherd confirm the overall distribution of silver and show areas of silver concentration (Fig. 5.39, p. 136). The surface also contains sodium and chlorine, showing

that salt was a principal active agent in the parting process. Lead is also present in the surface (4.3%), recalling Agatharchides' description of the cement used by the Egyptians which contained both tin and lead.[7] The surface composition is unlike the gold melting sherds and derives from the reaction products of the parting process.

The silver content of the gold flakes ranges from 1.5%, through 21–27%, to 41.8%. This spans the range from purified gold composition to very base electrum. Analysis shows the majority of the flakes have a high silver content and do not appear to have been purified, despite being on the parting vessel sherd. However, two flakes do have a low silver content, one in particular is very low, which suggests these flakes at least have been purified. This apparent contradiction suggests that the flakes are either from different batches, which implies reuse of the parting vessel, or that the original Pactolus alluvial gold was very variable in composition, about which there is little information. Clearly, the parting vessel was broken before the majority of the grains had been through the parting process.

THIN SECTION

In cross-section, the ceramic matrix is very similar to that of the gold-melting sherds, as exemplified by 44399U. It was only moderately fired, having linear porosity between the glassy filaments, and has a wide size range of quartz grains (see p. 161). The ceramic does not appear to have been heated above its original firing temperature during the parting process. On the surface of the sherd is a thick silver chloride-rich layer (variable c. 80 μm thick) (Fig. 5.40). The very top surface appears less bright on the micrographs and is less rich in silver. The porous ceramic has absorbed silver chloride, which appears as a fine precipitate within the sherd body nearer the surface, and the glassy filaments in this region also appear to be speckled; there seems to be a textural change due to the parting reaction (Fig. 5.41). Further into the sherd the matrix appears physically unaffected by the surface chemistry (Fig. 5.42) although, analytically, silver chloride is present throughout.

Analyses In section, the surface of the sherd contains Ag, Fe, Pb, Na, Mg*, Al, Si, Cl, K, Ca, Ti, P*. The bright top surface layer in the SEM micrograph (Fig. 5.40) contains the most silver (55.3%), which progressively decreases to about the middle, after which the silver content remains steady (c. 2.5%) through to the outside of the sherd. Sodium and chlorine are also present throughout. Lead is only present on and in the surface layers. The presence of both silver chloride and sodium chloride prove that this

Fig. 5.38 SEM detail micrograph of two gold flakes on the surface of parting-vessel sherd 44401Z.

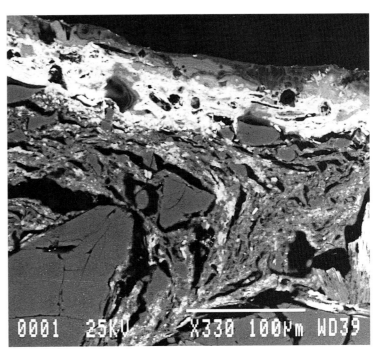

Fig. 5.40 SEM micrograph of the thin section through the surface of parting-vessel sherd 44401Z, showing the surface layer of silver chloride.

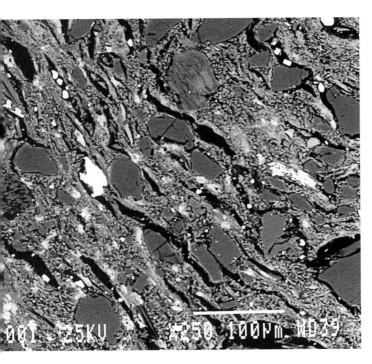

Fig. 5.41 SEM micrograph of the thin section below the surface of parting-vessel sherd 44401Z, showing the speckled texture of the sherd matrix due to silver chloride precipitation and the effects on the glassy filaments within the ceramic.

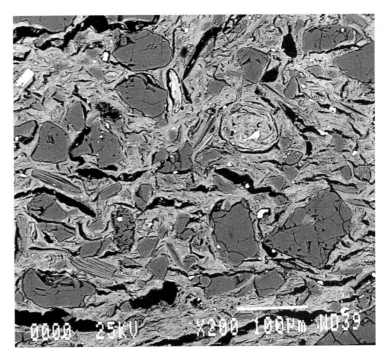

Fig. 5.42 SEM micrograph of the thin section within the core of the parting-vessel sherd 44401Z, away from the effects of the surface reactions, and showing the normal ceramic matrix, in which silver chloride is present.

sherd was part of a parting vessel. The distinctive purple colour is partly due to the silver chloride and partly due to the mobility of the iron salts and is also an indicator of the parting process.[8]

Sherd 45671S (Fig. 5.43, p. 137)

This is an irregular trapezoid-shaped sherd. The fabric is red throughout and was medium fired. The concave side has a sandy- and slightly purple-coloured and finely textured, consolidated surface. The sherd appears to have been part of a larger object when used. There are no gold flakes on this sherd.

ANALYSES

Surface analyses revealed the following elements: Ag, Cu, Fe, S, Na, Mg, Al, Si, Cl, K, Ca, Ti, Mn, P. It has a relatively high calcareous content. The presence of Ag, Cu, Na and Cl suggests that it is from a parting vessel rather than a melting sherd. The absence of gold flakes suggest that the original vessel might have contained foils rather than gold dust.

Sherd 46849Y (Figs 5.44, p. 137, 5.45 and 5.46)

This is a nearly square parting sherd with parallel turning lines of the original thrown pot. The core is pink/red in colour where exposed by a break. This colour indicated the possibility for the sherd being from a parting vessel. The surface is blackened but not vitrified or bloated, and shows no signs of overheating. The concave surface is two thirds blackened with a brown area between. The broken edges show that the sherd was part of a larger object when in use.

SEM examination shows the finely textured surface with no vitrification and the surface appears similar to the exposed core (Fig. 5.45). There are various tiny particles with a metallic appearance on the surface (around 10–20 μm diameter) which were not seen optically (Fig. 5.46).

ANALYSES

Surface analysis in different areas show similar results with the presence of silver, sodium and chlorine. The following elements are present: Ag, Cu (two of four), Fe (high), Na, Mg, Al, Si, Cl, K, Ca (high), Ti, Mn, P. The exposed pink core of the sherd also contains Ag, Na and Cl showing impregnation of the sherd by these elements characteristic of the parting process. The metallic particles are of three kinds. Two particles are of gold alloy, one of which is almost pure gold (Ag 1.4%) and the other of which is impure (Ag 23%, Cu 0.3%) and similar to an alluvial gold flake. Other particles are silver-rich, both with and without chlorine, but contain no gold. Finally, there are lead-rich particles which have no silver. The presence of these particles of very different composition on the same sherd is most puzzling and may suggest limited reuse. Overall, the combined presence of silver, sodium, chlorine plus the pink core, the non-vitrified surface, the tiny metallic particles which are not spherical, and thus have not been molten, show similarities to the parting sherd 44401Z, which suggests that sherd 46849Y is also from a parting vessel.

Characterisation of the furnace bricks

Furnace brick 20, 47668S (Figs 5.4, p. 133, and 5.47 to 5.49)

A thin section was made from this irregular-shaped fragment from a parting-furnace brick.

Some of the exposed surfaces appear to be clean and relatively new thus exposing the original internal core of the brick. The brick is red/pink colour throughout with some coarse white, cream and grey inclusions. The structure is clearly low fired[9] with some larger porosity that may well have originated from an organic binder which has burned out.

The thin section shows the structure very clearly, there being very little visible binding glassy phase between the clay and mineral components (Fig. 5.47; see also p. 160). Essentially it is a mud brick. The porosity is quite high, estimated to be around 30% between the larger grains when seen at ×200 magnification. The structure shows a variety of angular, crushed (?) mineral grains of various sizes, plus mica. The overall structure is seen at low magnification in Fig. 5.48.

ANALYSES

Analysis of the cross-section at various places gave similar results. In particular, there is a substantial silver content (3.6%) together with sodium and chlorine. Other elements present are Fe, Mg, Al, Si, K, Ca, Ti, P. In cross-section, there are precipitated grains of silver chloride. By contrast, the parting elements, sodium and chlorine, which are found absorbed throughout the section and impregnate the structure, show no physical form. The composition and colour of the furnace brick are similar to the parting sherd 44401Z, which is to be expected as they are of the same basic fabric and have both been exposed to the parting process. The whole structure and atmosphere of the working parting furnace with its parting vessel were clearly permeated with volatile silver chloride. The porous nature of the vessel allowed reaction products to move away from the gold and to be absorbed into the furnace walls from which the silver could be extracted by cupellation.

Fig. 5.45 SEM micrograph of parting-vessel sherd 46849Y, showing the surface has a similar appearance to the exposed core.

Fig. 5.46 SEM micrograph of a gold-rich (Au 77%, Ag 23%) grain on the surface of parting-vessel sherd 46849Y.

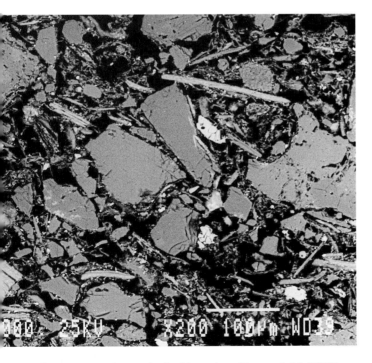

Fig. 5.47 SEM micrograph of a thin section of furnace brick 47668S shown in Fig. 5.4, p. 133. The image shows the mud-brick texture with very little binding glassy phase.

Fig. 5.48 SEM micrograph of furnace brick 47668S in cross-section at low magnification (×10), showing the mineral-grain temper.

Discussion

It is very noticeable that coarse cooking pots make up almost 100% of the ceramics used for the refining of gold, although they make up only about 5% of the total ceramics excavated from Lydian Sardis,[10] showing deliberate material selection of coarsewares for metallurgical purposes. Earlier reports[11] noted fragments of heat-distorted and bloated hydriae and other vessels of a finer ceramic body, and suggested these were the parting vessels. However, the majority of these have no metallic traces or evidence of metallurgical processes and presumably are from vessels that became involved in ordinary domestic conflagrations.

Refractory sherds with gold globules

The sherds in this group (apart from three) all had some gold globules trapped in the bloated or vitrified surfaces. Figure 5.5, p. 133, shows a typical sherd with a bubbly surface, which is clearly the result of strong heating from above, and gold globules are trapped in the once sticky glazed surface. This particular sample had many globules, that range in size approximately from 10 to 250 μm in diameter. Other sherds have fewer globules, but the overall range of size is extreme, 3–300 μm. Sometimes the globules appear in groups, and elsewhere they are spread around the sherd surfaces.

The globules are clearly derived from once-molten gold, demonstrated by their spherical shape and dendritic structures (Figs 5.22 and 5.26). They are thought to originate from melting small quantities of purified gold for assay.

There is little published data on the composition of the alluvial gold from the Pactolus, but unpublished analyses on a few samples show about 15–30% silver and perhaps a little copper. Table 5.1 shows the composition of gold globules analysed on the surfaces of the sherds. Most of the sherds have globules whose composition indicates purified gold with silver contents mostly of less than 5% and many globules are fully pure.

A minority of sherds have globules of very impure gold (e.g. sherd 45680Q) with a silver content of around 40%, which is well above the composition of the Pactolus argentiferous alluvial gold. However, it is very close to the values obtained by the analysis of the early Lydian electrum coins in the British Museum and in some of the Sardis foils analysed by NAA.[12] Further, some of the impure globules are associated with lead containing bismuth. This shows that the gold is not impure gold straight from the Pactolus. Together this evidence strongly suggests that old electrum coins were being melted down prior to being hammered into foils for parting.

One sherd, 45670U, has 11 pure gold globules and also one globule with 24.7% silver. Sherd 45681Z (sherd IV) also has electrum globules amongst purified globules. The presence of globules of gold with very different compositions poses a problem. One would expect any previous residual gold globules to immediately melt into a subsequent batch and homogenise with this gold. There are two possibilities. Perhaps the earlier gold globule was completely fused within the bloated sherd surface and protected from the subsequent melt(s), but was subsequently exposed due to physical loss of glaze. Alternatively, only very small quantities of gold may have been melted on the sherds, for purposes of assay, such that the different samples did not physically cover the same area.

This evidence of gold of very different composition on the same sherd suggests, as one might expect, that the gold-melting sherds were used many times, presumably until broken or unserviceable.

Sectioned gold globules

Impure gold globules were chosen because the results of their analyses would be more informative than those from the pure gold globules and additionally would enable any surface enrichment during 2.5 millennia of burial to be measured. Impure gold globules from two sherds were removed, mounted in resin and polished in cross-section. Table 5.2 shows the compositions of these globules. Those from sample 45684T are about 200 μm in diameter and the silver value of about 18% is similar to the in-situ surface values obtained on other globules on this sherd. Analysis across the sectioned gold also shows minimal variation between surface and core indicating that surface enrichment from burial has resulted in less than 1% reduction in the silver. A globule with a high-silver content (40%) from sherd 45680Q is seen in Fig. 5.23. This is also about 200 μm in diameter. Discrete enriched zones containing lead and bismuth were detected, which appear white in the micrograph. This is significant as these two elements are often associated with ancient silver but would not normally be present in silver from placer gold.

Early Lydian electrum coins contain more silver than Pactolus alluvial gold. Therefore, it appears that the first electrum coinage was made of the alluvial gold from the Pactolus plus extra silver. Thus the gold was debased to a fixed standard.[13] What we may be seeing in the electrum globules that have a high silver content and traces of lead and bismuth, is evidence for the addition of silver from another source, e.g. argentiferous galena, to make the original debased electrum coinage. Thus the extent of gold refining at Sardis goes beyond simply processing alluvial gold from the Pactolus.

It is suggested elsewhere (see p. 202) that all the scrap gold was cupelled prior to cementation and thus the lead could come from that operation. However, these impure globules would seem to be the remnants from an assay sample taken prior to the cupellation, as the metal also contains about 2% copper.

Surface analysis of bloated sherds

Table 5.1 shows typical surface analyses of the sherds. These show an iron-rich fabric which typically contains some phosphorus, sodium and chlorine. It is most probable that these elements are from the residues of the parting charge carried over with the foils. Calcium also tends to be high on some surfaces and white grains of calcite (?) are frequently seen. The regular presence of this calcite suggests that lime was added to flux the melting.

Cross-section of bloated sherds

Sherds 44399U and 45677R have typically bloated surfaces and contain some gold globules. Examination in the SEM showed that both sherds have continuously vitrified, bloated surfaces only on their concave sides (Fig. 5.11, sherd 44399U). Clearly, the sherds had been heated strongly from the upper concave surface for a limited time. This is consistent with the sherds being part of a receptacle heated from above to melt the gold. Petrographic study indicates that the temperature of the vitrified surfaces was well over 1100 °C to cause the observed bloating, and that this temperature was held for only a short time (p. 161).

The interpretation is clear. The sherds used for melting the purified gold have been strongly heated, but only from above, for the minimum time needed to melt the gold. The melting sherds were probably heated on the ground in some of the shallow depressions, also used for cupellation, by a small charcoal fire over the top.

Function of the gold-melting sherds

However, this does raise questions of whether the melting of what inevitably must have been small amounts of gold on pot sherds would have been sufficient to process the quantities of gold necessary for the Lydian coins and other large-scale production for jewellery. Perhaps we are seeing evidence of only a small part of the whole Sardis gold-refinery operation and these sherds were for melting gold for touchstone assay (see Appendix 5, p. 247). Theophilus[14] states that sherds of earthenware cooking pots were to be used for the fire refining of silver.

The bloating layer on the sherd surfaces is often fractured through, showing that the melting sherds must have been significantly larger when used. The curvature of the sherds suggests a large-diameter vessel. The fact that the sherds are heated directly from above suggests the open form of a sherd rather than of a complete and thus closed vessel. Some sherds with evidence of gold melting incorporate rims or the remains of handles, showing clearly that they are sherds from common cooking vessels. The common Lydian cooking pot has the same petrographic fabric as the melting sherds (see p. 158). This interpretation presupposes that sufficient broken domestic cooking vessels were available to supply the production requirements of the gold refinery. However, in the absence of any specialised crucibles there appears to be no alternative explanation. Perhaps the local production of coarseware cooking vessels was so prolific and cheap that new vessels were broken for the purpose, too.

Refractory sherds with lead-rich surfaces and no gold globules

These sherds are of the same fabric as the gold-melting sherds. They are characterised by highly glazed areas, which are often thick and a variety of colours – particularly black, but sometimes cream or brown. The glaze is only on the concave side on which they were heated and which formed the reaction surface. The glazed regions are crazed and they all contain high lead (as oxide) and various amounts of silver and copper, and some contain small white calcareous grains. These are likely to have been deliberate additions of lime to flux the operation. The surface chemistry is very different from the gold-melting sherds and none has any gold globules, and the sherds were therefore not used for melting gold. The glaze is often broken through at the edges of the sherds showing that they were of a larger size when used. The bodies of the sherds, as seen at the fractured edges, are all oxidised red in colour.

The thick lead-rich glaze and the absorbed silver and copper (globules of silver chloride are present on one sherd) suggest that these sherds have been used for cupellation purposes, but the small size of the sherds suggests that the cupellation was for assay.

Parting-vessel sherds

There is a predominance of sherds for melting gold, but only a small number of the actual parting vessels that are central to the parting process. This is understandable because they would have been recycled to extract the silver absorbed during purification, by cupellation, and it is perhaps significant that many burnt but unvitrified sherds were excavated from the north-east dump that also contained the litharge cakes (see p. 86).

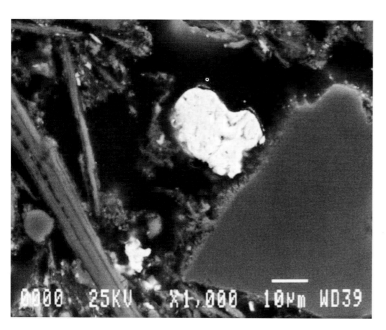

Fig. 5.49 SEM detail micrograph of an area of the sectioned furnace brick 47668S (×1000), showing the precipitate of silver chloride within the open structure.

Three sherds, 44401Z, 45671S and 46849Y, stand out amongst the whole assemblage as being different from the melting sherds, although the fabrics of all the sherds are broadly similar (see p. 158). These three sherds have similar characteristics to each other but sherd 44401Z is the most distinctive (see pp. 122–4) The principal distinguishing features of this sherd are:

- The colour of the sherd is slightly pink to purple.
- There are tiny flakes of gold or electrum on the surface, not spherical globules.
- The surface is loaded with silver chloride, and residual sodium.
- The sherd fabric is impregnated with silver chloride and sodium.
- The surface is not bloated by intense heat.
- There is no thermal gradient recorded through the sherd fabric.

The purple colour in the convex surface layers is probably due to the silver chloride, and the pink colour of the rest of the sherd results from the effect of chloride attack on the iron-rich refractory fabric.

Origin of the parting sherds

Although only three sherds were identified as being from parting vessels, they were clearly part of larger vessels. Two of the sherds, 44401Z and 45671S, have essentially the same fabric and thickness as the domestic coarseware cooking pots. Furthermore, some later descriptions actually specify the use of coarseware cooking pots[15] and the Lydian cooking pots do fit rather well into the contemporary cementation furnaces (Fig. 5.3). These pots had all of the properties required for an efficient parting vessel:

- They were large enough to hold a good volume of the parting-charge mixture and gold foils or grains.
- They were porous, which allowed air and the vapours of water and silver chloride to pass into and through the coarseware fabric, thereby removing silver from the region of the purifying gold.
- They often had handles for manipulation.
- The coarse-quartz temper imparted thermomechanical stability for long-duration heating.
- They were common and presumably cheap, with no specialised fabrication requirements.

Furnace bricks

The whole of the excavated furnace structures are pink in colour as a result of the absorption and chemical interaction of the salt vapours with the iron-rich fabric, as noted above. The distinctive colour is characteristic of salt parting which results in mobility of the iron salts.

A typical furnace-brick fragment (Fig. 5.4, p. 133) was removed from one structure and was mounted and polished in cross-section for SEM examination (see p. 124). The petrographic structure shows significant differences from the cooking vessels. There are large mineral grains as temper in a generally coarse-grained and very-low-fired body (Fig. 5.48). The low-fired body of the mud brick shows that the working temperature of the furnace was low, which is consistent with the requirements for the parting process and recovery of purified gold.

The brick body in cross-section contains 3.6% silver along with significant amounts of sodium and chlorine (Table 5.1). The extraordinary amount of silver makes the furnace an extremely rich silver 'ore'. Presumably, the Lydians recognised this and reprocessed the old furnace bricks to extract the silver, hence the heap of bricks awaiting reprocessing (see p. 86). Because of their low firing, these bricks would be relatively easy to crush for smelting.

The brick body illustrated in Fig. 5.49 at high magnification in the SEM, shows precipitated grains of silver chloride (white).

Summary of the parting process

The parting process involved surface chemical reactions

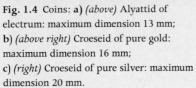

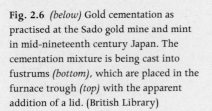

Fig. 1.4 Coins: **a)** *(above)* Alyattid of electrum: maximum dimension 13 mm; **b)** *(above right)* Croeseid of pure gold: maximum dimension 16 mm; **c)** *(right)* Croeseid of pure silver: maximum dimension 20 mm.

Fig. 2.6 *(below)* Gold cementation as practised at the Sado gold mine and mint in mid-nineteenth century Japan. The cementation mixture is being cast into fustrums *(bottom)*, which are placed in the furnace trough *(top)* with the apparent addition of a lid. (British Library)

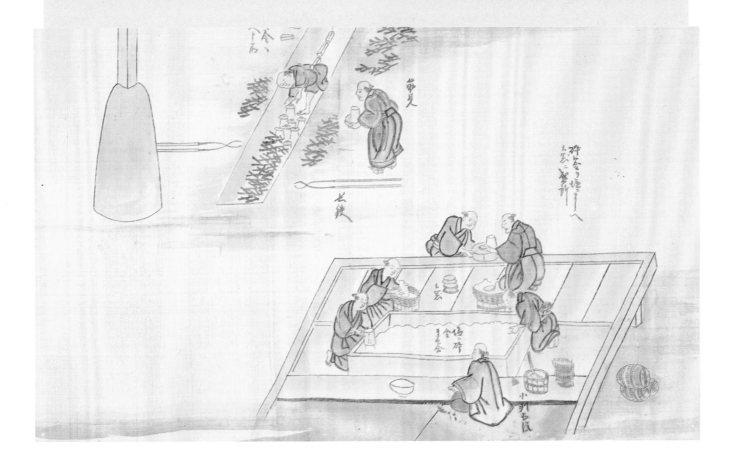

Fig. 4.20 Test trench in cupellation area A.

Fig. 4.28 Furnace area B: furnaces 1, 2 and 3.

Fig. 4.39 *(above)* Detail of coarse potsherd with gold globules.

Fig. 4.44 *(above right)* Granule of alluvial gold, first thought to be a large dribble (see Appendix 1, sample 30, p. 218).

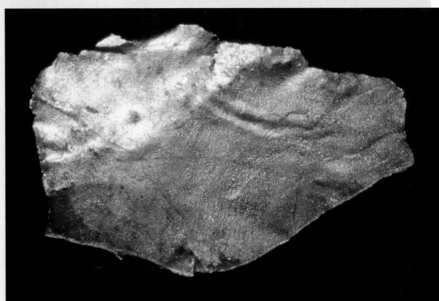

Fig. 4.46 Gold foil (see Appendix 1, sample 11, p.216).

Fig. 4.50 Oblique view of the scoriated and mended tuyere (see Appendix 2, item 1, p. 223).

Fig. 5.1 Sherd 45680Q: gold-melting sherd.

Fig. 5.2 Sherd 44401Z: parting-vessel sherd.

Fig. 5.3 Parting furnace with reconstructed cooking vessel put in the position of a parting vessel.

Fig. 5.4 Furnace brick 47668S.

Fig. 5.5 Sherd 44396P: gold-melting sherd with vitrified surface.

Fig. 5.6 Sherd 44398W: gold-melting sherd with vitrified surface.

Fig. 5.8 Sherd 44399U: gold-melting sherd with vitrified surface.

Fig. 5.13 Sherd 45673Z: gold-melting sherd with vitrified surface.

Fig. 5.17 Sherd 45676T. This sherd has no surface bloating and appears more like a parting sherd. There are no gold globules in the surface.

Fig. 5.30 Sherds 44400Q: three sherds with lead-rich vitrified surfaces.

Fig. 5.31 Sherd 45672Q: sherd with lead-rich vitrified surface.

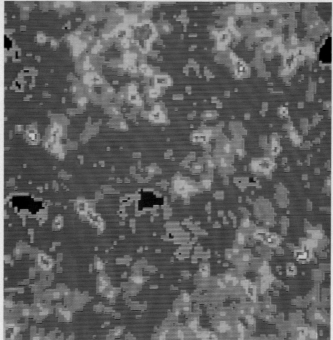

Fig. 5.34 *(top)* Sherd 45678P: sherd with thick surface layer rich in lead and silver.

Fig. 5.39 X-ray distribution map for silver on the surface of the parting-vessel sherd 44401Z. Magnification × 250: 25 mm = 100 μm on image.

Fig. 5.43 Sherd 45671S: parting-vessel sherd, but with no gold flakes on the surface.

Fig. 5.44 Sherd 46849Y: parting-vessel sherd with a few gold particles on the surface.

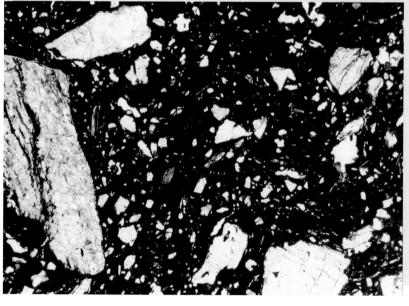

Fig. 6.1 Micrograph showing the fabric (fabric 1) of sherd 44401Z, thought to be from a parting vessel. This is the typical fabric of the parting-vessel sherds. Plane-polarised light, width of field *c.* 4.5 mm.

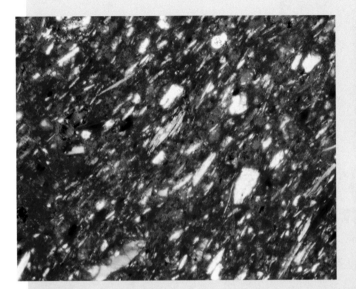

Fig. 6.2 *(above left)* Micrograph showing the fine fabric (fabric 2) of sherd 46849Y, thought to be from a parting vessel. This is the only parting-vessel sherd found to be made in this fabric. Crossed polars, width of field *c.* 1 mm.

Fig. 6.3 *(above)* Micrograph showing the distinctive fabric (fabric 3) of the so-called bread trays (sherd 45690X). Crossed polars, width of field *c.* 4.5 mm.

Fig. 7.1 PGE inclusions in a Lydian gold coin.

Fig. 9.1 *(above)* Refined gold foil sample 1A. Its surface has an extremely granular appearance.

Fig. 9.2a Refined gold foil sample 2A. Its surface has a granular appearance

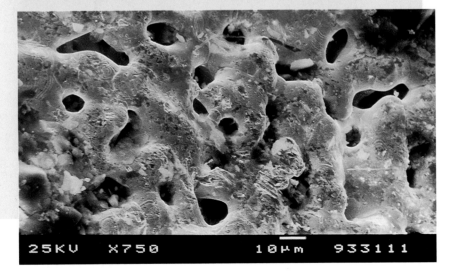

Fig. 9.2b SEM micrograph of the original surface of sample 2A, showing extensive porosity due to the removal of all of the silver during the parting process.

Fig. 9.3 *(above)* Gold foil sample 3A. Its surface is partially covered with thick, pinkish contamination. (Surface contains 4.7% silver.)

Fig. 9.4 Gold foil sample 4A. Its surface is partially covered with thin, transparent, pinkish contamination. (Foil contains 17.3% silver.)

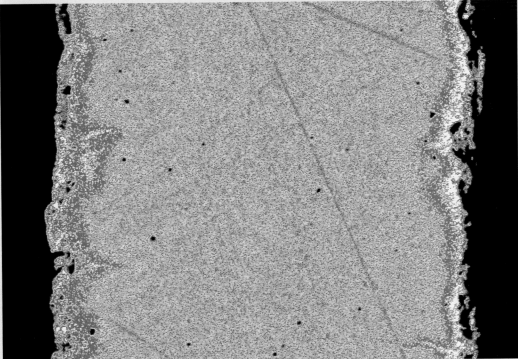

Fig. 9.5 *(top)* Gold foil sample 28A. Dark and buff-coloured contamination on the surface. The foil is appreciably thicker than the others and contains 22.8% silver at the surface.

Fig. 9.25 False-colour SEM micrograph of a section through foil G-11-2, showing initial attack at surface and grain boundaries after 30 minutes at 500 °C.

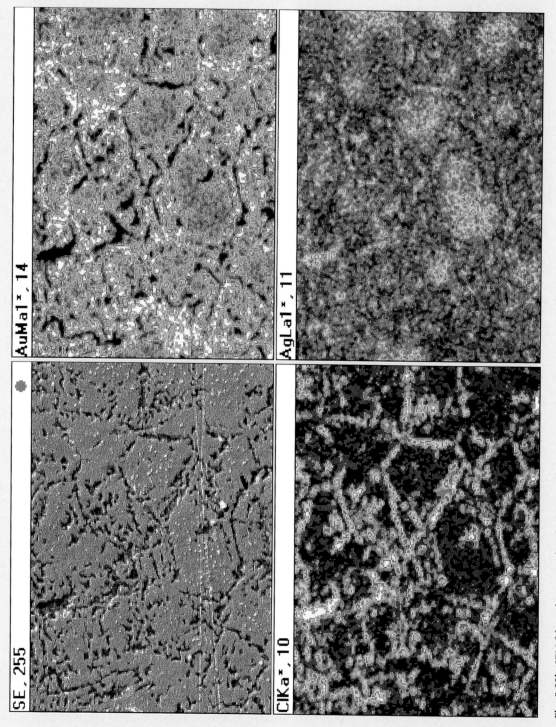

Fig. 9.33b SEM false-colour x-ray distribution maps of cube G-13-3 for gold, silver and chlorine.

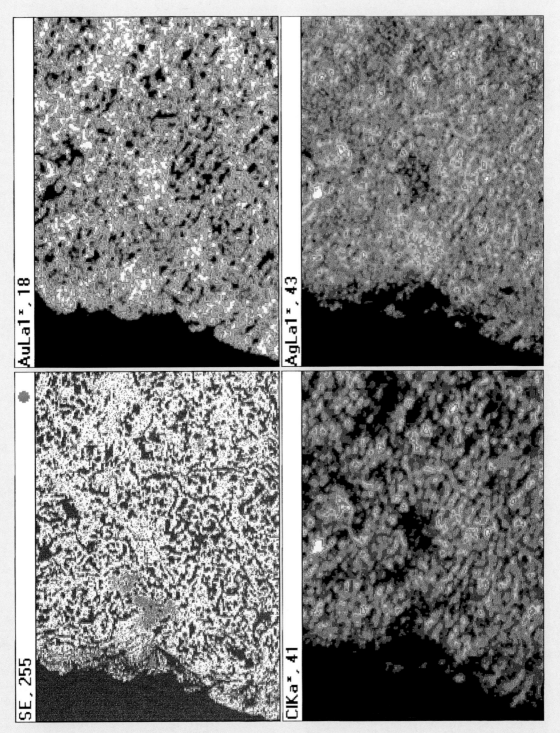

Fig. 9.34 SEM false-colour x-ray distribution maps of cube G-13-4 (after 60 minutes at 700 °C) for gold, silver and chlorine.

A1.10 **A1.10** The pieces of gold foil comprising sample 43.

Fig. A2.7 Three sherds of glazed and vitrified pottery: **a**) neck of a cooking pot, heavily glazed on the inside; **b**) rim of tableware dinos, heavily glazed and vitrified; **c**) cooking-pot sherd showing abrupt transition from vitrified to plain surface.

and thermally activated solid-state diffusion within the metal. During the process, silver in the surface of the impure gold reacted with chlorine ions and ferric chloride, and the resulting silver chloride volatilized thus reducing the silver content of the surface. Silver from deeper within the metal diffused towards the surfaces where it reacted with the chlorides. The reaction occurred most rapidly along the grain boundaries, leading to the formation of a network of grain-boundary porosity as discussed below and on pp. 182 and 187. Silver chloride melts at about 455 °C and loss by volatilization is appreciable at elevated temperature.[16] The absorbed silver chloride found inside the Sardis parting sherds and furnace bricks clearly shows its mobility.

The impure gold foils (or grains of alluvial gold) would have been mixed with salt and probably an inert carrier, such as brick dust, and put into the parting vessel. This vessel would have been put into the furnace and run under oxidising conditions for a period of time. It is estimated from the relatively low vitrification of the sherds that the temperature of the furnace was low, certainly no higher than about 800 °C, thus keeping the parting process well below the melting point of the impure gold (see p. 160). Another reason why the temperature of the parting process must not have been too high would be to prevent the vitrification of the charge in the presence of excess sodium. Vitrification or sintering would slow the reaction down and prevent extraction of the foils from the spent charge, and also render the parting vessel useless.

After the parting process, the purified gold foils would have been separated from the spent charge and the grains would have been recovered by washing. The spent charge would still have been rich in silver chloride. More of the silver chloride would have been in the parting-vessel walls and the furnace walls, while some would have been lost to the air.

EXAMINATION OF GOLD FOILS AND DUST

Gold fragments from the Archaeological Museum at Manisa

The gold from the PN site is described in Appendix 1 (pp. 215–20), and many of the gold foils from the parting process have been examined at the Istanbul Technical University and are reported and illustrated on pp. 184–7, and in Figs 9.1 to 9.5, pp. 139–41. Figure 5.50 illustrates the characteristic porous surface of one of these foils (sample 8A), which are around 60–100 µm thick. The compositions of these foils range between 45% silver and pure gold.

From the inventory of gold samples at the Manisa Museum and the subsequent list of 55 gold specimens sent for analysis by the Archaeometry Center at the Bosphorus University, five gold fragments were selected, mounted in resin and polished for microscopy and analysis at the British Museum.

Table 5.3 Analysis of polished sections of Sardis gold foils and pieces from the Archaeological Museum, Manisa

Sample	Analysed point	Au%	Ag%	Cu%	Cl%	Analysed point	Au%	Ag%	Cu%
16A (porous)						grain centre	99.6	0.4	–
19A (porous)	edge area	88.6	11.4	–	–	grain centre	89.0	11.0	–
		89.8	10.2	–	*				
26A (some pores)	edge area	68.9	30.3	0.8	–	grain centre	68.3	31.0	0.7
	core area	67.0	31.7	1.3	–				
30A (solid)	core area	83.3	16.2	0.5	–				
33A (some pores)	core area	82.0	18.0	–	*	grain centre	82.7	17.3	–
						grain centre	78.3	21.7	–
						grain point	85.9	14.1	–
						grain point	84.4	15.6	–
						grain point	84.5	15.5	–

* Small amounts of chlorine detected.

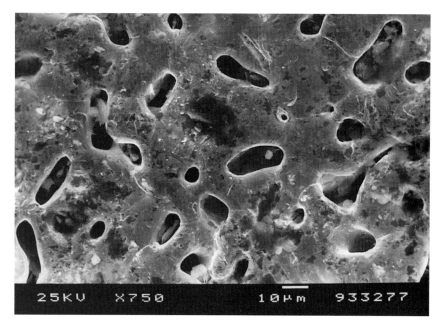

Fig. 5.50 SEM micrograph of the surface of gold foil 8A from the Sardis excavation, showing the characteristic porosity associated with foils that have been through the purification process within parting vessels in the furnaces.

SEM examination of the Manisa Museum's polished gold samples

The analytical results are shown in Table 5.3. The three foil samples, 16A, 19A and 26A, were mounted and polished as a section parallel to the plane of the foil.

Sample 16A: foil

Sample 16A, shown in Figs 5.51 and 5.52 (detail), is considerably thicker than the usual Sardis foils (which are around 60–100 μm thick). However, the microstructure of this sample is very informative with regard to the physical diffusion mechanism of the purification process. The porosity has the appearance of being in two forms, channels along grain boundaries and pores that are apparently closed within grains. However, in three dimensions, the porous structure is an interconnecting sponge network which allowed the reactive chemicals and reaction products to have access to a convoluted surface. Physical measurements show that from the centre of any solid grain the maximum distance to a free surface of a pore is around 6 μm, which is thus the maximum

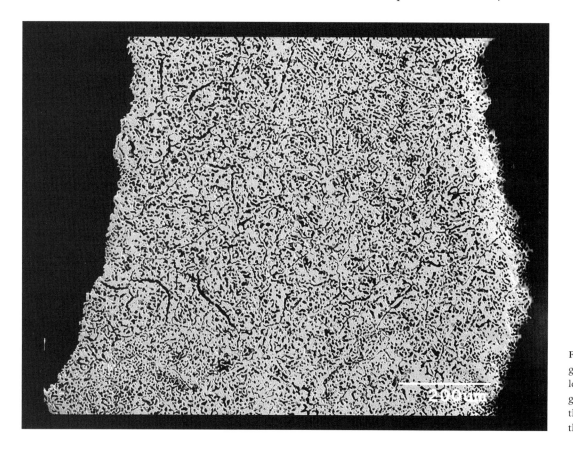

Fig. 5.51 SEM micrograph of a polished longitudinal section of gold sample 16A, showing the characteristic porosity throughout.

distance silver atoms must have diffused in order to react with chloride ions at a free surface. This gold fragment has clearly been subjected to the parting process.

ANALYSES

Analyses of various solid grains within the porous body show almost pure gold (Au 99.6%, Ag 0.4%). The homogeneity within the grains shows that the process was essentially complete.

Sample 19A: foil

The sample is relatively thick (c. 300 µm) (Fig. 5.53 and detail Fig. 5.54). It is homogeneously porous throughout, similar to sample 16A, but perhaps with a less well developed interconnecting network of pores, although the mean distances from grain centres to free pore surfaces are similar to those in sample 16A. Comments concerning the porosity and diffusion processes of sample 16A also apply to sample 19A.

ANALYSES

Analyses were made of various solid grains from the edge to the core. In all cases, the gold was found still to be impure with around 11% silver still present. (NAA analyses by Gordus reported 11.9% Ag; see Table 7.1, p. 170.) The

sample has clearly been subjected to the parting process and the alloy must originally have contained much more silver than is found now, judged simply by the volume of porosity throughout due to loss of silver. There does not appear to be any coring of the grains. This can be explained if heating continued long after the reactants had been exhausted, which homogenised the partly purified alloy.

Sample 26A: fragments

At low magnification, this sample consists of a mixed collection of fragmentary foils (Fig. 5.55). Their thicknesses vary between 20 and 300 µm. The thicker pieces show a very interesting intermediate purification structure with discontinuous porosity predominantly along larger grain boundaries and not uniformly distributed across the sample area (Figs 5.56 and 9.18, p. 193). The large grain size is interesting. As the grain growth cannot continue once the porosity network has been established, this suggests a relatively long time at elevated temperature before the porosity network became established.

ANALYSES

One of these pieces was analysed by Gordus (see Table 7.1, p. 170), who found 10.4% silver and 0.7% copper. Analyses here were made at various points from the edge to the

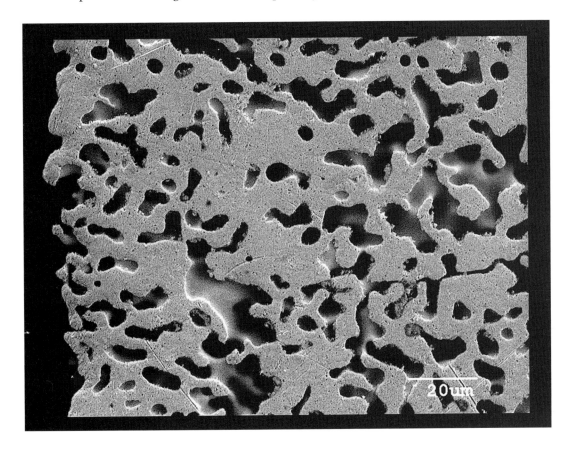

Fig. 5.52 SEM detail micrograph (×750) of a polished longitudinal section of gold sample 16A, showing the porous network penetrating the gold.

centre of one of the pieces. In all cases, similar analyses were obtained, the gold containing about 31% silver and 1% copper; clearly this was not the same fragment as that analysed by Gordus. The homogeneity confirms the evidence of the large grain size, that overall this was a process of long duration. Once again, the homogeneity also shows that the heating continued long after the reactants had been exhausted.

Sample 30A: natural granule of alluvial gold

The sectioned sample appears as an irregular-shaped, elongated piece which is quite narrow at the middle: 3.5 by 0.3 mm at the narrow point (Figs 5.57 and 4.44, p. 131; see also Appendix 1, p. 218). Two other smaller irregular pieces (the smallest is 0.3 mm across) are conjoined to the larger piece. Unlike the other samples, the pieces are not porous. The larger piece has some apparent cracks that lead inward a short distance from the surface but do not appear to be associated with corrosion; however, on close examination the cracks look more like folds in the metal. Moreover, the gold pieces are not rounded and therefore do not appear to be a 'dribble' that was once molten, as suggested in the original description. The irregular shapes and textured outlines of the pieces look more like the alluvial grain in Fig. 5.64.

ANALYSES

Area analysis of the cross-section of the large piece shows it to contain Au 69.6%, Ag 29.8%, Cu 0.6%, a typical composition for unrefined native gold. A small lump on the side of the large piece has a different composition, but still well within the range of natural gold: Au 83.3%, Ag 16.2%, Cu 0.5%. Probably these are two grains of alluvial gold stuck together over time. The non-porous nature of the sample shows it has not been in the parting process, and it has not been molten. No PGE inclusions were observed on this granule.

Sample 33A: fragment

This is an unusual-shaped piece with the appearance in section of a cross or starfish with three complete 'arms' and one broken one (Fig. 5.58). The body and 'arms' are largely porous at grain boundaries. The outlined grains appear equiaxed, showing that the gold is now annealed due to the temperature and time of heating in the parting process. The innermost regions of the 'arms' are slightly less porous than the edges (Fig. 5.59). The porosity is similar to that of samples 16A, 19A and 26A. Clearly, this sample has been in the parting process but the distribution of the porosity suggests that the process was not complete.

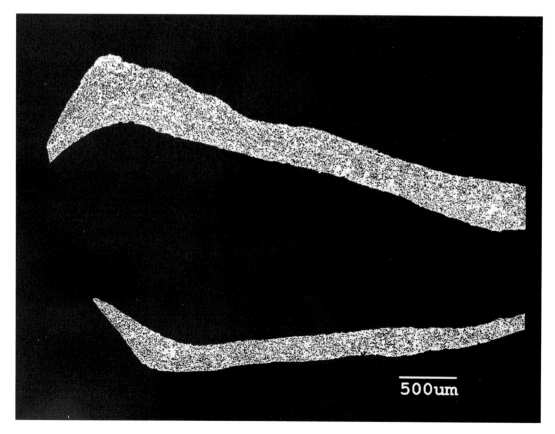

500um

Fig. 5.53 SEM low-magnification micrograph of a section through gold sample 19A.

ANALYSES

Area analysis of the cross-section shows that the sample contains about 18% silver. Detailed analysis of individual grains within the section showed only small variations in composition, indicating that the purification process had reached a reasonably stable situation, presumably at the exhaustion of reactive agents, although heating continued to homogenise the composition. The greater porosity at the surface shows that the parting process was still in progress when the removal of silver ceased, but the heating clearly continued to homogenise the composition. The only explanation is that the heating continued long after the active agents of the cement had become exhausted.

Summary

Sample 30A is very different from the other samples in that it has no grain-boundary porosity. This is a sample of primary, unprocessed alluvial gold.

The other four samples are all hammered foils or fragments. Three of them are of impure gold with substantial residual silver contents. Samples 19A, 26A and 33A all have varying degrees of grain-boundary porosity, showing that they too have been in the parting process but have not proceeded to completion. Sample 26A shows evidence of the beginnings of grain boundary penetration of the reactants and of silver loss. It is therefore seen arrested at an early stage in the parting process. Sample 33A has more pronounced porosity at the surfaces than in the interior, as one would expect in a piece where the parting was in progress.

These pieces of gold alloy thus show a range of purification stages. The fact that several of these pieces are still impure suggests that the reactants were exhausted and purification could not reach completion, but homogenisation of the silver content shows the heating continued long afterwards. This further implies that the operators were not aware of the exhaustion of salt.

Only sample 16A approaches full purity and also has the characteristic intergranular porosity which occurs when gold–silver–copper alloys are subjected to the parting process.

Diffusion mechanism in the parting process

The Manisa Museum's gold samples represent different stages in the purification process and the mechanism of the process can be observed from the unprocessed alluvial

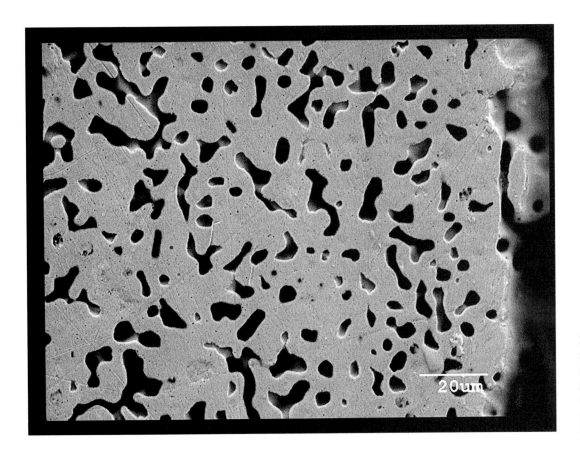

Fig. 5.54 SEM detail micrograph (×750) of a polished section through gold sample 19A, showing the porous grain-boundary network penetrating the gold.

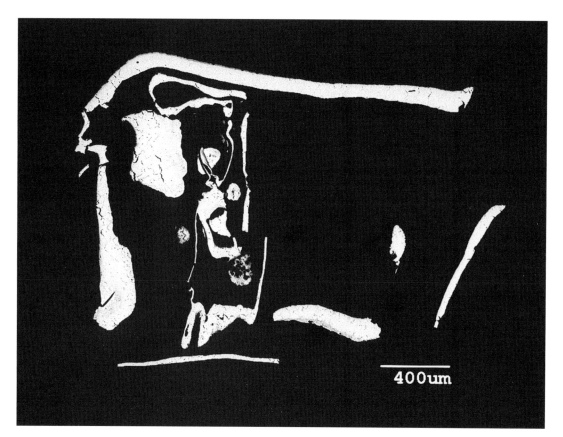

Fig. 5.55 SEM low-magnification micrograph of the collection of mixed scrap foil pieces of sample 26A.

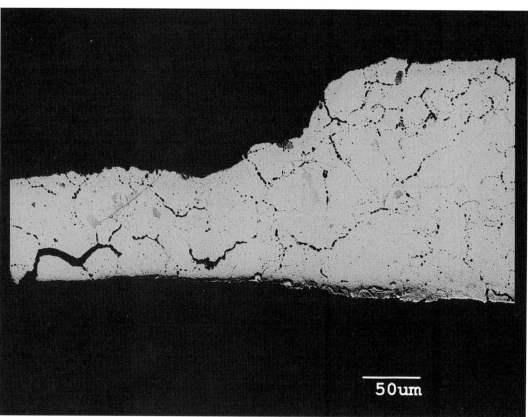

Fig. 5.56 SEM detail micrograph of a thick foil in sample 26A, showing the grain-boundary porosity of an early/intermediate stage of purification which appears to have stopped. The grain growth shows the high temperature was held after exhaustion of the reactants.

grains to fully purified gold. (See also pp. 187 and 200.)

The cementation parting process involves surface chemical reactions and thermally activated solid-state diffusion within the metal. The chemical reaction between the silver in the gold alloy and the active parting agents, most likely chlorine or ferric chloride released from the parting charge, occurs initially at the gold-alloy surface but appears to progress quickly inwards along grain boundaries to form pathways which allow reactants to penetrate the core of the alloy. Diffusion gradients soon form between the silver-depleted grain-boundary zones and the silver-rich core of the metal grains. As silver is lost from the interior, an open sponge-like network forms, creating more reaction surfaces. This is clearly seen in sample 33A (Fig. 5.58). In this way, purification becomes a three dimensional process rather than simply just progressing from the exposed outer foil surfaces. The open structure speeds the purification process, but leaves the gold in a weakened state.

The parting process proceeds as long as sufficient reactant (in this case sodium chloride) remains in the parting vessel, or until the gold is purified.

The purified gold grains

A sample of gold grains, or dust, from Sardis (now in the Archaeological Museum at Manisa) was studied for both surface morphology and composition by examination in the SEM at the British Museum. The exact find location of the grains is unclear but they are assumed to be associated with the gold refinery.

The sample-number prefix, KZP, relates originally to Kemal Ziya Polatkan, a former director of the Archaeological Museum at Manisa. Ten grains were selected from the small vial in which the gold dust is kept.

Surface examination of the gold grains by SEM

The grains have a variety of shapes and sizes with maximum dimensions between 250 and 1000 μm. The grains are clearly not molten globules of gold, such as are found on the coarseware sherds. Examples of these grains are shown in Figs 5.60a to 5.63a. Optically the surfaces of the grains were all matt from the microscopic surface textures. The grains are generally more flat than bulky and the edges and exposed surfaces are rounded from geological transport and abrasion. Some grains have thinner edges folded over. In the SEM at high magnification, the surfaces show fine textures, voids, irregular porosity and tiny protuberances (Figs 5.60b to 5.63b). These are the typical physical charac-

teristics associated with grains of natural alluvial gold.[17] However, although they have the gross morphology of natural grains, very high magnification suggests that they have been treated.

The grains are much larger than those found on the parting vessel sherds (which presumably became trapped because they were small), but placer gold does have a wide range of grain sizes. One grain was polished in cross-section and shows macroscopic porosity near the surface and microscopic porosity internally (Fig. 5.64).

Analyses

Analysis shows that all of the grains are of pure gold, thus confirming that they have passed through the parting process. The significance of this is that it shows beyond doubt that at least some of the placer gold was purified directly without being first converted into foils. Although unusual in later historic accounts of parting, the earliest detailed description of the process, given by Agatharchides,[18] states that the washed granules were placed straight into the parting vessel. The traditional Japanese process, as discussed by Gowland,[19] also used powdered (granular) gold.

The grains do not show the characteristic surface porosity found on the many gold foil fragments that have clearly been through the parting process. It is unclear, at this time, why this is so, although it could be explained by the small size of the grains allowing the natural surface microtexture to provide the necessary reaction surfaces.[20] The alternative hypothesis that the alluvial grains are of untreated gold is belied by the composition of the other grains of Pactolus gold, which, in common with alluvial gold elsewhere, have a silver content typically lying between 10 and 30%, as exemplified by sample 30A.

CONCLUSIONS

Detailed scientific examination of the refractory remains and gold at the Lydian gold refinery at Sardis has provided an insight into the main production processes.

Bulk gold for purification, probably from recycled scrap jewellery and electrum coins, was made into thin foils in order to make a large surface area to volume ratio which enabled the parting process to work efficiently. The purified foils show characteristic vacancy diffusion porosity both on the surface and within the metal, even where the gold is relatively thick. Grain-boundary diffusion is the key to fast removal of silver from the core metal. Residual reaction products from the parting process show that

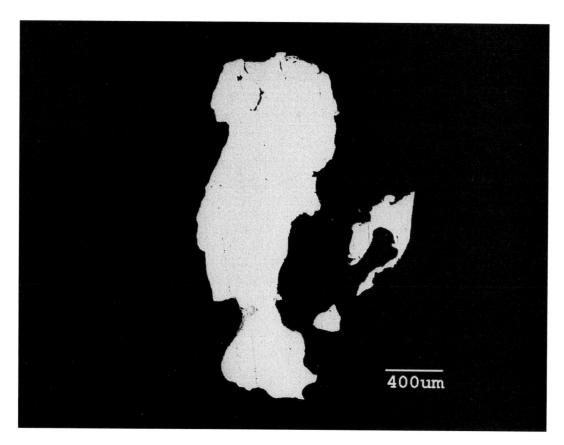

Fig. 5.57 SEM low-magnification micrograph of sample 30A, which is an irregular-shaped gold-alloy piece that does not appear to have been purified.

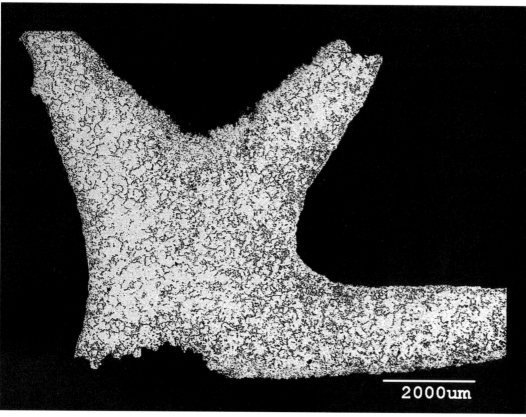

Fig. 5.58 SEM micrograph of a section of sample 33A. The gold-alloy fragment shows the extent of the network of grain-boundary porosity in this partly purified sample (18% Ag left).

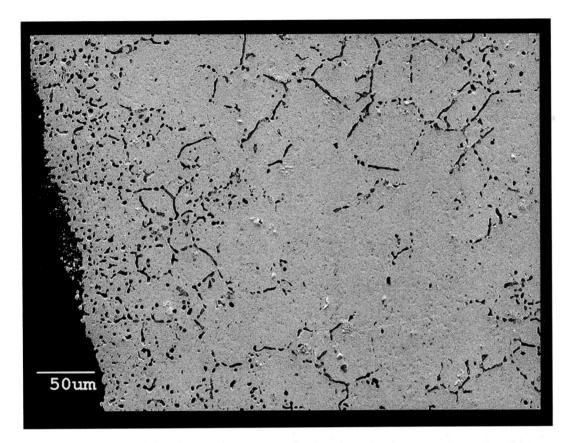

Fig. 5.59 SEM detail of sample 33A, showing the partly formed grain-boundary porosity of this partly purified alloy.

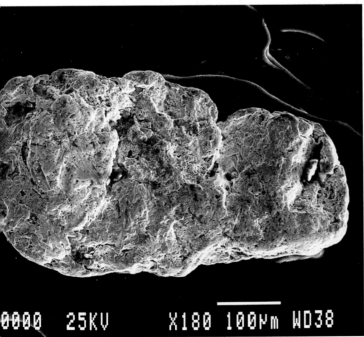

Fig. 5.60a SEM low-magnification image (×180) of a gold grain.

Fig. 5.60b SEM detail micrograph (×2200) of the grain in Fig. 5.60a, showing the porous surface texture.

Fig. 5.61a SEM low-magnification micrograph (×190) of a gold grain.

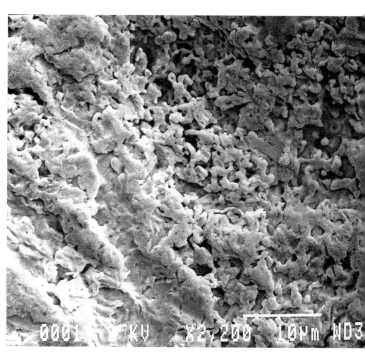

Fig. 5.61b SEM detail micrograph (×2200) of the grain in Fig. 5.61a, showing the porous surface texture.

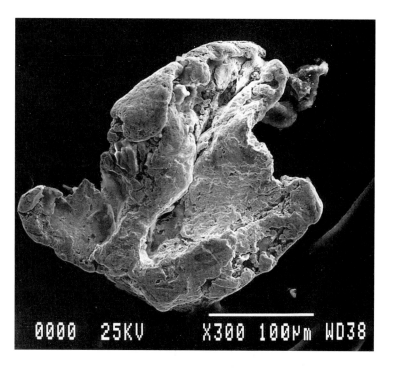

Fig. 5.62a SEM low-magnification micrograph (×300) of a small gold grain.

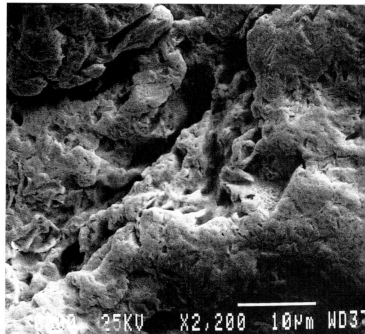

Fig. 5.62b SEM detail micrograph (×2200) of the grain in Fig. 5.62a, showing the porous surface texture.

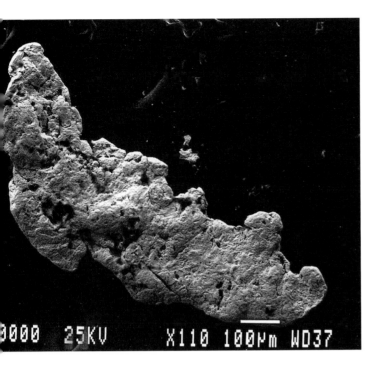

Fig. 5.63a SEM low-magnification micrograph (×110) of a gold grain.

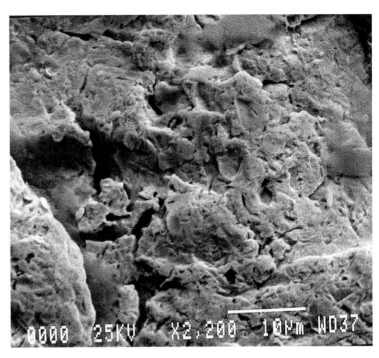

Fig. 5.63b SEM detail micrograph (×2200) of the grain in Fig. 5.63a, showing the porous surface texture.

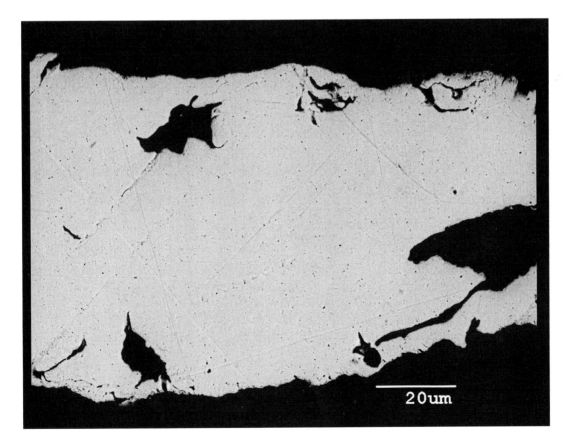

Fig. 5.64 SEM micrograph (×1200) of a cross-section through a gold grain, showing macroscopic porosity near the surface and microscopic internal porosity.

common salt (sodium chloride) was the chief reactive agent. The flakes of gold on the surviving parting sherds shows that the freshly mined alluvial gold was also treated directly in the parting vessels.

The petrographic evidence shows that the majority of the gold-melting sherds, as well as two of the three parting vessels, are of the same fabric as the common coarseware cooking vessels. The open texture of the ceramic is ideal for the parting process, allowing the continuous passage of water vapour and air *to* the reactants and of silver chloride *away from* the gold. Vitrification textures of the sherds and furnace bricks showed evidence of only moderate working temperatures of about 800 °C. High levels of silver (as chloride) were absorbed into the refractory vessels during the parting process and the small number of surviving sherds suggests that they may well have been recycled for extraction of silver, by a process involving cupellation. At Sardis, there are many small hearths and flat-sided tuyeres, both straight and right-angled, suitable either for melting gold on the sherds or for cupellation.

The furnace bricks are coarse grained with large mineral-grain temper and are very friable. They are unvitrified, once again showing that the furnaces were run at a relatively low temperature. The bricks are heavily impregnated with silver chloride absorbed from the parting process, and would have been routinely reprocessed to recover the silver.

The pink colour of the furnaces and the parting vessel sherds are characteristic of salt parting and the effects of the chlorides on the iron-rich clays. It also shows that the furnaces operated in an oxidising atmosphere.

Approximately half of the examined coarseware sherds had been used for melting relatively small quantities of gold. Analytical evidence from the gold globules trapped on the sherds shows that both purified gold and unpuri-

fied gold were melted, including what were probably earlier electrum coins for recycling. The bloated surfaces of these sherds show that they were heated from the top for a short period of time at temperatures of about 1150 °C during the gold-melting process. The concave fragments were suitable as containers for melting the purified gold foils and grains, heated from above. The gold would melt quickly on the insulating sherd. The sherds could only have contained small quantities of gold, suggesting they were used for melting samples of the gold ready for assay by touchstone.

Around a third of the coarseware sherds examined had significant levels of lead and silver in the vitrified surfaces, indicating their use in cupellation. Once again, the very small quantities that could be processed on the sherds suggests the cupellation was for assay.

The function of two of the sherds, 45676T and 45683V, are uncertain.

Many of the sherds had high calcium contents on the concave surfaces and small white calcareous lumps, suggesting crushed lime may have been added as a flux.

The samples of gold foils were in a range of states of purification. All had grain-boundary porosity to a greater or lesser extent. To some degree, the greater the extent of porosity the purer was the gold, which is to be expected if the gold to be treated was of reasonably constant composition. One granule of untreated natural gold alloy was studied, and a collection of granules of pure gold that, presumably, had been treated.

The purified gold foils were weakened after parting and, together with the purified granules of freshly mined gold, would have been consolidated by melting to form ingots. The lack of purpose-made crucibles at the PN site suggests this activity was performed elsewhere.

Notes

1 Cowell et al. (1998).
2 Bayley (1991a).
3 Bayley (1991a).
4 Meeks (1988).
5 Maniatis and Tite (1978–79).
6 Burstein (1989, pp. 65–6). See also p. 34.
7 Burstein (1989, pp. 65–6). See also p. 34.
8 Bayley (1991a).
9 See pp. 159–60 and 205–6 for results of temperature determinations.
10 Andrew Ramage, personal communication.

11 Ramage (1978a).
12 Cowell et al. (1998).
13 See Table 7.3, p. 171.
14 Hawthorne and Smith (1963, p. 96).
15 Boussingault (1833), quoted in Percy (1880, p. 390).
16 Mellor (1923, pp. 190–1).
17 Hallbauer and Barton (1987).
18 Burstein (1989, pp. 65–6).
19 Gowland (1917–18). See p. 49.
20 Reed-Hill (1973, pp. 254–93).

CHAPTER 6

Scientific Examination of Some Ceramic Materials and Samples of Litharge

A.P. Middleton, D.R. Hook and M.S. Humphrey

Introduction

A selection of ceramic materials excavated at Sardis has been examined using thin-section petrography, x-ray diffraction (XRD) and scanning electron microscopy (SEM) with elemental analysis using an energy-dispersive x-ray analyser (EDX) (see p. 101). A series of litharge samples from the cupellation process were chemically analysed using inductively coupled plasma spectrometry (ICP) and atomic absorption spectrometry (AAS) (Table 6.2). The aim of the investigation has been to characterise various fragments, which have been associated with several aspects of the gold refining and related processes thought to have been carried out at Sardis in the sixth century BC. The various stages in producing the refined gold and in recovering silver have been considered elsewhere and will not be described in detail here. The related activities are thought to have fallen into three stages: the purification of gold (i.e. removal of silver and any copper) by parting; the smelting of the silver-rich debris from the parting process with (added) lead; and the recovery of silver from the argentiferous lead by cupellation.

Materials and methods

The samples examined are summarised in Table 6.1. They include fragments of putative parting vessels, sherds believed to have been used for (re)melting gold, tuyere and furnace brick (all thought to relate to the gold-refining process), and also slagged furnace linings, tuyeres, burned bread trays and litharge cakes (all thought to relate to the silver-recovery process). In addition, some sherds of pottery and of unused bread trays (from domestic contexts) have been examined for comparative purposes. These various items are illustrated and described more fully on pp. 72–156.

Samples were removed from the majority of the ceramic fragments and prepared as thin sections for examination of their clay fabrics using a petrographic microscope. Some fragments were prepared as polished thin sections for examination in a scanning electron microscope, equipped with an energy-dispersive x-ray analyser for elemental analysis. The various techniques applied are indicated in Table 6.1. It was possible to observe the degree of vitrification of the ceramic fabrics of those samples examined by SEM. By comparison with the vitrification textures of small fragments of coarse pottery, refired under oxidising conditions at known temperatures (700–1100 °C) in a laboratory furnace, it was possible to draw some inferences regarding the temperatures reached and the duration of the various processes.[1]

Additionally, the litharge cakes were analysed for a range of major and trace elements using a combination of ICP[2] and AAS.[3] Seven samples (which had been identified previously using XRD as being principally of litharge, PbO, and cerussite, $PbCO_3$) were crushed and homogenised. Twenty-five milligrams of each sample were accurately weighed into high-purity graphite crucibles (Johnson Matthey C007220/UF-4S), followed by 100 mg of Specpure 100 lithium tetraborate. Batches of crucibles were heated in a furnace at 1050 °C for 15 minutes, to convert the samples into small glassy beads. (Some of the more volatile elements may have been lost during the fusion process, e.g. As, Cd, Hg, S, Sb.) In each case, the whole crucible was then plunged into 4% nitric acid (the solution being stirred to aid dissolution of the glass bead), made up to 50 ml with de-ionised water and filtered into polythene bottles. (Some silica may have been lost from the solutions due to its low solubility. Silica has therefore been quoted by difference in the results presented in Table 6.2.) The resulting solutions were analysed for 24 elements by ICP, using a specially developed analytical procedure on the ARL 2410 instrument at the Natural History Museum (London). In addition to the Sardis samples, duplicate samples of four commercial standard materials were also analysed, and duplicate synthetic standards were run after every eight unknowns, acting as drift monitors. Silver and gold were measured using AAS on a second batch of solutions made up using hydrofluoric acid and *aqua regia*. The results obtained using both techniques are summarised in Table 6.2. They have precisions of approximately ±2% to ±5% relative for the major elements (i.e. present at levels above 10%), ±10% for minor elements present between 1 and 10%, deteriorating to ±50% at the respective detection limits.

Observations

Petrography

Examination of the thin sections using a petrographic microscope allowed the nature of the clay and aplastic inclusions in the fabrics to be characterised. It was also observed that the clay matrices of some sherds had been vitrified and sometimes bloated (see pp. 159–61 for further discussion). The different fabrics are described below, and the assignment of sherds to the various fabrics is summarised in Table 6.1.

Fabric 1

This fabric (Fig. 6.1, p. 137) is characterised by common to abundant, poorly sorted aplastic inclusions, which range from silt/fine sand to quite coarse grains several millimetres in diameter. These inclusions are generally angular and predominantly of quartz or quartzose rocks (e.g. meta-arenites, metasiltstone, chert). A few fragments of schistose rock were observed in some sections. The silt/fine sand includes rare opaque grains and a few grains of moderate birefringence. The clay matrix is generally micaceous (muscovite is also present as a component of some of the lithic inclusions), and frequently dark coloured (red-brown) and/or isotropic. This fabric includes most of the refractory materials, including putative parting sherds, (re)melting sherds, bricks from the parting furnaces, tuyeres and linings from the lead-smelting furnaces; the coarseware pottery is also in this fabric.

Although included in this fabric group, the two fragments of the furnace linings are generally coarser and less homogeneous than the other materials, especially when observed in the hand specimens. The fragment of lining 45687Y is heavily slagged and vitrified; the vitrified material includes angular quartzose fragments (silt grade to coarse particles several millimetres across), which was initially thought to represent the refractory residue from the partial melting of the furnace wall itself, but see further discussion on pp. 167 and 208. The variability in the material of the furnace linings is exemplified in this fragment, in which the vitrified material appears to grade into regions of very micaceous clay which are almost devoid of inclusions coarser than silt grade. This fine-grained, unvitrified material is rather similar to fabric 2 described below.

Some evidence of straw temper was observed on tuyeres 7 and 8 (see p. 224).

Fabric 2

This is a fine fabric (Fig. 6.2, p. 138) with common silt but only sparse grains of fine sand (less than 0.25 mm diameter). The fabric is micaceous, with the mica flakes exhibiting strong, preferred orientation parallel to the walls of the vessel. The clay matrix is birefringent and there is evidence that the clay platelets, like the flakes of mica, are preferentially aligned parallel to the sherd walls. The interior of the sherd (i.e. away from its inner and outer surfaces) is distinctly redder in colour and less translucent, apparently because of the presence of fine particles of iron oxide. Only a single sherd from amongst the refractory materials, 46849Y (thought to have been used as a parting vessel), has been assigned to this fabric. However, two of the three sherds of fineware pottery examined petrographically appear to be in a similar fabric. The third fineware sherd (29733X) may also have been made from a similar clay but it contains markedly less silt-grade material. As already noted above, there is some indication that fabrics 1 and 2 are closely related, the main difference being the presence of coarse-grained, predominantly quartzose inclusions in fabric 1. None of the fineware sherds exhibits the same gradation in redness as the probable parting sherd, 46849Y, suggesting that this effect may be related to its use in the parting process.

Fabric 3

The so-called bread trays are in a very distinctive fabric (Fig. 6.3, p. 138), quite different from those described above. This fabric is characterised by abundant subangular fragments of metamorphic rocks, including talc schist, phyllite and serpentinite. These fragments are typically elongate and may be up to 1 cm long. The matrix is largely made up of disaggregated flakes from the rock fragments; amphibole, quartz and feldspar are generally rare and

Table 6.1 Summary of samples examined and techniques applied.

Sample	Description	Petrography (fabric group)	SEM	ICP/AAS
44401Z	parting vessel	✓(1)	✓	
45671S	parting vessel	✓(1)	✓	
46849Y	parting vessel	✓(2)	✓	
44399U	remelting sherd	✓(1)	✓	
45676T	remelting sherd	✓(1)	✓	
45677R	remelting sherd	✓(1)	✓	
47910X	tuyere	✓(1)		
47232V	tuyere	✓(1)		
47668S	furnace brick		✓	
47911V	furnace brick	✓(1)		
45686P	slagged furnace wall	✓(1)		
45687Y	slagged furnace wall	✓(?1)		
47908W	bread tray [Apparently used for metallurgical process]	✓(?3)	✓	
47909U	bread tray [Apparently used for metallurgical process]	✓(3)		
45690X	bread tray [No evidence for use in any metallurgical process]	✓(3)		
45691V	bread tray [No evidence for use in any metallurgical process]	✓(3)		
44389Y	litharge cake			✓
44390Q	litharge cake			✓
44391Z	litharge cake			✓
44392X	litharge cake	✓	✓	✓
44393V	litharge cake			✓
44394T	litharge cake	✓	✓	✓
44395R	litharge cake			✓
47905R	coarse pottery	✓(1)		
47906P	coarse pottery	✓(1)		
47907Y	coarse pottery	✓(1)		
29731Q	fineware pottery	✓(2)		
29732Z	fineware pottery	✓(2)		
29733X	fineware pottery	✓(?2)		

present as fine sand-silt sized grains, although fragment 45691V contains a few larger (up to 1 mm) grains of feldspar and quartz. Opaque grains are sparse.

Scanning electron microscopy

Parting vessels
Observation and analysis of the polished section of sherd 44401Z in the SEM (Fig. 6.4) showed that one surface of the sherd is impregnated to a depth of approximately 100 μm by silver salts (primarily silver chloride). The fabric of this sherd (and also other sherds in petrographic fabric 1) is low

in calcium (CaO less than 5%) but not particularly refractory (total alkalies plus iron over *c*. 15% as oxides). The clay matrix of the sherd appears to be of a rather uniform texture, with a low level of vitrification throughout its thickness; comparison with the refired fragments and with other unpublished data on material of similar composition suggests that this sherd has been heated to a temperature of about 800 °C. This could, however, be an indication of the original firing temperature of the vessel rather than of the temperature of the parting process, and therefore further tests were carried out on a mud brick from one of the parting furnaces (see p. 160).

Table 6.2 Analyses of litharge cakes.

Sample	Ag*	Al_2O_3	(As_2O_5)	Au*	[BaO]	Bi_2O_3*	CaO	(CdO)	CoO	Cu	FeO	HgO	$[K_2O]$*	MgO	MnO_2	$[Na_2O]$*	NiO	P_2O_5	PbO*	(S)	(Sb_2O_5)	$\{SiO_2\}$	SnO_2	TiO_2	(ZnO)
44389Y	0.589	1.83	<0.04	<0.005	0.19	0.39	6.65	<0.005	<0.005	4.67	0.28	<0.009	0.64	0.40	0.015	0.88	<0.008	0.77	57.2	0.02	<0.02	25.4	<0.01	0.07	0.01
44390Q	0.016	1.30	<0.04	<0.005	0.18	0.11	4.06	<0.005	<0.005	3.71	0.97	<0.009	0.24	0.46	0.056	0.64	<0.008	0.40	74.5	0.02	<0.02	13.1	<0.01	0.18	0.02
44391Z	1.070	3.83	0.05	<0.005	0.26	0.04	4.97	<0.005	<0.005	10.1	1.46	<0.009	0.71	0.99	0.044	1.16	0.019	1.09	46.0	0.03	<0.02	27.9	0.04	0.24	0.02
44392X	0.029	0.81	<0.04	<0.005	0.18	<0.02	2.55	<0.005	<0.005	1.42	0.16	<0.009	0.32	0.28	0.007	0.65	<0.008	0.26	81.3	0.02	<0.02	12.0	<0.01	0.04	<0.01
44393V	13.0	2.83	<0.04	0.06	0.16	<0.02	2.14	<0.005	<0.005	1.28	0.76	<0.009	1.33	0.36	0.023	1.00	0.010	0.40	65.8	<0.02	<0.02	10.7	<0.01	0.15	0.01
44394T	0.036	1.44	<0.04	<0.005	0.16	0.09	3.21	<0.005	<0.005	7.86	0.12	<0.009	0.27	0.49	0.006	0.58	0.038	0.55	70.1	0.02	<0.02	15.0	<0.01	0.02	0.01
44395R	0.270	0.80	<0.04	<0.005	0.17	0.29	3.55	<0.005	<0.005	9.03	0.19	<0.009	0.47	0.34	0.014	0.68	<0.008	0.20	73.7	0.02	<0.02	10.2	<0.01	0.04	0.02

Figures are in weight per cent.

Analyses were by ICP except those marked * which were by AAS.

Braces { } indicate the silica content was calculated by difference. This should be regarded as a maximum, as the presence of unquantified carbonates could lower this figure significantly.

Brackets [ ] indicate possible contamination from glass vials / distilled water.

Parentheses ( ) denote some of the analyte may have been lost during fusion.

< denotes less than the value quoted (i.e. the detection limit).

The precision of the analyses is c. ±2% to ±5% relative for major elements present above 10%, ±10% for minor elements present between 1–10%, deteriorating to ±50% at the respective detection limit.

The observations made on a second putative parting-vessel sherd (45671S) were essentially similar to those on sherd 44401Z. That is, it has a silver-impregnated zone about 50–100 μm thick at one surface and shows an essentially uniform vitrification texture (with a little fine bloating) of the clay matrix, throughout its thickness. These observations may suggest that this parting-vessel sherd was exposed to a slightly higher temperature.

A third putative parting vessel, 46849Y, which is in a finer fabric (fabric 2), similar to the sherds of fineware pottery examined, was also examined in the SEM. Chemical analysis using the EDX showed that, despite its finer fabric, its chemical composition is rather similar to those of the other parting vessels just described. The micaceous clay matrix is again uniformly vitrified with a little fine bloating, once more suggesting exposure to a maximum temperature of about 800 °C. This sherd, like the other parting vessels, exhibits a surficial zone in which silver salts have diffused into the ceramic fabric, in this instance to a depth of about 200 μm. Smaller amounts of silver were also detected at greater depths (over 1 mm) into the body of the sherd.

Brick from a parting furnace

A large fragment (47668S) of one of the bricks from a parting furnace was examined as a polished section in the SEM. This revealed that the degree of vitrification of the moderately coarse fabric is uniformly low, grading from regions where there is no evidence for vitrification, to those with only limited fusion of the micaceous matrix. The unvitrified state of some parts of the brick offers support for the notion that the bricks used to build the parting furnaces were originally unfired (or very low-fired), and the generally low level of vitrification indicates that even during use, the brick (like the parting-vessel sherds) was exposed to a temperature no higher than about 800 °C. Because of the very low degree of vitrification, it is not possible to make more precise estimates of the process temperature, nor is it reasonable to attempt to estimate the duration of the heating process in the parting furnace from these observations.

It was also noted during the course of the SEM examination that there is extensive penetration of silver salts into the fabric of the brick from its hot face, to a depth of at least several centimetres (see Fig. 5.49, p. 128).

(Re)melting sherds

Sherd 44399U exhibits a gradation of vitrification texture through its thickness from a continuously vitrified and very coarsely bloated texture at the hot face, through to a continuously vitrified (non-bloated) texture at the cool

Fig. 6.4 SEM micrograph of a polished section of parting-vessel sherd 44401Z, showing the silver-rich zone (light coloured) at one surface of the sherd. Backscattered electron image. Scale bar = 100 μm.

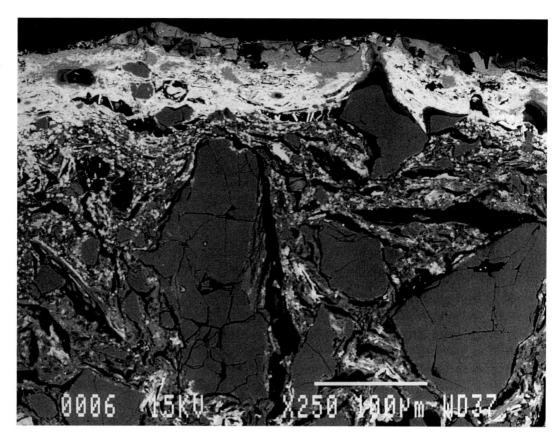

face (Fig. 6.5a–c). Comparison with the refired fragments and with other data suggests that the hot face of this sherd has been exposed to a temperature of about 1150 °C. Consideration of the very steep vitrification gradient across the sherd suggests that this temperature was maintained for only a short period, probably less than 30 minutes.

Another sherd (45677R) thought to have been used for (re)melting gold is less bloated at its hot face, but again exhibits a pronounced gradient through its thickness. The sherd contains somewhat less alkali and iron than 44399U and would therefore be expected to be more refractory. Taking this into account, the observations are probably consistent with this sherd also having been heated at its hot surface to a temperature of about 1150 °C. The steep vitrification gradient again suggests only a very short period of heating.

Bread trays

A polished thin section of sample 47908W was examined in the SEM (Fig. 6.6). It is heavily vitrified and has very coarse, bloating pores. There is a reasonably well-defined surface zone, up to about 0.5 mm thick, which contains approximately 10% lead oxide. Much of the vitrified body of the sherd is magnesium-rich, with a composition that is close to that of talc, though with around 7% alumina and low levels (less than 2%) of soda and lime. Other areas of the body have a somewhat higher mean atomic number and appear pale grey in Fig. 6.6. These are characterised by lower levels of magnesium (c. 10–15% as MgO), and rather more calcium and iron.

The highly vitrified and bloated texture throughout this fragment suggests that the bread tray has been exposed to a high temperature. However, it is probably not appropriate to make numerical estimates of either temperature or duration of heating because of a number of complicating factors. These include the conflicting influences of the distinctive magnesium-rich composition and the significant penetration of the fabric by lead oxide: the magnesia will make the fabric relatively refractory but the lead will act as a flux. Furthermore, macroscopic observation suggests that the burned end of the tray was heated from three sides, ruling out any meaningful observations of a thermal gradient, which would have allowed an estimate of the duration of the process to be made.

Litharge cakes

The results of the quantitative chemical ICP and AAS analyses of the litharge cakes are given in Table 6.2, and

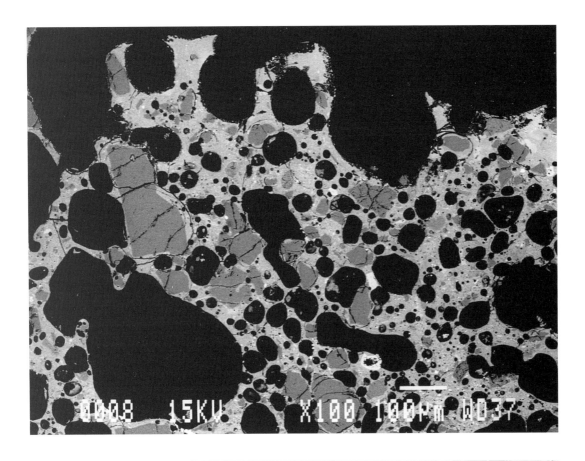

Fig. 6.5 SEM micrographs of a polished section, showing the variation in vitrification texture from the hot surface and into the body of (re)melting sherd 44399U.
a) *(facing page, top)* Very coarsely bloated texture at hot surface.
b) *(facing page, bottom)* Medium bloating, *c.* 3 mm from hot surface.
c) *(right)* Continuously vitrified texture with only fine bloating, *c.* 6 mm from hot surface. Backscattered electron images. Scale bar = 100 μm.

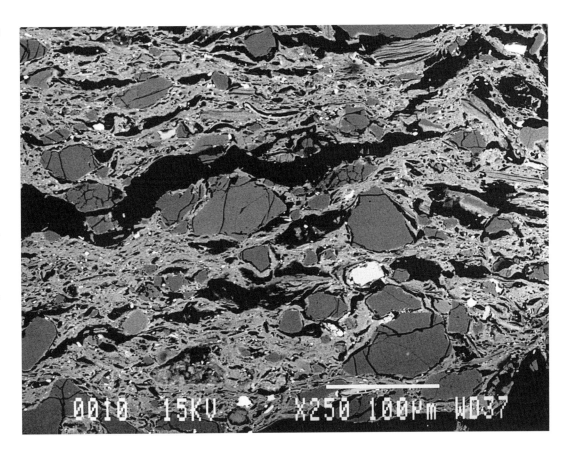

Fig. 6.6 SEM micrograph of a polished section of a fragment of a burned bread tray (47908W). Very coarse bloating of the vitrified fabric can be seen and also the enrichment in lead (lighter coloured band) at the surface of the fragment. Backscattered electron image, width of field *c.* 3.5 mm.

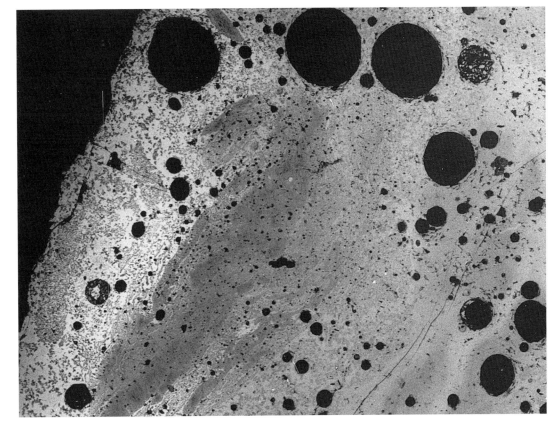

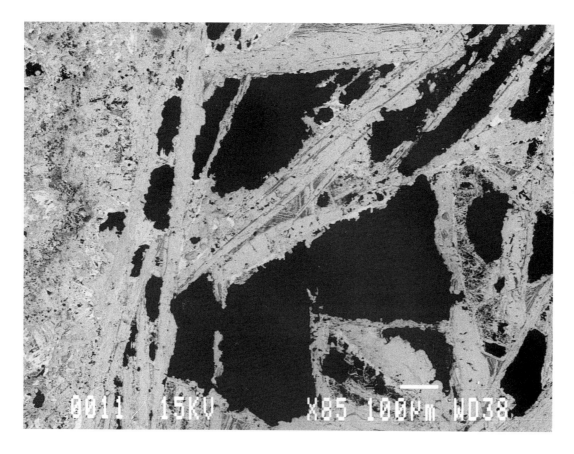

Fig. 6.7 SEM micrograph of a polished section of litharge fragment 44392X, showing an interlocking mass of prismatic crystals with interstitial areas enriched in copper and silicon (mid-grey areas). The black areas are voids in the section. Back-scattered electron image. Scale bar = 100 μm.

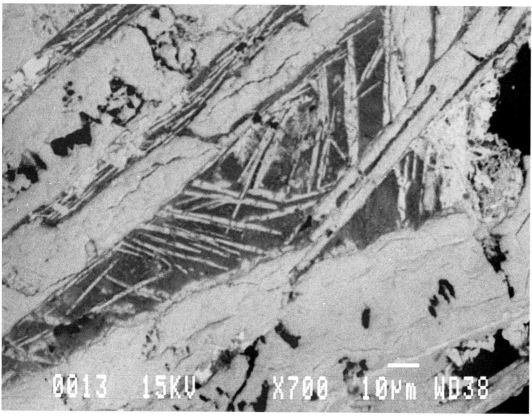

Fig. 6.8 SEM micrograph showing detail of the central area of Fig. 6.7. The mid-grey regions are enriched in copper and silicon. Backscattered electron image. Scale bar = 10 μm.

Fig. 6.9 SEM micrograph of a polished section of litharge sample 44394T, showing copper-rich globules (mid-grey). The darker grey areas (discrete grains and interstitial material) appear to be copper-bearing siliceous materials. Backscattered electron image. Scale bar = 10 μm.

Fig. 6.10 SEM micrograph of copper globules in a polished section of litharge sample 44394T. The small bright areas within and adjacent to some of the globules (arrowed) are rich in silver. The dark rims around the globules are siliceous. Backscattered electron image. Scale bar = 10 μm.

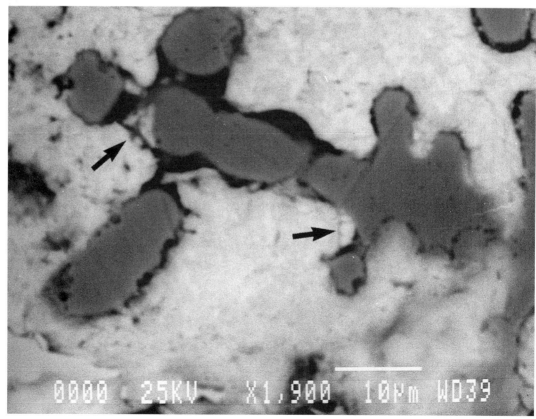

discussed more fully below. Fragments from two of the cakes (44392X, 44394T) were examined as polished sections in the SEM. Both consist mainly of what appears to be a mixture of lead oxide (litharge) and lead carbonate (probably formed naturally over time from the litharge by exposure to the atmosphere), consistent with the results of the XRD analyses. The textural relationships suggest that the litharge crystallised as an interlocking mass of prismatic crystals. A small proportion of other material, including silica and copper, is present as interstitial phases (Figs 6.7 and 6.8). Small (c. 20 μm diameter) copper-rich globules were observed in sample 44394T (Fig. 6.9). Analysis of these globules suggests that they are composed predominantly of copper and that they are argentiferous, containing approximately 3% Ag. In some areas, the silver has exsolved, producing very limited (c. 1 μm) areas which appear to be almost pure silver (Fig. 6.10). Silver was not detected in the lead-rich areas of the sample.

The samples of the litharge cakes analysed quantitatively using ICP and AAS were confirmed as being lead-rich (46–82% as PbO); the majority of the remainder is likely to be silicate material. Silver was detected in all the samples analysed, usually in the 0.01–1% range, although in one sample the silver content was found to be surprisingly high, containing 13% (44393V). This sample also contained the only detectable gold, 0.06%. Copper was present in all samples (in the range 1–10% when expressed as copper metal, although some is likely to have been present as oxides). The sample containing the highest level of copper also contained the only detectable arsenic and tin. Phosphorus was present only at relatively low levels, showing no obvious correlation with calcium, indicating that bone ash was probably not used in the metal processing. No mercury was detected in any of the samples.

Interpretation

Parting process

The petrographic observations have shown that with only a single exception amongst the albeit small number of sherds examined, the ceramics used in the gold-parting process (i.e. parting vessels, tuyeres, furnace linings, and (re)melting sherds) were made from micaceous clays heavily tempered with quartzose inclusions. Furthermore, these clay fabrics are essentially similar to those used for the local coarseware pottery. The single exception is sherd 46849Y, also interpreted as being from a parting vessel. This is in a finer micaceous fabric similar to two of the fineware pottery sherds examined; the aplastic inclusions are again predominantly quartzose but are mainly silt-grade with only sparse coarser-grade material. The significance of this particular sherd is not clear but on the basis of the small number of sherds examined, the use of such fineware fabrics in the parting process seems to be atypical.

Sardis is situated in an area of geologically recent sediments. Examination of a geological map[4] suggests that these sediments may have been derived by the erosion of the rocks of the Menderes Massif, which lies to the south of Sardis and which contains Mount Tmolus (see also p. 19). These strata are predominantly quartzose, including quartzites and schists, together with rare beds of marble. The quartzose nature of these rocks is consistent with the suggestion that derived (quartzose) sediments provided the raw materials for the domestic pottery and also for the majority of the metallurgical ceramics.[5]

The evidence from the vitrification textures observed in the SEM suggests that the sherds from vessels thought to have been used in the parting process have not been exposed to temperatures greater than about 800 °C. This may in fact reflect the original firing temperature of the pottery, rather than the temperature of the parting process itself. However, the observations do serve to set an approximate upper limit to the temperature of the process (see p. 205). Certainly, it would appear that it was not necessary for the parting vessels to be especially refractory. The relatively high porosity of the pottery, demonstrated by the diffusion of silver salts into the body of the ceramic, may have been a more important property than a highly refractory nature. This aspect is discussed more fully elsewhere (see pp. 180 and 203), but it is clear from the analyses that the porous parting vessels must have absorbed significant amounts of the silver liberated from the gold.

The (re)melting sherds, although apparently made from similar low-refractory clays, were clearly exposed to significantly higher temperatures (in the region of 1150 °C). This is close to the temperature at which the ceramic would be expected to lose its integrity due to melting. However, the observations suggest that this was probably not a problem because the heating was from one side only and the exposure to high temperature was limited to very short periods, so that melting of the ceramic was limited to the immediate surface of the sherd. In this context, it is perhaps pertinent to note that no ceramic forms recognisable as crucibles that might have been used for metal melting were found, and that no sherds in any obviously refractory fabric were identified from the Lydian material, although crucibles were recovered from the Roman layers immediately above the refinery.

Silver-recovery process

It is thought that silver was recovered from the parting-process debris in two stages. In the putative first stage, the silver-rich ceramics (i.e. broken parting vessels), fragments of the parting furnaces and any used cement (see p. 203 for a discussion on the probable nature of the cement) were smelted with (added) lead metal and/or recycled litharge, to produce argentiferous lead metal. Silver was then recovered from the argentiferous lead by cupellation.

Smelting

The observations on the two vitrified fragments (45686P and 45687Y), thought to have come from the lining of a smelting furnace, suggest that these were probably made from clay pastes essentially similar to those used for the local coarse pottery and for the parting-process ceramics. These two fragments were found in the north-east dump (see p. 86) with many pieces of litharge cake. It seems possible that the quartzose inclusions in the slaggy material adhering to the furnace lining were derived from the partial melting of the furnace walls. However, it may be that they represent the residue from the partial melting of the argentiferous ceramic waste which had been added to the furnace. Certainly, if the bulk of the silver-rich debris from the parting process was smelted with lead in order to recover the silver, it goes some way towards explaining the comparative rarity of sherds from parting vessels and of fragments of brick from the walls of the parting furnaces.

Cupellation

The so-called bread trays are rather enigmatic but it is thought that they may have been used to form a surround to the cupellation hearths. The highly vitrified and bloated texture of the burned bread trays examined in the SEM appears to be consistent with their use in the cupellation process; the presence of the lead-rich surface layer provides further support for this interpretation. Petrographic examination showed that similar bread trays from domestic contexts are, like the burned examples, in a very distinctive fabric, quite different from that of all of the other ceramic materials examined. In particular, the bread trays contain very little quartzose material and appear to be composed predominantly of talc schists and other magnesian minerals. This magnesium-rich composition was confirmed by the EDX analysis carried out on the fragment of a burned bread tray, examined in the SEM.

As noted already, Sardis lies in an area of Quaternary sediments and it seems likely that these were the source of the clays used for the local coarse pottery and for most of the metallurgical ceramics. Although these clays could pre-sumably have been used also to make the bread trays, the quartz-poor fabric and magnesium-rich composition of the bread trays suggest that the raw materials came from an area of ultramafic rocks (i.e. with no quartz): the nearest outcrop of such rocks lies about 25 km to the north-west of Sardis. Thus it appears that the bread trays (or the clay to make them) were brought to Sardis, and that the choice of ceramics in this talc-rich fabric, to be used in the cupellation process, was quite deliberate. This may have related to the particular properties of ceramics made from magnesian raw materials such as these. Searle[6] in his treatise on refractory materials commented that talc or steatite[7] could be added (up to 40%) to the clay used to make saggars, in order to improve their resistance to thermal shock. Another advantage is that ceramic bodies rich in talc or steatite may be vitrified without presenting a serious risk of physical distortion (as happens, for instance, with lime-rich bodies). Furthermore, Searle noted that crucibles containing talc or steatite, whilst not highly refractory, are often used for fusing caustic alkalies and other highly corrosive materials. Thus the bread trays, in the distinctive talc-rich fabric, may well have been selected for their resistance to thermal shock and more particularly for their resistance to the corrosive environment of the cupellation hearth.

Detailed examination and analysis of the two fragments of litharge (44392X and 44394T) in the SEM confirmed that these contain little material other than lead oxide and probably also lead carbonate produced naturally by the alteration of litharge. In particular, no phosphatic material, which might have indicated the use of bone-ash cupels, was detected. According to Percy[8] bone-ash was the traditional choice of material for cupels but Craddock[9] has noted that there is little evidence for the use of bone-ash cupels at primary metal production sites before the Middle Ages, and none for their use before the Roman period. However, the analysis of these litharge fragments from Sardis has provided no positive evidence regarding the raw materials used for the cupels. It seems possible that cupellation was carried out in the simple clay-lined hollows scooped into the ground surface (see pp. 81–3 and 208).

The observation that small quantities of copper, some of it metallic, are present in the litharge is at first sight surprising but similar observations have been made on litharge from several sites in Britain and elsewhere.[10] Moreover, consideration of the conditions required to reduce copper and lead suggests that the coexistence of litharge and metallic copper is quite possible.[11] Freshly mined gold contains little copper but the copper might have been present as an impurity in the gold scrap which was to be submitted to purification by the parting process. In this case, it would have been removed from the gold during a preliminary cupellation or during parting and, in the latter case, like the silver, much of this copper would have been deposited in the ceramics of the parting vessel, as exemplified by sherd 45671S.[12] The copper would then have been carried on into the cupellation process as an impurity in the argentiferous lead recovered by smelting the ceramic debris from the parting furnaces. Alternatively, copper may have been introduced as an impurity along with the lead and/or lead ore at the smelting stage. The origin of the copper is discussed further on pp. 202 and 205.

Summary

These studies have demonstrated that, with the exception of the bread trays, the ceramic materials used in the gold-refining and silver-recovery processes were, for the most part, made from local clays, similar to those used for the coarse pottery. In the parting process, the porous texture of these low-fired ceramics was probably a more important requirement than a highly refractory nature. Observation of the vitrification textures of the ceramics has permitted

some inferences to be drawn concerning the conditions of the parting process, with a maximum temperature no higher than about 800 °C being indicated.

Examination and analysis of the so-called bread trays, thought to have been used as surrounds to the cupellation hearths, have shown that this is a feasible interpretation of their function. The analysis has also indicated that these ceramics, which are in a distinctive magnesium-rich fabric, may have been deliberately selected for their relative resistance to the aggressive environment of the cupellation hearth. Information on the cupellation process and the nature of the material being cupelled has been derived from the chemical analysis of a series of litharge samples using ICP and AAS, and the examination of two of the fragments in the SEM. These results suggest that the charge contained a small amount of copper, which may have been present as an impurity in the precious metal submitted to purification, or was perhaps introduced during the silver-recovery process.

The significance of these various interpretations is discussed more fully, in the context of other observations, on pp. 200–11.

Notes

1 The progressive vitrification of the clay matrix of ceramic materials in response to increasing temperature is discussed by Maniatis and Tite (1981) and by Tite et al. (1982). For an example of the application of this approach to the determination of the temperature and duration of metallurgical processes, see Middleton (1991).

2 Hook (1998).

3 Hughes et al. (1976).

4 A geological map (1:500 000) and explanatory text published by the Mineral Research and Exploration Institute, Ankara (Dubertret and Kalafatçıoğlu 1973) were consulted.

5 Scott and Kamilli (1981) reported a petrographic study of a selection of pottery of various dates (sixth century BC to thirteenth century AD) from Sardis. They reached a similar conclusion regarding a probable local origin for pottery in quartz-rich fabrics.

6 Searle (1940).

7 Steatite is a compact, massive form of talc.

8 Percy (1880, p. 231).

9 Craddock (1995, pp. 228–9).

10 Bayley (1992c, p. 48); Bayley and Eckstein (1997). See also Rehren and Hauptmann (1995), Rehren (forthcoming) and references therein.

11 Temperature/pO_2 relationships for the oxidation of copper and lead were calculated using the approach described by Freestone et al. (1985) and based upon data given in Kubaschewski and Evans (1958). Metallic copper was reported to be present in the Roman cupellation-hearth material from Xanten, Germany, examined by Rehren (forthcoming).

12 See discussion on pp. 180–2 of the mechanism of parting using salt.

CHAPTER 7

Scientific Examination of the Lydian Precious Metal Coinages

M.R. Cowell and K. Hyne

Introduction

Coinages in electrum (sometimes called white gold),[1] pure gold and silver are attributed to the Lydians. It is generally accepted that they not only struck the first coins, in electrum, during the seventh century BC, but under Croesus (561–547 BC) they also introduced the first bimetallic currency system of pure gold and silver.[2] The earliest coins include the issues with a lion's head on the obverse and punch-marks on the reverse, and for their manufacture the Lydians made use of the local sources of native gold, particularly the alluvial deposits from the river Pactolus and other tributaries of the Hermus near Sardis. The gold refinery at Sardis is assumed to have been responsible for the production of at least the raw material for the pure gold issues and also, as a by-product, some of the silver. Recent work suggests that the Lydian electrum coins may not be composed exclusively of natural alluvial gold but are more probably mixtures with some added silver.[3] Nevertheless, they probably consist largely of unrefined material, whereas the pure gold issues consist of metal which has passed through a refinery system equivalent to that used at Sardis, if not at Sardis itself. The compositions of the Lydian precious metal coins are therefore relevant data for the interpretation of the Sardis gold-refining process, especially in relation to the gold found at the refinery itself.[4] The purpose of this chapter is to review and summarise recent and earlier work on the Lydian coinages.[5]

Previous analyses

Apart from the specific gravity measurements made on the coins catalogued by Head,[6] the first published analyses of the electrum royal types (lion's head right) are those reported by Kraay in 1958.[7] These analyses were made on coins in the Ashmolean Museum's collections using wavelength-dispersive x-ray fluorescence spectrometry (WDXRF) non-destructively, i.e. without surface preparation to remove potential surface enrichment.[8] Examples from several electrum coin series were analysed in this work and the Lydian issues (nine examples) are reported to contain 50–60% gold, 38–48% silver and 2–4% copper.[9]

Some of the same Ashmolean electrum coins and one gold coin (together with a few silver coins from the British Museum's collection and a selection of gold finds from Sardis, including gold foils prepared for the gold-refining process) were subsequently analysed by Gordus in 1970[10] using neutron-activation analysis (NAA) of streak samples; very good agreement was obtained with the previous XRF analyses of the coins.[11] For the streak method of sampling, the edge of the coin is rubbed against an etched quartz rod, which retains a small sample of the metal for subsequent analysis. If a preliminary streak is taken to remove any surface-enriched material, then this method may give more reliable results than completely non-destructive XRF, provided that the coin is not corroded and any surface enrichment is minimal. Gordus' previously unpublished analyses of electrum coins, and those of the gold foils, are reported in Table 7.1.[12]

One of the foils, sample 19A, was subsequently analysed by EDX in the SEM as part of this project[13] and the analyses are in good agreement (11% EDX and 11.9% NAA of silver). Sample 26A was also analysed by both methods but is in fact a group of fragments and it is clear from the analyses (31% EDX and 10.4% NAA) that different pieces were selected for analysis. Another of the items analysed by Gordus, sample 21A, described as a cut lump,

Table 7.1 Neutron-activation analysis of streak samples of the Ashmolean Museum's electrum coins and Sardis gold foils (from Gordus 1971).

Sample	Reference	Coin type	Denom.	Wt (g)	Au%	Ag%	Cu%
E-295/6	E.S.G. Robinson gift 1964	BMC Lydia 2/5	⅓ unit	4.70	51	46	3.4
E-297/8	E.S.G. Robinson gift 1962 (ex M&M sale 25, no. 469)	BMC Lydia 3/16	⅙ unit	2.38	52	46	2.2
E-337/8	E.S.G. Robinson gift 1969 (ex Bank Leu)	inscribed Alyattes	⅙ unit	2.37	60	38	1.9
E-339/40	E.S.G. Robinson gift 1968 (ex Bank Leu)	MN12, pl. I, 1	⅙ unit	2.36	51	46	3.6
E-341/2	E.S.G. Robinson gift 1964 (ex M&M 1955)	cf. Traite, pl. II, 3	1 unit	13.93	55*	42*	2.9
E-345/6	J.G. Milne gift 1924, O'Hagan sale no. 578	BMC Lydia 2/2	⅓ unit	4.65	54	44	2.1
E-347/8	Greville J. Chester gift 1892	BMC Lydia 2/2	⅓ unit	4.74	53	45	2.5
E-349/50	F.P. Weber gift 1906	BMC Lydia 2/2	⅓ unit	4.44	51	46	3.1
E-351/2	Keble College, Oxford	BMC Lydia 2/2	⅓ unit	4.72	51	46	3.1
E-355/6	Balliol College, Oxford	BMC Lydia 2/2	⅓ unit	4.74	53	44	2.2
E-357/8	Bodley miscellaneous	BMC Lydia 2/2	¹⁄₁₂ unit	1.17	55	43	2.1
24	light foil				64	33	3.1
19	light foil				86	11	2.6
26	light foil				89	10	0.7
20	light foil				57	42	1.2
21	cut lump				84	15	1.0

The gold, silver and copper figures are the mean of two measurements, precision typically ±1% absolute.

* Comparisons with XRF analyses of this coin suggest that the NAA figures for gold and silver were transposed in the original data table supplied. The figures shown here are the likely correct NAA analyses.

has been suggested to be a fragment of a coin.[14] However, its composition is completely different from the electrum coins and visual examination[15] has confirmed that it is not a coin fragment.

The streak method of neutron activation[16] was also used by Meyers[17] to analyse a larger number of Lydian silver coins in the British Museum's collection. A detailed examination of Lydian electrum issues, albeit on a small number of coins, was reported by Pászthory in 1980.[18] He examined three *hemihektai* (¹⁄₁₂ unit) coins, including one plated contemporary forgery, with non-destructive surface analysis by XRF, wet chemical analysis of the bulk metal, and energy-dispersive x-ray analysis in a scanning electron microscope (EDX–SEM) of a section through the plated coin. Specific gravity measurements were also made. The results (other than for the contemporary forgery) are shown below in Table 7.2.

The two electrum coins are similar in composition and there is reasonable agreement between the different methods of analysis and the specific gravity measurements. Pászthory has also subsequently reported the analysis of another early electrum issue.[19]

More recently, Avaldi et al.[20] and Vismara[21] report analyses, by non-destructive energy-dispersive XRF with measured and calculated specific gravity results, of a large number of electrum and gold coins attributed to Lydia and elsewhere. Vismara's results for Lydian issues are shown in Table 7.3.

A selection of coins from the British Museum's collections attributed to Lydia, or Ionia, were examined recently. These included a number of electrum issues of which almost half were of the royal types (lion's head right), mostly third staters, and some smaller denominations with geometric designs, similar to those found in the excavations on the site of the Temple of Artemis at Ephesus[22] (Fig. 1.4a, p. 129). Four gold coins were also analysed and nine silver coins, including both staters and half staters (Fig. 1.4b, c, p. 129). The coins were analysed using a combination of EDX analysis (in the scanning electron microscope) and energy-dispersive XRF on an abraded area on the edge of each coin. Specific gravity measurements were also made for comparison.[23] The sample preparation and microscopic examination of the coins revealed that three contemporary

Table 7.2 Gravimetric analyses of Lydian electrum coins (from Pászthory 1980).

Sample	Wt (g)	Au%	Ag%	Cu%	Pb%	Fe%	SG
1	1.32	47.82	46.96	2.58	0.24	2.54	12.76
2	1.26	45.97	41.30	1.78	0.43	9.65	12.74

Table 7.3 XRF analyses of Lydian electrum and gold issues (from Vismara 1993b).

Sample	Denom.	Wt (g)	SG meas.	SG calc.	Au%	Ag%	Cu%	Fe%
170	⅓ stater	4.77	13.10	14.50	61.5	35.5	2.4	0.53
171	⅓ stater	4.76	13.20	14.67	63.1	34.6	2.0	0.32
173	⅓ stater	4.73	13.10	14.96	65.7	32.8	1.2	0.28
174	⅓ stater	4.73	13.60	14.61	62.4	35.1	2.3	0.19
176	⅓ stater	4.70	13.10	14.53	61.6	36.1	2.1	0.30
177	⅓ stater	4.70	13.40	14.66	62.9	35.0	1.8	0.30
178	⅓ stater	4.70	13.60	14.82	64.5	33.5	1.8	0.19
180	1/12 stater	1.19	14.40	14.40	61.3	34.8	1.9	2.0
181	1/12 stater	1.18	14.60	14.34	60.7	35.3	1.9	2.1
182	1/12 stater	1.18	13.60	14.69	63.4	34.0	2.1	0.47
183*	1/12 stater	1.17	9.10	14.06	56.0	42.5	1.3	0.25
185	stater	10.76	19.00	19.03	98.5	0.5	0.2	0.78
186	stater	8.01	17.10	19.06	98.6	0.6	0.1	0.67

* This coin may be plated.

Table 7.4 Combined XRF and SEM analyses of Lydian electrum and gold coins from the British Museum's collection.

Sample	Registration	Type	Denom.	Wt (g)	SG meas.	Au%	Ag%	Cu%	Pb%	Fe%
46852	BMC14	royal	⅓ stater	4.66	13.42	53	45	1.6	<0.04	0.04
46854	1964, 13–3, 13	royal	⅓ stater	4.69	13.51	55	44	1.3	0.11	<0.04
46855	BMC7	royal	⅓ stater	4.73	13.51	53	45	1.6	<0.07	0.04
46856	BMC6	royal	⅓ stater	4.71	13.34	54	44	1.8	<0.04	<0.04
46857	1928, 4–4, 1	royal	stater	14.24	13.77	57	42	1.0	<0.04	0.09
47049	BMC2	royal	⅓ stater	4.72	13.76	55	43	2.0	0.13	0.12
47050	BMC9	royal	⅓ stater	4.73	13.48	52	46	2.3	0.12	0.11
47051	BMC11	royal	⅓ stater	4.70	13.61	54	44	1.7	<0.05	0.14
46863	BMC30	geometric	stater	10.65	19.04	99	0.7	0.2	<0.06	<0.04
46864	BMC31	geometric	stater	8.04	19.09	99	0.7	0.2	<0.06	<0.04
46865	BMC34	geometric	stater	8.06	19.09	99	0.6	0.2	<0.06	<0.04
46866	BMC35	geometric	stater	4.11	19.03	99	0.9	0.2	<0.06	<0.04

Precision: major components (over 10%) ±1%; minor components (1–10%) ±10%; trace components (<1%) ±20–50%.

Table 7.5 Combined XRF and SEM analyses of Lydian silver coins from the British Museum's collection.

Sample	Registration	Denom.	Ag%	Au%	Cu%	Pb%	Bi%
46869	BMC37	stater	99	0.92	0.68	<0.03	0.23
46870	BMC40	stater	100	0.07	0.28	<0.03	0.11
46871	1987, 6–49, 436	stater	99	0.33	0.09	0.30	0.18
46872	1948, 7–12, 10	½ stater	99	0.47	0.11	0.32	<0.01
46873	1948, 7–12, 11	½ stater	97	0.52	1.5	0.32	0.14
46874	BMC49	½ stater	99	0.44	0.36	0.19	0.12
46875	BMC50	½ stater	99	0.45	0.29	0.20	0.09
46876	BMC51	½ stater	99	0.61	0.39	0.23	0.09

Precision: major components (over 10%) ±1%; minor components (1-10%) ±10%; trace components (<1%) ±20–50%.

forgeries, plated with a foil of precious metal over a baser metal core, had been included in the batch to be analysed, two electrum and one silver denomination. The full quantitative results for the royal-type electrum and gold coins are reproduced in Table 7.4 and the silver coins are in Table 7.5. Some of the same silver coins had also been analysed by Meyers (see above) and the results for these are presented in Table 7.6 for comparison.

Electrum coins

Most of the analyses of Lydian royal-type electrum coins by Kraay, Gordus, Pászthory and Cowell are in reasonable agreement for the major alloying components even though a variety of techniques were used. The results reported by Vismara are, however, consistently higher in gold content than the others. In addition, the iron contents observed by Pászthory are also much higher than otherwise reported.

In the case of Vismara's results, the XRF analyses (and the specific gravities calculated from them) are not in good agreement with specific gravity measurements made on the same coins. Considering the technique used, non-destructive XRF of the whole coin, it is possible that there may have been some overestimation of the gold contents because of surface enrichment effects; it is significant that the calculated specific gravities are higher than those measured. Some differences between calculated and measured specific gravities are expected from porosity in the metal and surface deposits, both of which will lower the measured specific gravities. However, the magnitude of the differences in some cases is much larger than usually observed. One coin in particular (no. 183) shows such a large discrepancy that there is a possibility it may be plated.[24]

Regarding the iron content of the electrum coins, Pászthory reports up to 9% in the two coins he analysed and Vismara found similar amounts in some cases. Iron was determined in all the British Museum's electrum coins and found to be consistently low, rarely above 0.1%. It is not

certain why very large amounts of iron have been found in the analysis of some of these coins. The iron may be present in surface deposits and therefore preferentially detected when using non-destructive surface procedures but not by bulk-analysis methods. In any case, it is unlikely that the iron forms a deliberate component of the alloy.[25] If the iron content is ignored and Pászthory's results are recalculated, then the agreement with other analyses is improved.

Pászthory and Cowell both detected small amounts of lead in the electrum. About half the British Museum's electrum coins contained detectable lead, above 0.05%. Pászthory also implies the presence of unquantified traces of bismuth in the gravimetric analysis of the coins. The presence of lead and possibly bismuth have implications for the origins of the alloy being used for these coins.[26]

In summary, therefore, the Lydian electrum coins (based on the British Museum's analyses) contain about 54±2% gold, 44±2% silver and 2±0.5% copper together with traces of lead, up to 0.2%, and iron, in the range 0.1–0.2%. It is relevant to compare this with the composition of native gold, both from Sardis and elsewhere. Native gold extracted from the Pactolus is reported to contain about 17–30% silver.[27] This can be seen to be very different from the alloys used for the coins, which contain much more silver. Of course, a small number of analyses cannot hope to define the composition of all native gold available to the Lydians, although this result is supported by the analyses of flakes of gold alloy, apparently alluvial gold, on the surface of parting-vessel sherd 44401Z from Sardis[28] (see pp. 122–4) and is typical of a number of other native gold–silver alloys.[29]

The electrum coinage alloy is different from native gold in other respects. The gold contents of the coins are very consistent (a necessary requirement for coinage), with all but two of the British Museum's royal types within 1% of the average of 54%, whereas alluvial gold is usually more variable in composition.[30] The average 1.7% copper content of the coins is also not typical of native gold, nor are the small amounts of lead.[31] These factors suggest that the coinage alloy is at least partly artificial and may be best explained as a mixture of native gold and silver.

If the added silver had been refined by the usual process of cupellation then this would account for any lead and possibly bismuth since these are typical impurities in cupelled silver. The amounts of lead (0.05–0.1%) are consistent with the separate addition of part of the silver in the coins, corresponding to one fifth of the coin weight; the added silver was likely to contain about 0.2–0.3% lead (as in the silver coins) which would be diluted by a factor of about five when mixed with alluvial gold. On the other

Table 7.6 Neutron-activation analyses of streak samples of coins from the British Museum's collection (from Meyers 1983).

Registration	Denom.	Ag%	Au%	Cu%
BMC37	stater	98.8	0.91	0.31
BMC40	½ stater	98.2	0.57	1.20
1948, 7–12, 10	½ stater	99.3	0.52	0.24
1948, 7–12, 11	½ stater	98.1	0.50	1.40

hand, a wholly artificial alloy, with 45% added silver, would introduce about 0.2–0.3% lead in total, rather more than is seen in the coins. Furthermore, if the added silver was already alloyed with small amounts of copper then this would account for the presence of copper in the final alloy used for the coins. The implication is that the rather variable alluvial gold was debased down to a fixed level of gold using silver containing a little copper.[32]

Gold and silver coins

All the analyses of the Lydian gold and silver coins show them to be of very high purity with typically 98% or more of the principal metal. It may be assumed that the composition of the gold is representative of that metal which has come from the refinery and is a measure of the efficiency of the parting process. Gold of similar purity or better was found on the sherds used to melt the refined metal.[33] As for the silver coins, it cannot be determined, solely from the analysis of the coins, whether the silver was extracted from the silver-rich debris of the parting process or from completely separate argentiferous lead-ore deposits. There is no doubt that cupellation was practised at Sardis to recover silver from parting-process waste[34] but this by itself would clearly be insufficient to supply the needs of the silver coinage as a whole. The amounts of bismuth detected in the silver coins could only have been derived from lead but this could equally well be the lead used in cupelling the gold-refinery debris as well as that extracted from any argentiferous lead ore. Similarly, the gold content of the silver recovered by cupellation from the refinery is similar to that found in silver from other sources, as exemplified by the gold and silver content of the litharge fragment 44393V (identification number assigned by the British Museum: see concordance, Appendix 6, p. 251).

Platinum group element inclusions

White metal inclusions of platinum group elements (PGE) are frequently found embedded in the surfaces of both the Lydian electrum and the gold coins (Fig. 7.1, p. 138). Such inclusions are commonly found on ancient gold derived from placer deposits. From the work of Ogden[35] and Meeks and Tite,[36] these are known to be usually alloys of osmium, iridium and ruthenium with a range of compositions. Their occurrence in material from the Sardis excavations and associated metalwork is discussed by Meeks et al.[37]

The presence of PGE inclusions in the Lydian gold coins indicates that the inclusions were able to survive the gold-refining process and it is relevant to consider whether their composition has been changed by this processing. The PGE inclusions in some of the Lydian coins in the British Museum's collections have already been analysed by Meeks and Tite[38] and it seems that there are differences between the composition of these in the electrum and the gold coins. In particular, the inclusions in the gold coins generally contain more ruthenium (typically greater than 20%) and less iridium and osmium than the electrum coins. If the two coin types represent respectively the native gold (albeit diluted with silver) and the same processed gold, the implication is that the refining process tends to preferentially remove both osmium and iridium from the inclusions, with a proportionate increase in ruthenium. However, this interpretation is based on the analyses of a very limited number of inclusions in a small sample of coins and hence should be treated with caution. Nevertheless, some changes in the composition of the inclusions would be expected since the component metals are not unreactive.[39] For example, Bowditch[40] notes that in the high-temperature oxidising-refining process of cupellation, most of the osmium is removed along with substantial proportions of ruthenium and iridium; osmium also readily reacts with chlorine at temperatures above 750 °C. The subtle composition changes which seem to have occurred in the PGE inclusions on the gold coins suggests that they were subjected to a somewhat less reactive regime than this, especially as the parting process is believed to have been conducted over many hours or even days.

Conclusions

The Lydian electrum coinage has been shown to be of very consistent composition and it seems likely, from the data presented here, that it was produced from an artificial alloy. This may have been obtained by alloying native gold with additional silver, rather than refining followed by alloying, perhaps with the intention of producing a coinage alloy of consistent value. In the absence of a process to routinely refine gold on a large scale from the non-uniform native gold, the only option for ensuring a consistent gold content was to debase with silver down to a fixed level.[41] This assumes that the native gold could be assayed to determine the correct amounts of silver to add.[42] Clearly, however, this method of restricting the amount of gold allowed only limited control over the alloy. Once the gold parting process was fully developed, any alloy of gold and silver could be manufactured from the basic metals.

Notes

1 The definitions and distinctions of electrum and white gold are discussed by Healy (1980a). Healy considers that the term white gold should refer to naturally occurring gold–silver alloys, whereas the term electrum should be restricted to artificial alloys of this type. There is some support for this when the accounts of the ancient authors are considered, and so elsewhere in this volume electrum is defined specifically as being an artifical alloy. However, for specific gold artefacts, e.g. coins or jewellery, the origins of the alloy, which could range from wholly natural to wholly artificial, may not be definable from the analyses alone. In view of this, the term electrum has been used in this chapter to refer to all alloys of gold and silver (which invariably also include some copper) without making a judgement on their origins.

2 Herodotus I.94.1. See also pp. 17, 78 and 88.

3 See pp. 17 and 172–3; Cowell et al. (1998).

4 It should be noted, however, that none of the analysed Lydian coins came from the Sardis excavations. Indeed, no Lydian gold coins have been found in the current excavations (Buttrey et al. 1981, p. 2). The one possible fragment has been conclusively refuted by its composition (sample 21, see Table 7.1, p. 170, and p. 217).

5 Healy (1980a).

6 Head (1892 and 1901).

7 Kraay (1958).

8 The equipment was probably that described in Hall (1958).

9 Only concentration ranges were reported in the paper, but the authors are grateful to E.T. Hall for making available an unpublished list of individual analyses.

10 Gordus (1971), and personal communication. The results for Sardis artefacts, analysed at the same time, are reported in Waldbaum (1983, p. 186, Table V.6).

11 Gordus (1971).

12 One of the foils, sample 20A, with 42% silver, is similar in composition to the electrum coins; the others contain less silver although none is pure gold. For discussion and more recent analyses of some of these foils, see Table 5.3, p. 145, and p. 185.

13 See p. 147.

14 By Goldstein in Hanfmann (1983).

15 Andrew Ramage, personal communication.

16 Meyers et al. (1973).

17 Meyers (1983).

18 Pászthory (1980).

19 Pászthory (1982).

20 Avaldi et al. (1984).

21 Vismara (1993a, b).

22 British Museum excavations 1904–5.

23 Cowell et al. (1998).

24 The calculated and measured specific gravities are 14.06 and 9.1 respectively.

25 Coins containing this amount of iron would be attracted by a magnet.

26 Lead and bismuth were also found in globules of base gold found on one of the melting sherds, 45680Q, which contained 40% silver, suggesting it might have come from an old electrum coin that was being reprocessed in the PN refinery. See also p. 113.

27 Waldbaum (1983). See also pp. 100 and 201.

28 Meeks et al. (1996). See also p. 122.

29 Morrisson et al. (1987); Antweiler and Sutton (1970).

30 Antweiler and Sutton (1970).

31 Antweiler and Sutton's data for American gold sources, for example, shows that copper is usually below 500 ppm and lead below 100 ppm in placer deposits.

32 If this hypothesis is correct then the copper content of the added silver–copper alloy is calculated to be about 6–9%.

33 Meeks et al. (1996). See also Table 5.1, pp. 103–7.

34 Meeks et al. (1996). See also Table 6.2, p. 160, and p. 209.

35 Ogden (1977).

36 Meeks and Tite (1980).

37 Meeks et al. (1996). See also pp. 201 and 205, and Appendix 4, p. 241.

38 Meeks and Tite (1980).

39 See Appendix 4, p. 238.

40 Bowditch (1973).

41 As previously suggested by Wallace (1987).

42 The most likely technique is by touchstone. See Halleux (1985); Oddy (1993b); Appendix 5, p. 247.

CHAPTER 8

Replication Experiments and the Chemistry of Gold Refining

P.T. Craddock

Previous experiments and replications

Since the salt cementation process became redundant well over a century ago, it has attracted little recent theoretical consideration, no developmental research and, consequently, little practical experimentation.[1] Although the main reactions of all the materials involved in the salt cementation and related processes are recorded in the standard works on inorganic chemistry,[2] these were compiled long after the particular processes we are concerned with here had ceased to be used. Thus the chemistry of the reactions taking place under the specific conditions of the processes have not been described in detail. The few chemical treatises which do mention the processes in passing suggest that hydrogen chloride gas[3] was generated, which in turn attacked the silver or base metals in the gold.[4] Percy, in his excellent account of salt cementation, also believed that hydrogen chloride was the main active reagent, but did note the possibility that chlorine gas could also be involved.[5] This was a prescient comment in light of the experimental reconstructions of the process which took place in this century. Further, it was common experience in the ceramics industry that the traditional salt-glazing processes produced fumes of both hydrogen chloride and chlorine.[6]

It is proposed to describe here some of the experiments which were carried out while the salt cementation process was still operating in the eighteenth and early nineteenth centuries, and the later experimental reconstructions performed in the twentieth century, specifically to demonstrate the viability or otherwise of the descriptions of the process made by Agatharchides.[7] (This does not include the experiments associated with the present project, which are reported on pp. 187–8 and 195–9.)

In the latter category are the researches of Notton and of Hall.[8] The well-known experiments performed by Notton, although meticulously carried out, do not discuss the chemistry of the process at all. In this respect, the unpublished experiments performed by Hall, as part of a thesis on the more general topic of the analysis of antiquities, are much more useful, and provide the strongest evidence that chlorine was likely to be the main agent attacking the silver in the gold. A number of important and well-reported experiments carried out in the eighteenth and nineteenth centuries[9] augment Hall's work, and show that some of his conclusions should be qualified. From these experiments and the replications carried out as part of this project, it has been possible to work out the chemistry of the process. First, the experiments will be described.

The experiments of d'Elhuyar

In 1790, Don Fausto d'Elhuyar published a monograph on the theory of amalgamation, in which his experiments on salt cementation were recorded.[10]

In his first experiment, he took silver sheet, washed it with acid, and heated it with a mixture of cleaned and crushed quartz and common salt in the (clay) muffle of an assay furnace. Fumes of hydrochloric acid were driven off and the residue, when washed, contained no metallic silver but only silver chloride. He believed that this experiment showed 'conclusively that the vapour of hydrochloric acid powerfully attacks and dissolves silver'. In his next experiment, he calcined mixtures of common salt and Carrara marble, and of common salt and porcelain clay, and in both instances hydrogen chloride was generated. In his third set of experiments, which are of more direct value to our understanding and interpretation of the ancient processes, he tried using cements comprising mixtures of chlorides and sulphates: namely, common salt and gypsum, and common salt and barytes. Both mixtures resulted in the evolution of hydrogen chloride but at perceptibly lower temperatures than when using common salt alone. This is of some significance, as several of the ancient recipes, notably those of Pliny, specify the use of mixed cements containing both common salt and *misy* (ferric sulphate).[11] A lowering of the reaction temperature would have been of real value if, as the evidence from Sardis suggests, the process was carried out in antiquity at a low temperature compared with later, historically recorded processes.

D'Elhuyar found that a mixture of hematite and common salt also yielded hydrogen chloride, but that a higher temperature was required. He also heated some common salt in a clay crucible and kept it molten for an hour. Afterwards he noted that the remaining salt was acidic in solution, showing that acid had been evolved. Significantly, he noted that the salt which had been in direct contact with the crucible wall had been decomposed (catalytically) by the clay of which the crucible was made, and that some of the acid evolved had remained in the residual salt or attached to the sides of the crucible, which had been coated in clay. He was also clear that:

> Since it cannot be assumed that any of the alkali should have been converted into vapour and volatilized, it is not possible to account for the excess of acid in the product, otherwise than by supposing that the alkali separated had combined with another substance, which could be nothing else than the substance of the crucible.

Thus even these first experiments recognised that the process was of some complexity and that the actual material of the parting vessel participated in some manner in the chemistry of the process.

These experiments were reported by Percy, and prompted some experiments of his own which both confirmed and amplified d'Elhuyar's work.

The experiments of Percy

Percy[12] heated common salt 'at comparatively high temperature' in a crucible of fine-grained French clay and noted the evolution of an acidic gas that was quite distinct from salt vapour. On evaporation to dryness, he noted brown patches on the side of the crucible, which proved to be of ferric chloride. Percy correctly deduced that the silica of the crucible had been involved in the reaction between the salt and atmospheric moisture to produce hydrogen chlo-

ride, which had in turn attacked the iron salts in the clay of the crucible, forming volatile ferric chloride. This is significant as the first recognition of the involvement of iron salts in the reaction.

A mixture of rock salt and gypsum was then heated 'at strong red heat' both in an open clay roasting dish and on a platinum dish. In both instances, an acidic vapour was formed, which contained no sulphuric acid. The gypsum and rock salt were clearly not respectively pure calcium sulphate and sodium chloride, as after the second experiment the platinum dish had a yellow tint, which on analysis proved to be of iron oxide.

The experiments were repeated with rock salt but substituting Carrara marble and an artificial marble of precipitated calcium carbonate for the gypsum: that is, carbonates instead of sulphates. First, the salt and Carrara marble were reacted on the clay roasting dish, and evolved hydrogen chloride. Then a lining of marble or the precipitated calcium carbonate was made over the clay dish before the mixture was added, so that the salt was isolated from the ceramic. On heating, a vapour was produced that was only faintly acidic. Finally, the marble and the precipitated calcium carbonate mixtures with rock salt were heated on platinum dishes and no acid vapour was produced at all. This confirmed d'Elhuyar's idea and as Percy stated:

> The substance of the roasting dish was concerned in the production of the acid reaction of the vapour in those experiments.

Percy had previously stated that he believed that the vital component was the silica.

It should be emphasised that Percy noted that, although the production of silver chloride could be most simply explained by the action of hydrogen chloride alone, it was possible that some chlorine was also formed. This was because it was known that when silica and the chloride of an alkali metal or alkaline earth were heated to redness in air, a little chlorine was produced as well as hydrogen chloride. Experiments conducted in the twentieth century confirm that hydrogen chloride is generated when common salt is heated with either alumina or silica in temperature ranges from about 830 to 1000 °C for alumina and slowly from about 600 to 1000 °C for silica,[13] in the presence of some water vapour. In both these reactions, there was a subsidiary production of chlorine, but this must have been very limited under most conditions, and certainly the industrial conversion of hydrogen chloride to chlorine required the participation of other metals (see the Deacon process described on p. 180).

The presence of di- or trivalent metal salts strongly promotes the formation of chlorine from hydrogen chloride. In fact, as will be shown, the majority of the chlorine suspected by Percy to have been generated in his experiments probably owed its formation to small amounts of iron mineral impurities in either the refractory clays or the gypsum carrier, rather than to the silica itself.

The experiments of Boussingault

Another important set of experiments was carried out by the French chemist Boussingault in the 1820s whilst he was studying the gold-purification methods used in South America, where salt cementation was still in use.[14] The local refiners used parting vessels of porous earthenware, but Boussingault decided to experiment with impervious 'Cornish' crucibles (dense kaolin stoneware). However, although he used the same cement as the local refiners of one part common salt to two parts brick dust, strongly heated for periods up to 72 hours duration, only minimal quantities of silver were removed from the gold and, as he put it, 'to the great satisfaction of the local workmen' he was compelled to return to the old methods, using porous earthenware vessels.

Boussingault pondered on this failure. He speculated that the problem was the imperviability of the 'Cornish' crucibles, and that the absence of air might be the cause of his failure. Accordingly, he sought to test this by heating thin pieces of silver under a variety of conditions. One piece was placed in a small porcelain vessel filled with the cement of brick dust and common salt. This vessel was itself placed inside a larger vessel packed with charcoal dust, well tamped down, with every precaution being taken to exclude oxygen. Another identical piece of silver was placed in the same cement on an open cupel. (The composition of the cupels is not stated in these experiments, but it is assumed that they were of the usual pressed bone-ash; no clay seems to have been present.) After prolonged heating, the first piece of silver had lost little or no mass but the second piece had lost two thirds of its mass and the remaining cement was heavily impregnated with silver chloride.

He then demonstrated that water was also essential in the process by heating the silver in the usual cement to red heat in a porcelain tube in a current of dried air. The silver was unaffected.

Boussingault carried out further experiments, ostensibly to investigate the function of the air, but which incidentally showed that the material of the parting vessel also played a part in the process. A piece of the silver was covered with common salt, but without the brick dust, placed on a cupel and heated to dull red heat for 3 hours, but the silver was unaffected. Clearly, something in the brick was

also important. As the bricks were principally of clay, which itself contains aluminium silicates, he again took two pieces of silver and placed one in a cupel with a mixture of common salt and silica (SiO_2), and the other in a cupel with a mixture of common salt and alumina (Al_2O_3).[15] The cupels were heated for 4 hours at above cherry-red heat (*c*. 900 °C). After this, he found that the silver in the cupel with the silica cement had lost about one third of its mass and that the remaining cement had lost its salty taste and had vitrified.[16] The vitrification would have protected the remaining metal and this was almost certainly the reason why the silver was not completely reacted. The silver in the cupel with the alumina cement was completely consumed, and the spent cement was initially brilliant white but soon acquired a violet colour on exposure to sunlight, showing that silver chloride was present.

These experiments showed that silica was necessary if hydrogen chloride was to be generated. As it was already known that hydrogen chloride would not be generated by heating common salt in the presence of silica unless water vapour was present, it was inferred that the moisture in the air was essential.

Finally, Boussingault subjected pieces of silver to dry hydrogen chloride in porcelain tubes under a variety of conditions. When heated with dry hydrogen chloride alone, the silver rapidly formed a thin crust of silver chloride which prevented further reaction. When the silver was surrounded by alumina, the forming silver chloride was partially absorbed. Finally, salt was added to the alumina and on passing dry hydrogen chloride through the tube the silver was completely consumed. At the temperature at which these latter experiments were conducted, the salt, which melts at 804 °C, was molten and Boussingault believed that its function was to dissolve the silver chloride as it formed.

Although interesting in themselves, the latter experiments with hydrogen chloride are of less importance to the understanding of the salt cementation process. Boussingault's experiments were carried out at much higher temperatures than the traditional salt cementation processes. A still more significant difference was that his experiments took place under a moving current of hydrogen chloride and the reaction products, especially the reactive hydrogen, produced by the reaction between the hydrogen chloride and silver, would have been swept away, whereas in the enclosed environment of the real cementation process the hydrogen would only have diffused much more slowly. This is an important difference because the equilibrium of the reaction

$$2Ag + 2HCl \rightleftharpoons 2AgCl + H_2$$

lies firmly to the left unless the hydrogen is being continuously and rapidly removed. Thus, in a reasonably static atmosphere of hydrogen chloride, such as would have pertained in a parting vessel, the silver would be much more slowly attacked, if at all.

The experiments of Notton

J.H.F. Notton's work[17] was concerned primarily with testing the accuracy of Agatharchides' description of the salt cementation process, preserved in the accounts of Diodorus Siculus and Photios,[18] and also to find the purity of the gold treated by the ancient methods. Notton followed fairly closely the ancient methods as described by Agatharchides but with two significant divergencies.

- The gold alloy chosen by Notton to simulate the freshly mined impure gold that the ancient process would have treated, was a 9-carat white-gold alloy which contained only 37.5% gold, the rest being made up of silver and copper. The composition was said in his 1971 paper to be very similar to that of an ancient electrum coin from Mytilene, which had 55% silver and 7.5% copper. This was more copper and silver than would usually have been found in the gold treated in the ancient refineries.[19] The ancient process as described by Agatharchides treated gold straight from the mines of the Eastern Desert of Egypt, which would have contained variable but appreciable quantities of silver (typically in the range 10–30%) but very little copper or anything else.

- Notton's scorifiers and crucibles were heated in an electric laboratory muffle furnace, which inevitably would have had a very much drier atmosphere than that generated by the burning wood of the traditional cementation furnaces, and this is likely to have led to important differences in the reactive gases generated and the whole chemistry of the process.

In Notton's experiments, the gold alloy was in the form of thin sheets approximately 63 μm thick, which is almost exactly the same as the Sardis foils (50–60 μm). In his first experiment, the sheets were placed in a scorifier (open crucible) under a mixture of common salt and brick dust and heated to about 800 °C for about 8 hours, until no more white fumes of salt were observed. This treatment removed all of the copper and when operated at 780 °C raised the purity of the gold to 86%, and at 820 °C raised the purity to 93%. Operating at 900 °C was not successful because the salt evaporated too quickly from the open scorifier. The non-metallic residue was a vitrified slag stained blue by copper salts from the gold alloy, and the interior of the

scorifier was stained red, which was also ascribed to the copper. However, it seems more likely that the red coloration was in part due to the deposition of mobile iron salts from the brick dust.

Notton's next experiments followed the ancient description more closely, although still in a dry environment. They were carried out at about 800 °C in a crucible of sillimanite. This has the general formula $Al_2O_3.SiO_2$, and typically contains about 1% of iron oxide.[20] The cement and the metal were sealed in the crucible with a plug of alumina cement, although the body of the crucible was porous. First, the gold sheets were heated with common salt alone. After five days, fumes of salt ceased to emerge[21] and the gold content of the sheets was found once again to have risen from 37.5% to 93%. The salt had entirely disappeared and there was no residue left inside the crucible, conforming exactly to Agatharchides' description. Another important observation was that although the crucible was sealed, considerable quantities of silver chloride were recovered from the crystalline deposits of sublimed salt which had crystallised in the cooler parts of the furnace, recalling the very similar situation in the furnaces at Sardis.

Notton repeated the experiment but with brick dust added to the cement. The process worked perfectly well, but a greeny blue slag was formed on the parting vessel, from which the gold had to be chipped. This is contrary to Agatharchides' quite specific statement that the parting vessel was empty apart from the refined gold at the end of the process. Such slags are recorded by Bayley[22] in some of the medieval parting vessels she has examined, but were not found on any of the Sardis fragments.

Most procedures stress the necessity for water as the source of the hydrogen required in the formation of the hydrogen chloride. However, the atmosphere in Notton's laboratory furnace must have been very dry, and in these conditions the salt vapour is likely to have reacted directly with the iron salts in the sillimanite, producing ferric chloride vapour, which is in itself a powerful oxidising reagent, as well as generating chlorine. Thus in Notton's experiments at least, the ferric chloride could have been the principal reagent (see p. 181).

Agatharchides' description states that barley bran was added to the mixture and to simulate this Notton added charcoal. This had a deleterious effect and the gold content was raised to only 80%. This, as we now appreciate, was to be expected, as the process must be carried out in an oxidising atmosphere.

The additions of lead and tin to the cement, as apparently stipulated by Agatharchides, reduced the gold content of the refined product still further. The tin dissolved in the gold and was still present at the end of the process,[23] and the lead formed a glassy slag, which had not been absorbed by the sillimanite pot. Notton postulated that perhaps the pots used in antiquity would have been better able to soak up the molten lead oxide slag and that the function of the lead might have been to remove any siliceous material remaining in the gold. Notton was clearly dubious about the value, or the reality, of adding either tin or lead. Any remaining siliceous material would have been removed far more effectively when the refined gold was melted after cementation and any foreign matter would float and could have been skimmed from the surface.

Note that although the experiments were judged successful they only produced gold that was 93% pure at best. Notton suggested that repeat refining would be necessary to produce pure gold.

Notton's work demonstrated that the process described by Agatharchides was broadly viable (except for the additions of tin and lead, and of bran). His experiments have proved invaluable in the interpretation of the results of the scientific study of the materials found at Sardis.

The experiments of Hall

E.T. Hall's unpublished experiments were carried out both to establish the viability of the traditional salt cementation process and to determine the identity of the active agents attacking the silver.[24] In the first experiment, a gold alloy containing about 30% silver was used to simulate freshly mined gold. This metal was rolled to a thickness of about 1.27 mm: that is about 20 times the thickness of the average Sardis foil. Pieces of this foil were laid in a platinum crucible containing a cement of equal parts of common salt, brick dust and kaolin (pure white clay) and heated to 850 °C in an electric furnace. At this relatively high cementation temperature, the salt was molten and rapidly dissolved the forming silver chloride, thereby removing the silver from the surface of the gold. After set periods of time extending through 6 days, the gold was removed and analysed (Fig. 8.1). Under these conditions, the removal of silver from the gold followed an exponential curve with the gold being about 99% pure after 6 days. This compares with a purity of only about 93% after 5 days obtained by Notton at the lower temperature of 800 °C on a gold alloy initially containing appreciable quantities of copper in addition to the silver. Using x-ray fluorescence spectrometry, Hall was able to analyse sections cut from the gold sheets and establish that the silver had been removed uniformly throughout the thickness of the gold and not just from the surface.

Hall initially believed that the salt, clay minerals (alu-

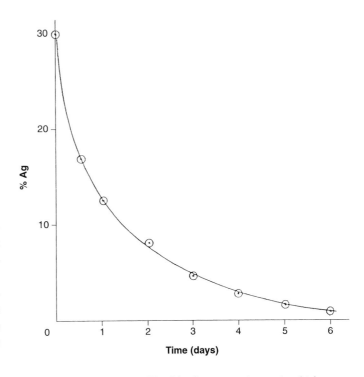

Fig. 8.1 Progress of E.T. Hall's gold-refining experiments, in which silver was removed from a 70% gold–30% silver alloy by cementation. (Based on Hall 1953, Fig. 65)

mina, Al_2O_3, and silica, SiO_2) and water reacted to produce hydrogen chloride. He noted, however, that hydrogen chloride only reacts very slowly with silver, and he suspected, as Percy had done before him, that chlorine was also being produced.

He set about testing this by a series of experiments in which a variety of mixtures of chemicals were heated to about 850 °C in a platinum crucible contained in a silica tube. Air was blown through the system continuously and the gaseous reaction products were carried away to be

Table 8.1 Summary of E.T. Hall's experiments to discover which were the active gases generated in the salt parting process (from Hall 1953).

Cement mixture	Equivalents of products produced grams per hour per gram $\times$ 10^{-4}	
	Chlorine	Hydrochloric acid
NaCl + Al_2O_3 (pure)	0.1	0.5
NaCl + SiO_2 (sand)	0.05	0.3
NaCl + kaolin	24.1	35.6
NaCl + brick dust	22.9	36.2
NaCl + kaolin + silver	0.05	28.1

identified and quantified. The products were blown through two wash bottles, one containing potassium iodide solution to absorb any chlorine and the other containing alkali solution to absorb any hydrogen chloride formed. The reactions are summarised in Table 8.1 and it can be seen that neither the mixtures of common salt with pure alumina nor those of common salt with silica in the form of sand, produced much gas.[25] However, when mixtures either of common salt with kaolin, or of common salt with brick dust were tried, there was a copious evolution of both chlorine and hydrogen chloride. When silver was added to the charge of common salt with kaolin, the recovery of hydrogen chloride was reduced a little but the chlorine was heavily reduced and the silver was entirely transformed to silver chloride. This strongly suggested that chlorine was the main reagent attacking the silver, not hydrogen chloride.

Hall believed that these experiments together showed that the water of crystallisation in the kaolin and brick dust was the vital ingredient necessary for the reaction to proceed, and that this water was missing from the silica or the alumina alone. He also noted that streaming moist air over the crucible did not increase the yield of chlorine or of hydrogen chloride. The experiments with wet air must have excluded mixtures of common salt, silica and alumina, because at high temperature common salt and water vapour react violently in the presence of both silica and alumina, producing hydrogen chloride, as Boussingault's experiments had shown over a century previously. This also suggests that the air blowing through the remainder of Hall's experiments must have been quite dry, and that the only water available was that contained in either the brick dust or the clay. If the air were very dry in the conditions of these specific experiments, then water from some other source (such as bound water) would have been necessary for the generation of hydrogen chloride. But when translated to the conditions of the traditional salt cementation process, the water of crystallisation contained in the solid reagents is unlikely to have played an important part. Burnt brick only rarely contains any bound water and clays quickly lose their bound water at temperatures between 500 and 600 °C. The medieval and post-medieval salt cementation process operated at a higher temperature than this. Thus it would be expected to have lost all the bound water within a matter of hours at the most, yet the process continued for days, long after all the bound water would have been released. In fact, the atmosphere in the cementation furnaces, under real working conditions with a fuel of burning wood, contained sufficient water vapour for the reactions to proceed, as noted by Percy.[26]

Active agents in the salt cementation process

The experiments outlined above enable the chemistry of the process to be established. Hall's work was especially important as it suggests that chlorine was the active agent in the attack on the silver, even though the reaction between pure salt and water vapour in the presence of alumina or silica at high temperature yields predominantly hydrogen chloride. This apparent chemical contradiction can be resolved by taking into consideration the particular chemical environment of the experiments. It is known that the presence of some metal salts promotes the conversion of hydrogen chloride to chlorine at high temperature in the presence of oxygen. The metal used with most commercial success was copper, and this formed the basis of the Deacon process for the production of chlorine.[27] First, the hydrogen chloride attacked the copper salts, forming cupric chloride. This then dissociated to form cuprous chloride with the evolution of chlorine:

$$2CuCl_2 \rightleftharpoons Cu_2Cl_2 + Cl_2$$

The cupric chloride can be regenerated by first oxidising the cuprous chloride to form copper oxyl chloride:

$$2Cu_2Cl_2 + O_2 \rightleftharpoons 2Cl-Cu-O-Cu-Cl$$

This then reacts with more hydrogen chloride to form cupric chloride:

$$Cu_2OCl_2 + 2HCl \rightleftharpoons H_2O + 2CuCl_2$$

This reaction also works for other metals that can exist in both the di- and trivalent states, including iron, and in fact processes were suggested utilising ferric chloride.[28] However, although chemically viable, these were not commercially practicable. Note that neither gold nor silver could have promoted the reaction. Gold is too unreactive readily to form a chloride and silver is univalent.

The reactions between hydrogen chloride and iron minerals are somewhat similar to those of copper.[29] In a hot oxidising atmosphere, the iron oxides present in the brick-dust carrier or the clay of an earthenware crucible would react with the hydrogen chloride, generated by the reaction between common salt and water vapour in the presence of silica, to give ferric chloride. At temperatures of between 700 and 800 °C, dry ferric chloride vapour will decompose into ferric oxide and chlorine. The oxide so formed could then react with further hydrogen chloride to give more ferric chloride. If water vapour is present, then the decomposition is more complex, yielding both chlorine and hydrogen chloride and transforming the ferric chloride into ferrous chloride, ferric oxide and the oxychloride FeOCl, recalling the reactions in the Deacon process:

$$2FeCl_3 + 3H_2O \rightleftharpoons Fe_2O_3 + 6HCl$$
$$FeCl_3 + H_2O \rightleftharpoons FeOCl + 2HCl$$
$$FeCl_3 + 3H_2O \rightleftharpoons Fe(OH)_3 + 3HCl$$

and

$$2FeCl_3 \rightleftharpoons 2FeCl_2 + Cl_2$$

The oxides, oxychloride and ferrous chloride so formed then react with more hydrogen chloride to reform ferric chloride and water vapour.

Note that ferric chloride vaporises at about 300 °C and thus at the temperatures of the cementation process it would be a vapour. It is extremely reactive and could also have reacted directly with the silver and any base metals in the gold. As suggested above, ferric chloride could have been the real active agent in the very dry environment of Notton's experiments, although not necessarily in the ancient processes described by Agatharchides, which Notton's experiments were supposed to be replicating.[30]

These reactions will only take place in oxidising conditions. Clearly, some water vapour is necessary to initiate the production of hydrogen chloride, and thereafter it would be necessary to maintain some water vapour. This would be no problem because, as Percy pointed out, the cementation furnaces were normally fired with wood, producing copious water vapour all around the porous cementation pots.

In Hall's experiment with common salt and silica, the silica was in the form of sand, which almost inevitably must have contained some iron minerals, and it might be argued that this should have promoted the reaction. However, if the conditions were too dry then no hydrogen chloride would have formed. Even if the conditions had been suitable, the reaction would still have been only temporary because the putative ferric chloride formed would have volatilised and been carried away to condense in a cooler part of the system. This was the principal drawback with some of the experiments of Boussingault and of Hall, who both used moving currents of gases to try to replicate the reactions of the salt cementation process. Vapour movement in the enclosed, but not completely sealed, environment of the cementation vessel would necessarily have been much more restricted.

Evidence from salt-glazing experiments

Some confirmation of this conjectured role of ferric chloride in the reaction is provided by a consideration of the chemical reactions which take place during the traditional ceramic salt-glazing process. In this process, common salt was introduced into the kiln with the stonewares which were being fired, typically at temperatures in excess of 1150 °C; the salt volatilised and the sodium ions attacked the silica of the clay, giving the distinctive salt-glaze appearance to the pots. The other products of this reaction were hydrogen chloride and chlorine, together with some unburnt carbon monoxide furnace gases,[31] and the environmental hazards of this led to the banning of the process, at least on an industrial scale. Meanley and Byers[32] conducted some experiments on the gases generated and came up with some surprising results. To summarise their experiments: they fired their kilns at a variety of temperatures from about 900 to 1200 °C in the presence of common salt in a moist oxidising atmosphere, sometimes with added water. The waste gases from the chimneys were sampled and analysed. Some hydrogen chloride was generated, but none of the experiments produced any chlorine at all. This apparently contradicts the experiments of Hall, and the common experience of the salt glazers. However, there were significant differences between Meanley and Byers' experimental conditions and usual practice. Their kilns contained a charge of porcelain ceramics with a low iron content[33] and the furnace walls were of insulating bricks specially treated to prevent glaze formation. In other words, there were few iron salts available, or any other metal salts which could promote the production of chlorine. This absence of iron salts was confirmed by the analysis of a white residue which had condensed on the inner walls of the chimney of their furnace, in which no ferric or ferrous ions were detected. If significant quantities of iron had been present anywhere in the system, then some volatile ferric chloride would surely have also condensed. These experiments seem to confirm, admittedly by negative evidence, the vital role of iron minerals in the production of chlorine during the salt cementation process.

This hypothesis, stressing the hitherto neglected role of the iron minerals in the salt cementation process, accommodates all the previous experimental work, as well as the descriptions of the earlier authorities on salt cementation as outlined on pp. 32–50 and 54–67. The process may be chemically summarised as follows.

Chemistry of the salt cementation process

The experiments suggest that chlorine is likely to be the main active agent in the salt cementation process, and that a trivalent metal ion, usually iron, plays a vital part in its production.

In order for the cementation reaction to proceed, the common salt must be in an oxidising atmosphere containing water vapour in the presence of ferruginous siliceous material, the latter forming either a component of the cement, typically brick dust, or the walls of the cementation vessel and furnace, provided these are of rough porous earthenware. On heating, hydrogen chloride would be formed first of all. This is most probably not the principal active reagent in the attack on silver and any base metals in the gold. Some would react with any iron oxides present, forming ferric chloride, which would be in the vapour phase at the usual cementation temperatures. This is extremely reactive in its own right, but also breaks down in air to form a mixture of hydrogen chloride, chlorine and iron oxychloride. The latter reacts with more hydrogen chloride, to reform ferric chloride and water, and so the process continues. Chlorine gas readily attacks silver to form silver chloride which could be absorbed in the salt surrounding the metal, or evaporate.

Thus, provided the system was kept oxidising with a small amount of water vapour, the process would continue until either the salt or iron minerals were lost from the system by evaporation. The bleached bricks forming the walls of the furnaces at Sardis show that ferric chloride vapour must have been prevalent throughout the atmosphere of the furnace system, and thus the loss of iron by evaporation from the parting vessel itself should not have been severe.[34] The water vapour in the atmosphere would be augmented by that from the burning fuel, made accessible by the open structure of the fabric of the parting vessel. The limiting factor for the duration of the process would thus be the consumption of the common salt from the cement that could not be augmented or maintained from outside the parting vessel.

Physical chemistry of the parting process

Even at the lowest feasible operating temperature of around 650 °C, the principal reagents postulated, chlorine and ferric chloride, would be in the form of a vapour. The examination of some of the foils from the Sardis refinery and from the replication experiments (pp. 145–51, 187–8 and 195–9) shows that the vapours initially attack the silver in the surface of the impure gold, removing it as silver chloride vapour, thereby creating porosity within the metal. The attack is not uniform but proceeds preferentially along the grain boundaries, as the atoms there will be in a more energetic state. SEM micrographs of both the ancient Sardis pieces and the foils from the replication experiments show that networks of intergranular porosity

were established throughout the metal (Figs 5.51 to 5.59, pp. 146–53, and Figs 9.26 to 9.32, pp. 195–9) through which the astringent vapours could permeate the metal and the resulting silver chloride escape. This is a relatively rapid process with the porosity network fully established through the metal before the silver content of the individual grains is seriously depleted.

Once the porosity network has been established, the parting process continues by the removal of silver from the pore surfaces, which are continually replenished with fresh silver diffusing from the grain interior. The rate of the diffusion, and thus the rate of removal of the silver, is a function of the temperature and of the distance that the silver atoms have to travel to reach a reaction surface. Thus the grain size is an important factor. That is, a smaller grain size brought about by extensive hammering facilitates the refining process by reducing the distance the silver has to diffuse. It is estimated that the silver atoms within the gold of the ancient foils examined here would not have had to travel more than 6 μm through the grains to reach a reaction surface.

The formula for determining the rate of the reaction is

$$t = x^2/D$$

where t is the processing time, $D = D_0 \exp(-Q/RT)$ is the diffusion coefficient and x is the diffusion distance. For example, at a temperature of 800 °C, with a diffusion distance of 6 μm, the time will be of the order of 12 hours to remove the silver, which accords well with the experimental evidence.

For most of the foils and the larger granules, as exemplified by sample 30A (p. 148), the total area of the internal porosity will be large compared with the overall surface area of the piece. In these cases, the diffusion distance is directly related to the minimum grain diameter, so the finer the grain size or the thinner (more flattened) the grains, the faster the depletion of the silver. Thus hammering definitely would help the process. Some grain growth will occur at the beginning of the process but should be stabilised once the porosity network has developed.

In these cases, the temperature will have been the main rate-determining factors although the thickness will still be of some significance. It is obvious that the thicker the gold, the longer the porosity network will take to form, and once formed, the longer it will take for the active vapours and silver chloride to diffuse through it.

For the small flecks of impure gold, such as those on parting sherd 44401Z (p. 122), and for thin hammered foils, the surface area of the whole piece can be large compared with the overall grain area and thus is likely to be paramount (as shown in Fig. 9.28, p. 196). Thus for the flecks and thin foils up to about 40 μm in thickness (for a grain size of about 12 μm in diameter), the overall thickness would be of greater importance; the rate-determining factors would be the thickness and temperature. This is reflected in the microstructures of the experimental gold in the foils, where the surface area is large compared with the area of the internal porosity. The overall pattern of silver depletion is towards the foil surface, whereas in the cubes,

where the internal porosity is large compared with the surface area, the overall pattern of depletion is towards the porosities.

These considerations show that it would have been much more efficient to have converted the granules of impure mined gold into thin foils of uniform thickness and small grain size before treatment. This was the practice in Europe from the Middle Ages onwards, although apparently not so in antiquity.

Notes

1   The only experiments known to the present author from the twentieth century are the consciously historical replications of Hall (1953), Notton (1971 and 1974), Healy (1974) and Szabó (1975). The experiments of Notton and of Hall are described on pp. 178 and 179 respectively.

2   Mellor (1923, pp. 190–1; 1935, pp. 71–3; 1961, p. 992); Thorpe and Whiteley (1939, pp. 48–9).

3   The usual term used in the nineteenth-century reports of the experiments given here was hydrochloric acid gas. The modern equivalent term, hydrogen chloride, will be substituted in this chapter except in the direct quotations.

4   Liddell (1926, Vol. IV, p. 286), for example.

5   Percy (1880, p. 397).

6   Starkey (1977, p. 660); Tudball (1992). See also p. 181.

7   Burstein (1989, p. 66). See also p. 34.

8   Notton (1971 and 1974); Hall (1953).

9   Reported in Percy (1880, pp. 390–7).

10   d'Elhuyar (1790, Vol. II, p. 200).

11   *Pliny: Natural History*, 33.84 (Rackham 1952, p. 68). See also p. 35.

12   Percy (1880, pp. 395–7).

13   Clews (1925); Clews and Thompson (1922). Summarised in Mellor (1961, p. 922).

14   Boussingault (1833). His detailed and valuable description of the traditional cementation process is given on pp. 66–7.

15   Note these experiments took place in South America and so it is most unlikely that Boussingault had access to pure materials.

16   But see Hall's experiments discussed on p. 179.

17   Notton (1971 and 1974). See also p. 34.

18   Burstein (1989, p. 66).

19   In fact, Notton had unwittingly created an alloy similar to that made up for the graduation procedure, where the gold–silver alloy was diluted with more silver or copper, as described in sources such as Theophilus and Ercker (see pp. 39 and 64). The procedure is still used in modern acid-parting assay processes (see Appendix 5, p. 250, note 15). At Sardis, the freshly mined gold would have contained only small amounts of copper, and it is likely that the majority of the copper in the recycled electrum coins or other scrap gold was removed by cupellation prior to cementation. See p. 201 for the likely composition of the gold from the Pactolus.

20   Lefond (1975, pp. 729–30).

21   Notton noted the coincidence with the five days mentioned in Agatharchides' description.

22   Bayley (1985; 1991a, b; 1992a, b, c).

23   Note that tin was found on the gold globules on the Sardis melting sherds 45673Z, 45675V, and 45680Q (p. 112), and on one of the foils, sample 29A (p. 186). Tin was also detected on the surface of melting sherd 45673Z (see p. 112). Lead was detected on the surface of parting-vessel sherd 44401Z, and in the sectioned globule from 45680Q. (Note: the numbers with a letter suffix used here to identify the sherds were assigned by the British Museum. See concordance, Appendix 6, p. 251.)

24   Hall (1953).

25   But note the experiments of Boussingault reported on p. 177.

26   Percy (1880, p. 392).

27   Thorpe and Whiteley (1939, pp. 48–9); Oxland (1845).

28   Thorpe and Whiteley (1939, p. 49).

29   Mellor (1935, pp. 71–3).

30   Rose (1915, p. 288) notes that 'ferric chloride chloridises silver with great energy at red heat'.

31   Starkey (1977, p. 66); Tudball (1992).

32   Meanley and Byers (1996), and see correspondence and replies in the following two issues of *Ceramic Review*.

33   The actual ceramic body used in Meanley and Byers experiments had about 0.65% of Fe_2O_3, suggesting that there is a practical threshold level. Analyses of traditional salt-glazed stonewares suggest that the bodies had about 2% of iron (expressed as iron oxide, FeO), but the applied glaze often contained rather more iron (Tite 1991). It was common experience that ferruginous clays glazed much more rapidly than ordinary stoneware bodies (Searle 1929, pp. 136–8). This could be because the reaction creating the sodium ions responsible for the glaze, expressed as

$$2NaCl + H_2O \rightleftharpoons Na_2O + 2HCl$$

is driven to the right by the removal of the hydrogen chloride by reaction with the iron salts, thereby promoting the production of sodium oxide. Additionally, the increased reactivity could be also be due to the iron salts promoting the vitrification of the surface of the ceramic, which is then better able to take the glaze (Singer and Singer 1963, p. 599; Salmang 1961, p. 325).

34   This mobility of iron minerals in ceramics has been noted elsewhere. Matson (1971) reported that ceramics made from saline clays such as those from Mesopotamia, or ceramics used in the production of salt, frequently displayed bleaching due to the loss of iron as ferric chloride.

CHAPTER 9

Examination of the Sardis Gold and the Replication Experiments

A.E. Geçkinli, H. Özbal, P.T. Craddock and N.D. Meeks

The scientific examination of the foils and other fragments of gold from the excavation of the PN refinery, now housed in the Archaeological Museum at Manisa, is presented and discussed, together with the interim results of the examination of gold treated in a series of refining replication experiments. The outcome of this work has an important bearing on the interpretation of the ancient processes.

EXAMINATION OF THE GOLD FROM SARDIS

The gold foils had been previously studied by Goldstein and analysed by Gordus[1] but their re-examination was essential for four reasons:

- To extend and check the previously unpublished analyses performed by Gordus.
- To carry out a more thorough search for other metals in the gold, especially for lead, which the previous NAA method could not detect.
- To examine and record the structure of the foils (Figs 9.1 to 9.22).
- To carry out a thorough check for any platinum group element (PGE) inclusions in the gold. Previous surveys, by Goldstein and latterly by Craddock and Meeks in the Manisa Museum, using the small binocular microscope from the excavation's conservation laboratory, had failed to find any. Their presence or absence was of such crucial importance to the interpretation of the ancient processes[2] that it was considered essential to make a more detailed examination of the foils with the better microscope facilities of the Department of Metallurgy and Materials Engineering of the Istanbul Technical University.

Gold samples selected for examination

Thirty-six small samples, selected and labelled by the excavators, were received from the Archaeological Museum at Manisa. Of those, 26 were in the form of foil fragments, three were lump pieces, five were in the form of minute globules (Fig. 9.21) and two comprised many grains of gold dust (Fig. 9.23). With the exception of the gold dust, the gold studied came from the 1968 excavation season at the PN refinery. The origin of the gold dust is uncertain, but it is believed to have come from the refinery (see p. 100).

The selected pieces of gold retain their original sample numbers, but because in many cases a single sample number covers more than one piece, those pieces examined here and at the British Museum are identified by the suffix A or B.

Methods

All of the samples were first carefully examined by binocular microscopy to observe and record the colour, texture and other characteristics such as adhering surface deposits. The major part of this investigation was carried out by scanning electron microscopy (SEM), using a JEOL 330 instrument and energy-dispersive x-ray microanalysis (EDX) to characterise the surfaces of the samples, both metallographically and chemically. In addition, some further analyses were performed by energy-dispersive x-ray fluorescence specifically to check for lead and bismuth, but none was detected. The detection limit for most of the elements sought in this work was 200 ppm. In order to preserve the original surface structure of the samples, no cleaning or surface preparation whatever was carried out. The microanalyses were peformed on areas of the samples uncontaminated by surface deposits of soil or spent cement. Five of the more problematic samples – 16A, 19A, 26A, 30A and 33A – were mounted longitudinally and polished for detailed microanalysis and metallographic examination, both in Istanbul and London (see pp. 146–51).

Results and discussion

Macroscopic examination

Examination of the samples under a binocular microscope showed that the samples have the following features (Figs 9.1 to 9.6a).

- There is evidence that all of the foils examined had been through the cementation process either in part or, more rarely, to completion.
- The colour changes from sample to sample, ranging from rich gold to lighter shades.
- Foils that have a rich gold colour tend to have extremely granular surfaces: samples 1A (Fig 9.1, p. 139), 2A (Fig. 9.2a, p. 139) and 7A (Fig. 9.7).
- The foils are generally curved or folded, and often have several tears: samples 1A, 3A, 9A, 16A, 20A, 22A, 23A, 25B and 26A (Figs 9.1 and 9.3, pp. 139–40).
- Some foils have definitely been cut *after* their removal from the refining process (samples 3A, 4A and 6A), and the small lump, sample 21A (Fig. 9.6a), has been cut on three sides.
- Generally, the purified foils, weakened by the porosity, have jagged edges where they have fractured: samples 4A, 6A, 7A, 13A, 16A and 28A (Figs 9.4 and 9.5, pp. 140–1).
- Both of the two longest foils (samples 26A and 29A), about 1.5 cm long, are curved over into a tube.
- The foils generally have thin, pinkish (the colours range from orange to purple) contamination on their surfaces, although some of them have thicker layers: samples 5A, 7A, 8A, 11A and 23B (Fig. 9.22). A few foils have buff-coloured contamination (sample 20A). Some of the foils (samples 3A, 4A, 6A, 9A, 13A, 16A, 20A, 28A and 29A) and the globules (sample 28B) additionally have black spotty stains on their surfaces. A thin, milky-white to transparent contamination was also noted on some of the samples.
- One sample (30A) has the appearance of being a molten dribble, although on examination it turned out to be a composite natural granule of unrefined native gold (see Fig. 5.57, p. 152).
- No carbonate deposits were detected on the surfaces of the samples.
- The foils are usually between 50 and 60 μm thick, but some (such as foils 4A, 28A and 29A) are approximately 100 to 120 μm thick, and foil 19A is approximately 30 μm thick.
- No metallic PGE inclusions were observed on the surface of any of the samples examined. See Appendix 4, pp. 242–3, and p. 202 for the significance of this.

Scanning electron microscopy study

Analysis of the gold samples

Semiquantitative EDX analysis of the surface of the gold shows general agreement with the NAA results obtained previously by Gordus on streaks taken from the gold.[3] The microanalytical results indicate that the general silver content of the samples lay between 0 and 35%, and the copper contents of the metal did not exceed 2%. In general, the gold with high silver content tended to have the highest copper content. No other elements were detected in the metal (but note the tin on the surface of 29A). In particular, lead was sought by x-ray fluorescence, looking for evidence of cupellation, but was not detected in any of the samples.

The major characteristics of the surfaces of gold samples are recorded in Figs 9.6b to 9.22, revealing the characteristic porosity of the treated samples. The surface as well as the bulk of the gold foils, seen in cross-section, show either extensive or partial porosity, which resulted from the parting process (Figs 9.2b, p. 139, and 9.7 to 9.20). The porosity extends along the grain boundaries and triple points from the surface. The amount of the porosity appears to be inversely related to the silver content of the sample (cf. Figs 9.2b and 9.7 of samples 2A and 7A, both with no detectable silver at all, and Figs 9.19 and 9.20 of samples 28A and 14A with 22.8% and 28.9% silver respectively). Another observation shows the correlation between the porosities and the chemical composition, in which a high density of *fine* porosity correlates with a still high silver content and probably represents the initial stages of the parting process. After the parting process had gone on for a long time, the porosities became larger and sometimes coalesced with each other as more of the silver was removed. Therefore, if the silver content of the native gold was initially high, at the end of the cementation process the amount of porosity in the now-purified gold would have increased extensively. A semiquantitative estimate of the initial silver content was made from a study of the porosity on samples 16A, which is fully purified, and 19A, which is only partially purified (see micrographs in Figs 5.51 and 5.52, pp. 146–7, and 9.12). The value for the original silver content for sample 16A comes out to about 35% and that for the partially purified sample 19A at

21.3%. As the latter piece still contains 11% silver, this makes the original content about 32%.

Samples which initially had a large grain size retain a low density of fine porosity, such as can be observed in sample 26A (Fig. 5.56, p. 150), and a high silver content (about 30%). Thus the initial grain size of the sample before the parting process is yet another parameter which affects the kinetics of purification.

Development of the porosity during the parting process may be the cause of the observed crinkles on the thin foils. During the initial stages of the process, the high temperatures can cause grain growth before the porosity network becomes established, which inhibits further growth by separating the grains. The granular appearance of the surfaces of the purified gold foils, such as that observed on samples 1A, 2A and 7A, is the surface manifestation of grain-boundary attack. The structure of those samples represents their condition at the end of the parting process.

At the end of the cementation, the extensive porosity of the purified foils renders them very weak, and the foils would have to be melted to produce a more consolidated form. The numerous globules of pure gold, such as sample 12A, and the many globules adhering to some of the melting sherds are from this stage of the process or from assay. Melting the gold at the end of the treatment would have an additional purifying effect in that any solid impurities from the cement sintered to the gold surfaces would tend to float and could be skimmed off. All the globules examined have a dendritic structure, as exemplified by sample 25A (Fig. 9.21) of unpurified gold.

Sample 30A consists of two conjoined granules, the smaller of which has 16.2% silver and 0.5% copper. The granules display no porosity, and were revealed to be granules of impure gold still in their native, untreated, condition.[4]

Analysis of the solid contamination on the gold samples

Microanalysis of the pink contamination which still adheres to the surface of the gold samples, shows an iron-rich deposit of clay particles, also containing some sodium, silver and chlorine. It is most probable that the latter elements are from the residues of the parting charge, still adhering to the gold. The contaminated surface on foil sample 29A also contains tin, recalling its presence on the gold globules on three of the melting sherds.[5] Its presence on a partially refined foil strongly suggests that it was present in the parting vessel, either as a deliberate addition to the cement[6] or as a component of a putative base-metal contamination in the gold undergoing purification.

The structure of the pinkish contamination clearly shows that the particles are firmly adhering to the surface but that there is no vitrification (Figs 9.16b and 9.22). The salt would promote the sintering of any clay materials, such as brick dust, in the cement, and the condensation of the volatile silver chloride vapour could also help to cement material to the surface of the foils.

Leaching of the iron as the volatile ferric chloride from the parting vessels and furnace bricks gives rise to their distinctive pink coloration (Bayley, personal communication). (See Figs 5.4, p. 133, 5.43 and 5.44, p. 137.) The occurrence of the same coloration in the material on the foils suggests it was formed during the parting process rather than as a corrosion deposit formed during burial. It should be noted that the ordinary clay in the nearby bed of the Pactolus stream is a light claret, not the pink colour observed on the gold.

Refined gold dust

The morphology of the particles of gold is shown in Fig. 9.23. These are irregular flattened flakes with rounded edges, having the characteristics of placer grains. Microanalysis of these gold particles shows that they are of pure gold, except one with 4% silver. However, although they are pure, suggesting that they have been refined, the surface texture and porosity are markedly different from that observed in the treated foils.[7]

Surface enrichment during burial

A selection of the partially purified foils, samples 19A, 26A and 33A, were mounted in the plane of the foil surface and polished to remove the surface layers. The analysis of the polished interior surface of sample 19A was similar to that of the original surface, but the other two samples showed a marked increase in the overall silver content, suggesting that surface enrichment had taken place. The interior grains are homogeneous in composition, showing that when the foils left the furnace they were of uniform composition throughout and that the present observed surface enrichment is likely to have occurred during burial. All the samples show extensive porosity (Figs 9.11b, 9.12b and 9.13).

The cut lump, sample 21A, with a dendritic structure and porosity (Fig. 9.6b) and the granule of unrefined gold, sample 30A, both relatively thick samples, were checked in cross-section for evidence of surface enrichment. Surface enrichment was observed on sample 30A.[8]

PGE inclusions

No PGE inclusions were detected in any of the sample surfaces examined, suggesting that the gold may well have been previously cupelled. See Appendix 4, p. 242, and p. 202.

REPLICATION EXPERIMENTS AND EXAMINATION OF THE EXPERIMENTALLY TREATED METAL

Form and examination of the test pieces

In order to understand the conditions under which the porosity observed on the gold from Sardis had developed, a series of experiments was conducted at Bosphorus University. The results are summarised in Table 9.1. Gold in two very different forms was made for the investigation: rolled foils (about 1 cm wide) of gold that contained 20% silver and 1.0% copper; and cubes of a cast alloy containing 63.1% silver (which is much higher than that found in the alluvial gold from the Pactolus), 36.6% gold and 0.3% copper.

The experimental foils were prepared by casting the alloy, followed by rolling down to a thickness of 60 μm by an experienced silversmith. The foils were then cut, each having an area of about 1.5 cm². The bulk samples were cast and cut into cubes with sides of approximately 3 mm.

These test pieces were treated to a simulated parting process in a covered alumina crucible containing a 1:1 mixture of salt and brick dust, heated in an electric oven. The experiments were performed at temperatures between 500 and 750 °C for periods of between 30 minutes and 13 hours (see Table 9.1 and Figs 9.24 to 9.32).

The test pieces were then examined by SEM in the Technical University, Istanbul, and their surface topography recorded (Fig. 9.24). A selection of the pieces were then sectioned and analysed by SEM/EDX at the British Museum.

Mechanism of the removal of silver

The results, presented in the Table 9.1, are extremely interesting as they demonstrate the difference in the actual mechanism of the removal of the silver between the two types of gold: worked thin foils with about 20% silver and cast cubes with about 63% silver. In both types, the initial attack was at the surface (Fig. 9.25, p. 141, foil section), then the attack proceeded preferentially down the grain boundaries, relatively rapidly producing a network of porosity throughout the gold (Fig. 9.26, cube section). This porosity effectively increased the surface area for the

Table 9.1 Summary of results of gold-refining experiments.

a) 30 minutes at various temperatures.

Sample	Position of analysis	Au%	Ag%	Cu%	Comments
G-11-2, 500 °C	section edge	92.4	7.3	0.3	
G-11-2, 500 °C	centre of foil	78.8	20.2	1.0	unchanged
G-11-3, 600 °C	section edge	95.3	4.5	0.2	
G-11-3, 600 °C	centre of foil	79.0	20.0	1.0	unchanged
G-11-5, 700 °C	section edge	100.0	0.0	0.0	
G-11-5, 700 °C	centre of foil	87.3	12.2	0.5	part pure

b) 750 °C for various times.

Sample	Position of analysis	Au%	Ag%	Cu%	Comments
G-7-1, 30 min	section edge	99.4	0.6	0.0	
G-7-1, 30 min	centre of foil	88.8	10.7	0.5	part pure
G-7-2, 60 min	section edge	99.6	0.4	0.0	
G-7-2, 60 min	centre of foil	99.0	1.0	0.0	nearly pure
G-7-3, 120 min	section edge	100.0	0.0	0.0	
G-7-3, 120 min	centre of foil	100.0	0.0	0.0	pure
G-7-4, 180 min	section edge	100.0	0.0	0.0	
G-7-4, 180 min	centre of foil	100.0	0.0	0.0	pure
G-7-5, 240 min	section edge	100.0	0.0	0.0	
G-7-5, 240 min	centre of foil	100.0	0.0	0.0	pure

c) Bulk samples at different temperatures.

Sample	Position of analysis	Au%	Ag%	Cu%	Comments
G-13-1, 500 °C, 30 min	section area	38.5	61.1	0.4	little overall change
G-13-1, 500 °C, 30 min	grain centre	40.3	59.2	0.5	slight purification
G-13-1, 500 °C, 30 min	grain boundary	51.2	48.2	0.6	some change
G-13-2, 600 °C, 30 min	section area	40.4	59.0	0.6	overall change
G-13-2, 600 °C, 30 min	grain centre	38.8	61.2	0.0	very little change
G-13-2, 600 °C, 30 min	grain boundary	51.3	48.7	0.0	some change
G-13-3, 700 °C, 30 min	section area	53.1	46.9	0.0	some overall change
G-13-3, 700 °C, 30 min	grain centre	40.0	60.0	0.0	slight purification
G-13-3, 700 °C, 30 min	grain boundary	65.0	35.0	0.0	some change
G-13-4, 700 °C, 60 min	section area	61.4	38.3	0.3	overall change
G-13-4, 700 °C, 60 min	grain centre	55.2	44.8	0.0	purification in progess
G-13-4, 700 °C, 60 min	grain boundary	71.5	28.5	0.0	change
G-13 original unpurified composition	section area	36.6	63.1	0.3	

attack by the salt vapours. However, on the experimental foils, 60 μm thick, the grain size is quite large and thus the area of the external surface remains relatively large compared with the surface area of the porosities (Figs 9.27 to 9.29), and the minimum distance the silver has to diffuse in the gold to reach a reaction surface is governed by this factor. The rate-determining factor for the experimental foils is thus their thickness, as expressed by

$$t = x^2/D$$

where t is the processing time, $D = D_0 \exp(-Q/RT)$ is the diffusion coefficient and x is the diffusion distance. The activation energy, Q, for silver self-diffusion in gold–silver solid solutions, is given as 40–41 kcal/mole.[9]

Figure 9.30 shows the x-ray line scan of gold and silver contents across foil G-11-3 (30 min at 600 °C), showing the diffusion gradient of silver from the surface of the gold alloy. By contrast, Fig. 9.31a shows the situation in the relatively thick cube G-13-4, where the surfaces of the interior grains, exposed by the initial attack along the grain boundaries (60 min at 700 °C), have a relatively large area compared with the overall external surface. Therefore, the grain size will also be an important rate-determining factor. Silver chloride, the main reaction product of the cementation, has to permeate along the porosity channels at the grain boundaries out to the surface, and fresh supplies of the reactants have to diffuse in. Thus the thickness is still of some importance even for the bulk samples.

Figure 9.32 shows the x-ray line scans of the gold and silver contents across bulk specimen G-13-1 (30 min at 500 °C), showing local grain-boundary silver depletion/gold enhancement.

The treatment of a foil for 1 hour at 750 °C removed the bulk of the silver, but the treatment of a cube for 1 hour at the, admittedly, somewhat lower temperature of 700 °C still left over half of the silver in the gold.

As was to be expected, the gold whilst under treatment is heterogeneous in composition, with its content highest at the surfaces both of the foil and the bulk specimen (Figs 9.25 and 9.30) and also near the grain boundaries for the bulk specimen (Fig. 9.32). Conversely, the silver content is highest at the centres of the grains (Figs 9.33a and 9.33b, p. 142). This is also reflected in the chlorine concentrations, probably present as silver chloride, which in both G-13-3 and G-13-4 (Fig. 9.34, p. 143) are highest at the grain boundaries.

These concentration differences are important, as three of the ancient foils (samples 19A, 26A and 33A) each had the typical grain-boundary porosities but still contained an appreciable amount of silver.[10] However, in marked contrast to the experimental samples discussed above, the silver content of all three of the ancient samples was homogeneous across the grains. The only explanation for this is that the parting process had ceased, possibly due to the exhaustion of salt, but the heating had continued for a considerable time, enabling homogenisation to take place.

The G-11 experiments show clearly that there is little or no reaction below 600 °C, beyond a very limited surface attack, as one would indeed expect from the very low vapour pressure of either the salt or ferric chloride at this temperature. Otherwise, the experiments, G-11-3 at 600 °C, G-11-5 at 700 °C and G-7-1 at 750 °C, show that the reaction rate increases with temperature, as predicted by the kinetics of the reaction.

Conclusions

Overall, these experimental samples produced structures which complement the structures observed in the ancient foils and help to explain the process. The surface encrustations of salty brick dust were likewise reminiscent of those observed on some of the Sardis foils.

There is a major difference in the composition between the ancient foils and the experimental samples. The internal grains of the three ancient foils exposed in the longitudinal sections were all homogeneous in composition, unlike the present replications, which were quite heterogeneous, with higher silver contents towards the centre of the grains. This suggests that the ancient foils had been held at elevated temperatures for long enough for the metal to homogenise after the cessation of the actual parting.

Thus, although the work is still at a preliminary stage, the replication under known conditions has been of enormous help in the interpretation of the structures observed in the ancient gold samples.

Notes

1 See Table 7.1, p. 170.
2 Goldstein (1977). See also p. 202 and Appendix 4, p. 242.
3 See Table 7.1, p. 170.
4 See Fig. 5.57, p. 152.
5 Sherds 45673Z, 45675V and 45680Q, p. 112. (Note: the numbers with a letter suffix used here to identify the sherds were assigned by the British Museum. See concordance, Appendix 6, p. 251.)
6 As mentioned in the accounts of Agatharchides (Burstein 1989), recounted in Diodorus Siculus III.12–14 (Oldfather 1935,

pp. 115–23). See also pp. 34, 35 and 178–9.
7 See also p. 151.
8 See also p. 148.
9 Wise (1964, p. 103, Table 8–3). Based on the data from Hall's experiments (see p. 179), a somewhat lower figure for the activation energy of 36–36.5 kcal/mole is obtained. This is because of the contribution from the grain-boundry diffusion in the porous structure, which has a much lower activation energy.
10 See pp. 147–8 and Table 5.3, p. 145.

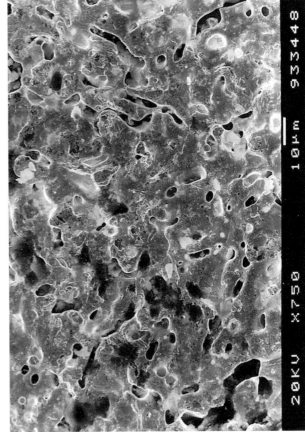

Fig. **9.6b** SEM micrograph of sample 21A, showing porosity developing along the dendrite arms.

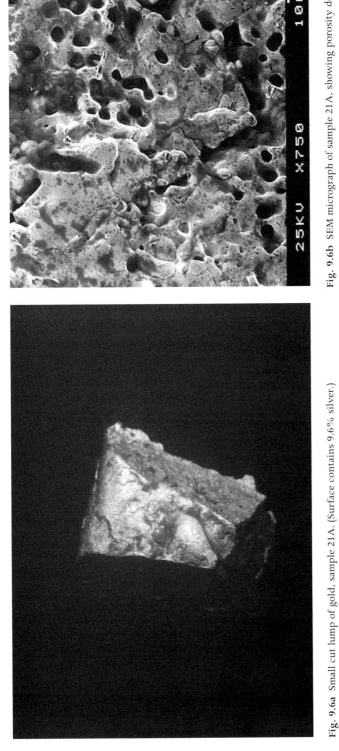

Fig. **9.6a** Small cut lump of gold, sample 21A. (Surface contains 9.6% silver.)

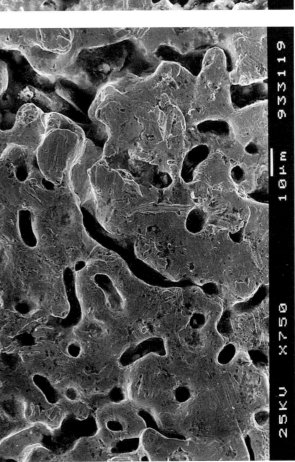

Fig. **9.8** SEM micrograph of the surface of refined-gold foil sample 23B.

Fig. **9.7** SEM micrograph of the surface of refined-gold foil sample 7A, showing extensive interconnected porosity due to the removal of all the silver.

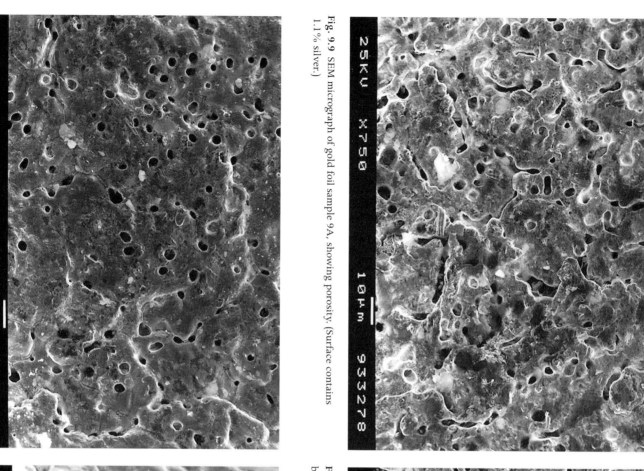

Fig. 9.9 SEM micrograph of gold foil sample 9A, showing porosity. (Surface contains 1.1% silver.)

Fig. 9.11a SEM micrograph of the structure of foil sample 20A. (Surface contains 7.2% silver.)

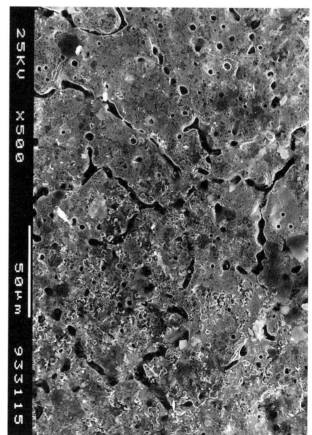

Fig. 9.10 SEM micrograph of gold foil sample 13A, showing porosity along the grain boundaries. (Surface contains 2.4% silver.)

Fig. 9.11b SEM micrograph of the fractured edge of foil sample 20A, showing extensive interior porosity.

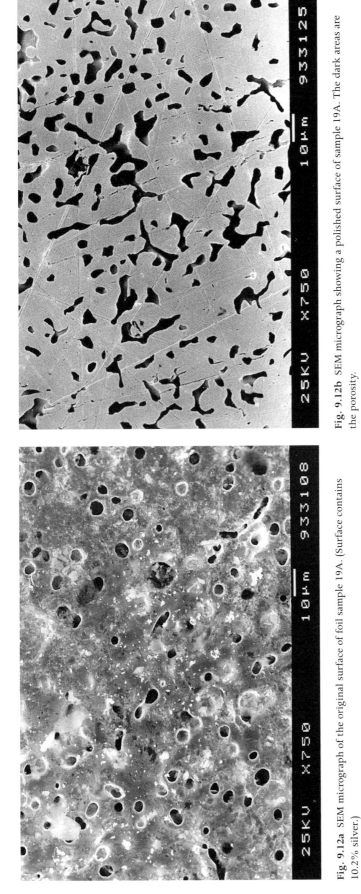

Fig. 9.12a SEM micrograph of the original surface of foil sample 19A. (Surface contains 10.2% silver.)

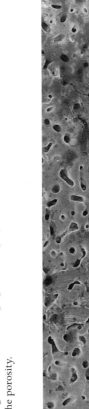

Fig. 9.12b SEM micrograph showing a polished surface of sample 19A. The dark areas are the porosity.

Fig. 9.14 SEM micrograph of gold foil sample 18A, showing the fine distribution of porosity. (Surface contains 16.4% silver.)

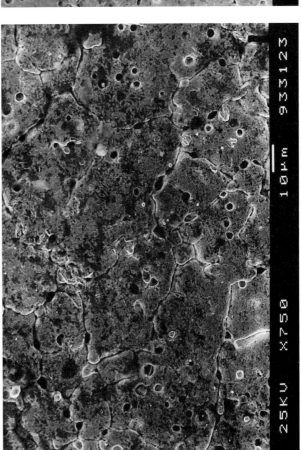

Fig. 9.13 SEM micrograph of the surface of gold foil sample 33A, showing porosity at the triple points and grain boundaries. (Surface contains 10.3% silver.) See also Fig. 5.59, p. 153.

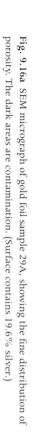

Fig. 9.15 SEM micrograph of gold foil sample 4A, showing the fine distribution of porosity. (Surface contains 17.3% silver.)

Fig. 9.16a SEM micrograph of gold foil sample 29A, showing the fine distribution of porosity. The dark areas are contamination. (Surface contains 19.6% silver.)

Fig. 9.17 SEM micrograph of the surface of light-coloured gold foil sample 20B, showing recrystallization structure. (Surface contains 22.1% silver.)

Fig. 9.16b SEM micrograph of the contamination on sample 29A. The sample contains silver, chlorine and tin in addition to iron, calcium, silicon, potassium and aluminium.

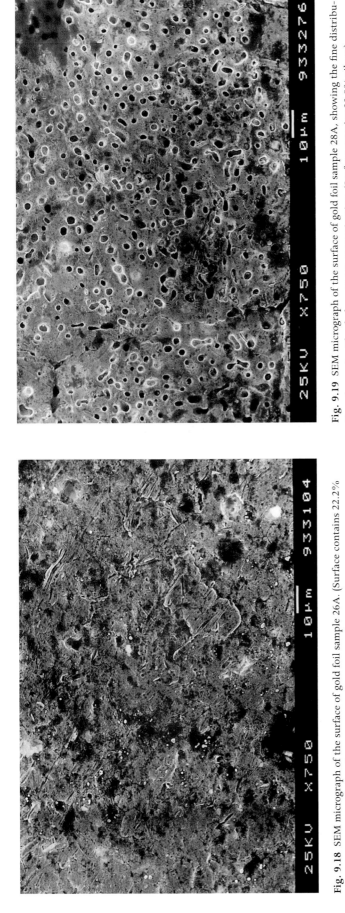

Fig. 9.18 SEM micrograph of the surface of gold foil sample 26A. (Surface contains 22.2% silver.) See also Figs 5.55 amd 5.56, p. 150.

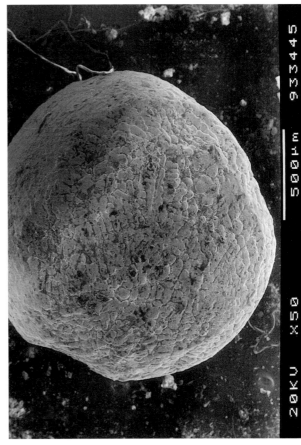

Fig. 9.19 SEM micrograph of the surface of gold foil sample 28A, showing the fine distribution of porosity. The dark areas are contamination. (Surface contains 22.8% silver.)

Fig. 9.20 SEM micrograph of the surface of the small lump, sample 14A, showing equiaxed grains and fine porosity at triple points. (Surface contains 28.9% silver.)

Fig. 9.21 SEM micrograph of gold globule sample 25A, showing its dendritic structure. (Surface contains 30.3% silver.)

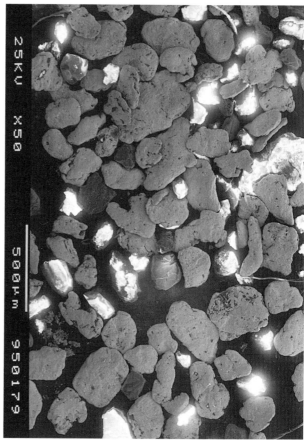

Fig. 9.22 SEM micrograph of a detail of the contamination on foil sample 11A.

Fig. 9.23 SEM micrograph of refined alluvial gold grains.

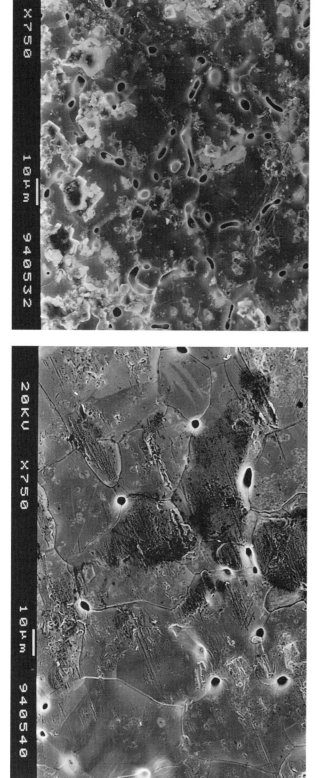

Fig. 9.24a SEM micrographs of the surface structure of the experimental foil after 30 minutes at 500 °C. (Sample contains about 2.5% silver.)

Fig. 9.24b SEM micrographs of the surface structure of the experimental foil after 13 hours at 750 °C. Note the much larger grain size. (Sample contains no detectable silver.)

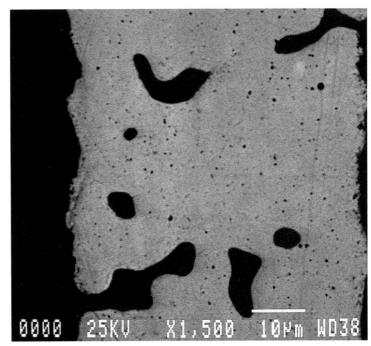

Fig. 9.26 SEM micrograph of a section through cube G-13-2, showing a network of grain-boundary porosity developing after 30 minutes at 600 °C.

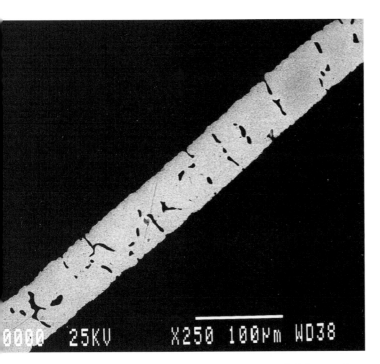

Fig. 9.27a SEM micrograph (×250) of a section through foil G-7-2 after 1 hour at 750 °C.

Fig. 9.27b SEM micrograph (×1500) of a section through foil G-7-2 after 1 hour at 750 °C.

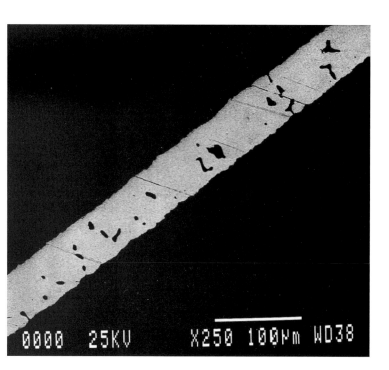

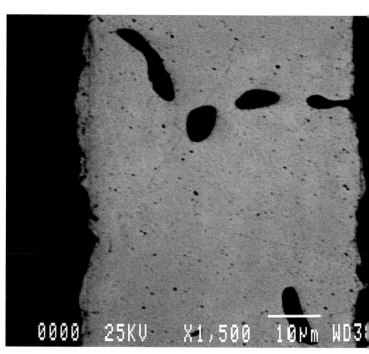

Fig. 9.28a SEM micrograph (×250) of a section through foil G-7-3 after 2 hours at 750 °C. Note the porosity is decreasing.

Fig. 9.28b SEM micrograph (×1500) of a section through foil G-7-3 after 2 hours at 750 °C. Note the porosity is decreasing.

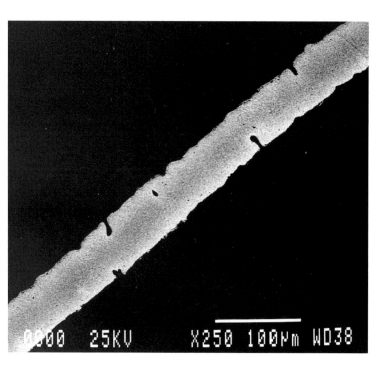

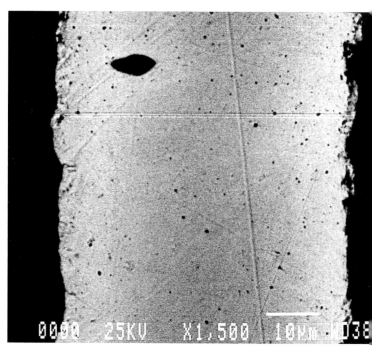

Fig. 9.29a SEM micrograph (×250) of a section through foil G-11-5 after 30 minutes at 700 °C.

Fig. 9.29b SEM micrograph (×1500) of a section through foil G-11-5 after 30 minutes at 700 °C.

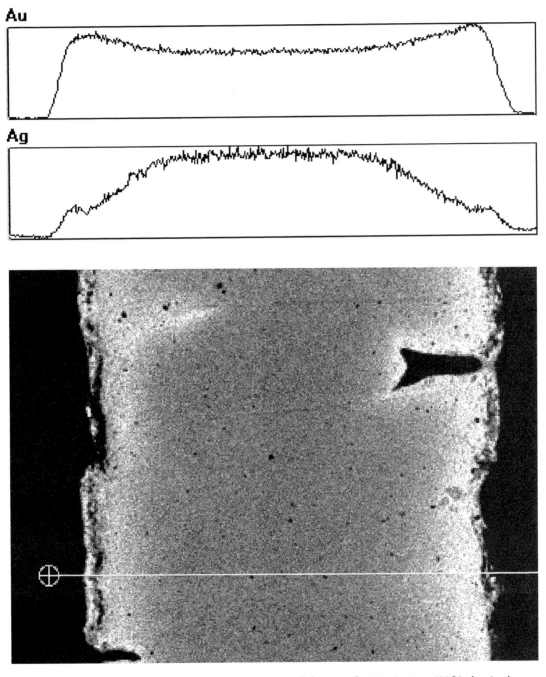

Fig. 9.30 SEM x-ray line scan of gold and silver contents across foil G-11-3 after 30 minutes at 600 °C, showing loss of silver and corresponding increase in gold from surface regions.

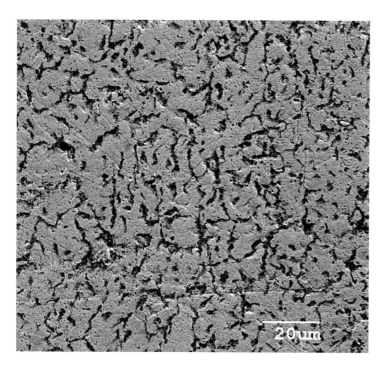

Fig. 9.31a SEM micrograph showing the porous bulk across a section of cube G-13-4 after 60 minutes at 700 °C.

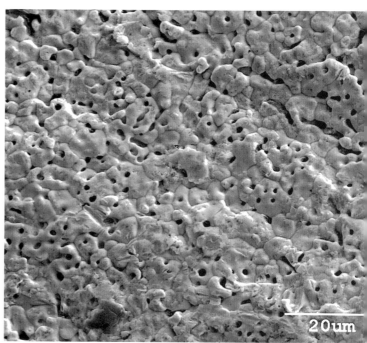

Fig. 9.31b SEM micrograph of the surface topography of cube G-13-4 after 60 minutes at 700 °C.

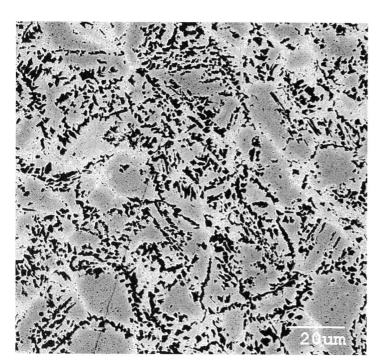

Fig. 9.33a Backscattered electron composition image of cube G-13-3 after 30 minutes at 700 °C. The dark grain centres are unpurified, i.e. silver rich. The light grain boundaries are depleted in silver, i.e gold enriched and purified. (Fig. 9.33b is on p. 142.)

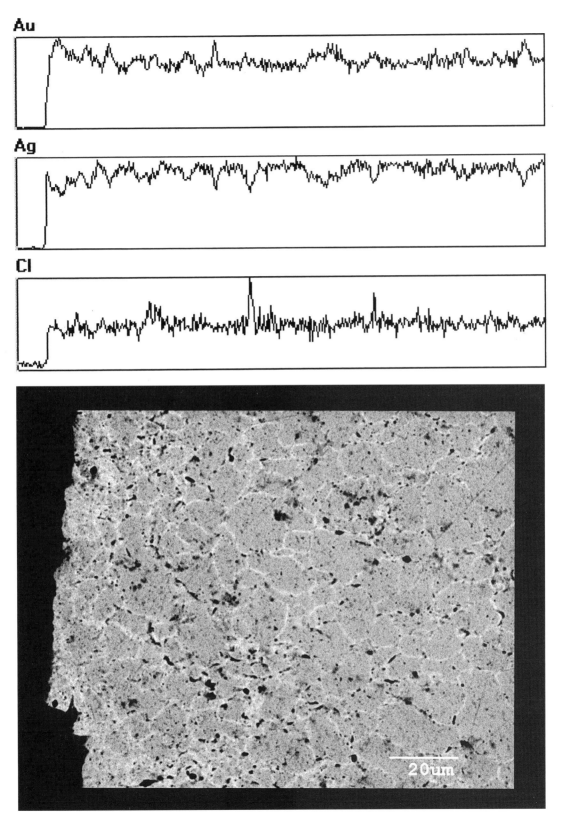

Fig. 9.32 SEM x-ray line scan for gold, silver and chlorine across bulk sample G-13-1 after 30 minutes at 500 °C, showing grain-boundary enrichment but little overall depletion.

CHAPTER 10

Reconstruction of the Salt Cementation Process at the Sardis Refinery

P.T. Craddock

Introduction

From the previous chapters it is possible to draw together a coherent and detailed, although not completely unambiguous, description of the gold-refining processes pursued at Sardis. These reconstructions are based on the material excavated from Sardis and their scientific study, coupled with the evidence from the analytical study of the contemporary Lydian coins, the historical evidence of the traditional salt cementation processes from around the world, and the replication experiments.

The metallurgical finds from the PN area, dating to the Lydian period, are at first sight disparate. They include the cementation furnaces and small burnt hollows as surviving features. From the same archaeological contexts, although not in direct association, there are a variety of small finds related to metallurgical processes. This material includes the foils and other pieces of gold and the litharge cakes. There are also fragments of identifiable metallurgical ceramics, namely the tuyere fragments and two fragments from a smelting furnace. Together with these are various burnt sherds of coarsewares and some finewares, and the so-called bread trays that have clearly been used or reused in metallurgical processes.

This is a disparate assemblage, and it could be argued that a whole range of metallurgical processes was practised at the PN site, each process represented by some of the material recovered from the excavations. However, we believe that these fragments can all, with the exception of the majority of the burnt finewares, be incorporated into the reconstruction of the gold-refining and related silver-cupellation processes. The whole interpretation of the site is based on this assumption and clearly it must be justified.

The main evidence for believing that the finds at the refinery belong to one set of operations is the unusual range of material found. Most metallurgical operations that involve heat – typically smelting, the refining and alloying of base metals, and casting – are performed at high temperature and produce considerable quantities of distinctive waste. This would include slag, together with the typical refractories, notably slagged and vitrified furnace linings from smelting operations,[1] fragments of the crucibles for melting the metal, and mould fragments if metal was being cast.[2] At the Lydian PN site, there was almost nothing one would expect from such a typical metallurgical installation, with the exception of the two small fragments of furnace lining which were slagged and vitrified.[3] Otherwise, the absence of typical materials, such as slag, true crucibles or mould fragments, demonstrates that this was not an ordinary workshop and, further, that no ordinary metalworking operations seem to have taken place there – not even the manufacture of the coins for which it is believed that the bulk of the refined gold was intended. Further evidence is given by the spatial distribution of the finds: the gold foils tended to be found in the vicinity of the structures interpreted as parting furnaces, and the globules of gold, together with the tuyere fragments and the fragments of vitrified coarsewares, were found close to the burnt hollows. Thus it seems that it is safe to assume that all of the metallurgical remains found at the PN refinery do indeed belong to the refining and cupellation processes (Fig. 10.1).

Source of the metal treated at the refinery

The geological studies carried out at Sardis in recent years[4] suggest that the gold originates in the conglomerate at the

Fig. 10.1 Reconstruction of the principal operations carried out at the Sardis gold refinery.

base of Mount Tmolus, with little evidence for primary gold sources further up the Pactolus and the valleys of adjacent streams, where they cut through the metamorphic rocks of Mount Tmolus. Thus the gold is likely to have been derived from secondary placer deposits reconcentrated by the two streams, probably within the immediate vicinity of Sardis itself. The presence of the PGE inclusions in the electrum and gold coins also suggests that the majority of the gold, if not all of it, must have been derived from secondary sources. Thus the present geological evidence is in accord with the ancient descriptions which firmly state that the gold came from the Pactolus river.

The surviving evidence indicates that two forms of impure gold were treated at the refinery: particles and granules of native alluvial gold, and the hammered foils (Fig. 10.2).

The presence of flakes of impure gold on two of the three parting sherds (44401Z and 46849Y) and of granules of refined gold shows that the freshly mined gold was refined in its natural particulate state, as was the practice at the Egyptian mines described by Agatharchides some centuries later.[5] The range of composition of the impure gold from the Pactolus has not been established with any degree of certainty. The major impurity in gold from placer deposits generally is silver, typically between 5% and 30%, and a little copper, usually below 1%.[6] The few analyses that have been reported suggest silver compositions lying between 17% and 24%.[7] These figures are supported by the composition of the flecks of gold on the parting sherd 44401Z, and by the analysis of gold sample 30A, which was found to comprise two unrefined grains, containing about 30% and 16% silver respectively.

The gold would have been separated and collected mechanically by washing. From classical antiquity there is no evidence[8] for the treatment of finely crushed gold ores with mercury to extract the gold. Clearly, at Sardis, the majority, if not all, of the freshly mined gold was treated directly, still in its original form.

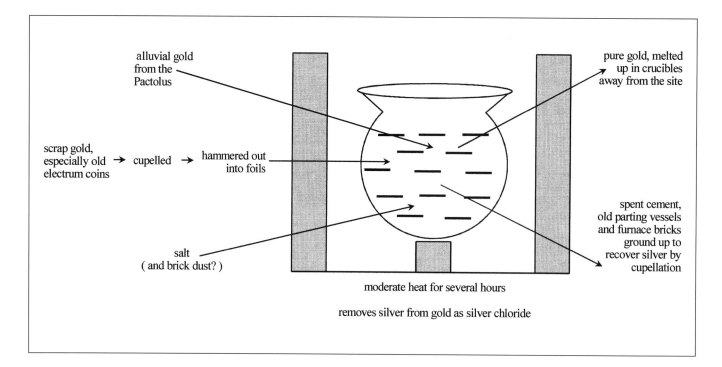

Fig. 10.2 Diagrammatic representation of the principal operations carried out at the Sardis gold refinery.

The other form in which the gold was treated at the Sardis refinery was as foils. The purpose of the hammering was to create a very thin piece of metal with a small grain size, both of which greatly facilitated the solid-state cementation parting process (see pp. 187–8). Scrap gold would have been treated after being hammered out into foils, and this is likely to be the origin of the majority, if not all, of the metal found in the foils at Sardis. The composition of foil 20A is very similar to the alloy used for the Lydian electrum coins, which suggests that the refinery was also engaged in the refining of old electrum coins, and presumably other scrap gold. In addition, one of the large dribbles and the globules on the surface of one of the melting sherds also have this composition.[9]

It is likely that the recycled impure gold was cupelled to remove any base metals present prior to being hammered into foils.[10] The evidence for this at Sardis is twofold: the presence of large quantities of copper in the litharge cakes,[11] which is not present on the surfaces of the admittedly small number of parting-vessel sherds recovered, and the total absence of PGE inclusions in the gold foils (see p. 187). If, as seems likely, the old electrum coins made up at least part of the precious metal undergoing refining, then the foils which had been made from them should also contain the inclusions, unless they had been

previously cupelled as the first stage in the refining process, which would have removed them (see Appendix 4, p. 238). There is a little direct positive evidence for such cupellation at this stage: one of the gold–silver globules on melting sherd 45673Z is associated with lead, those on sherd 45680Q contain lead in cross-section, and parting-vessel sherd 44401Z has lead on its reaction surface.

Parting vessels

The early sources often specify that the parting vessels were to be of earthenware, and a very high proportion of the ceramics from the Sardis refinery that were used in the metallurgical processes are indeed of earthenware, typically the coarseware Lydian cooking pots.[12] (See Figs 4.31, p. 89, 5.2, p. 132, and 5.43, p. 137.) The reasons for this preference over the otherwise prevalent finewares is that the coarsewares have an enchanced ability to absorb thermal shock,[13] as well as satisfying the requirement for the parting vessels to have permeable walls through which both water vapour and air could move, maintaining the conditions necessary for the reaction to take place. If, as now seems probable, there was no brick dust or clay present in the parting vessels (see p. 204), then the walls of a vessel itself would have had to perform the function of promoting the formation of the reaction gases (hydrogen chloride, chlorine and possibly ferric chloride) both by supplying the necessary chemicals, especially the ferric oxides, and by providing suitable sites physically for the reactions to take place. For both of these, the rough open surface of a permeable coarseware would be necessary. Very few parting-vessel sherds were retained from the excavations (see Figs 4.31, 5.2, 5.43, 5.44 and 6.4) but large numbers of the distinctive burnt but unvitrified sherds were found in both the north-east and north-west dumps, where, together with the fragments of bricks from the furnaces, they were awaiting processing to recover the quite large amounts of silver they contained.

Because so few sherds of the parting vessels were retained, it is difficult to be sure of their shape. The fabric of two[14] of the three candidates for parting vessels are identical to that used for the Lydian cooking pots, and their curvature suggest diameters of the order of 20 to 30 cm, similar again to the cooking pots. These globular pots fit very conveniently into the furnaces, and even the handles would have been very useful for lifting the pots whilst still hot from the furnaces, at the conclusion of the process (see Fig. 5.3, p. 132). The cooking pots have round bases and it might have been rather precarious to balance them on the central pedestal of the cementation furnace. Therefore, the vessels intended specifically for use as parting vessels may have had flat bases, although the junction between the wall and base would have created a weak point, especially vulnerable in a thermal process. No evidence was found for flat bases. Alternatively, there may have separate supports such as ring bases or wedges, for which once again there is no surviving evidence.

Sherd 45676T (see p. 112) has the appearance of having come from a parting vessel but does not have any gold flakes or evidence of silver salts in the surface. Thus if it is from a parting vessel then it was never used. The presence of pure gold flakes *within* the ceramic body of the sherd suggests the vessels were made near to where gold was being handled, most likely in the vicinity of the PN refinery itself.

There is also no evidence for lids on the parting vessels, but all of the ancient and most of the traditional descriptions specify them, often with detailed instructions of how they are to be securely sealed with a lute of clay.[15] Note that the lid need not necessarily have been turned on the wheel to a precise shape, but could have been little more than a seal of clay, similar to the plug of alumina which Notton used in his experiments. Bayley found that some of the Romano-British parting vessels at York had bodies of a recognised ceramic type, sealed with 'thin sheets of luted clay', and the Anglo-Saxon parting vessels from Lincoln were of a standard Saxon fabric, known as Crambeck ware, and 'a handmade lid in a different (?local) fabric was added to the dishes and smoothed down to seal it onto their outer surfaces'.[16] A similar arrangement could

well have been employed at Sardis. The principal function of the lid would have been to stop the salt from dispersing too quickly by evaporation and to limit the silver loss, also through evaporation. Their survival is unlikely as they would normally have been processed along with the parting vessels to recover the silver they contained.

It was initially believed that the high silver content in the furnace walls at Sardis indicated that there were no lids to prevent the escape of the silver, but Notton (see p. 178) describes similar silver deposits on the furnace walls from his sealed parting vessel. Also, the parting-vessel sherds are impregnated throughout with silver chloride, which clearly could pass easily through the walls and into the bricks of the furnaces. Thus, on balance, it seems likely that the Sardis parting vessels would have been sealed.

Cementation furnaces

It seems most likely that the surviving remains of the furnaces (Figs 4.24 to 4.30, pp. 84–8) represent substantially what was originally there and that they were not as complex as those illustrated in *De Re Metallica*, for example (Fig. 3.1, p. 60). It is likely that the parting vessels were fired singly, sitting upon the central pedestal with the fire all around them, thus ensuring even heating from all sides. It could be argued that the central pedestals were the supports for putative grates, such as are shown in Fig. 3.1. These, if they had existed, would most probably have been of ceramic, either clay bars such as are often found at the sites of pottery kilns, or perforated grate bricks such as those recovered from some pottery kilns and metallurgical installations, notably the zinc-distillation furnaces at Zawar.[17] However, none was found at Sardis. Iron grates could be postulated but these are rarely attested from ancient kiln sites. A grate of iron would additionally have been very prone to attack by the vapours of the cementation process, although this did not deter their use in the post-medieval period. Perhaps the most telling evidence against there once having been grates is the absence of ledges running around the furnace walls on the inside to support a putative grate at the height of the pedestals.

The entrance in the front of each of the Sardis furnaces would have enabled the ashes to be periodically raked out whilst keeping the fires burning without interruption.

Cementation mixture

There are two aspects to consider here: the active chemical agents (chloride, sulphate or nitrate) and the putative carriers such as brick dust or clay.[18] Here, there are problems of finding surviving physical evidence for active soluble chemicals in mixtures that had only a transient existence. Once again, great reliance has to be placed on the ancient sources.

Of the three active agents given above which are mentioned in the ancient sources, the strongest evidence is for chlorides. Chloride impregnation has been found throughout the furnace walls and in the parting-vessel sherds, and the same walls and ceramics have the distinctive coloration brought about by the mobility of iron minerals in the ceramic after the formation of the volatile ferric chloride. Silver chloride is frequently encountered on the surface of the gold foils (see p. 186). Common salt is the only active agent recorded by Agatharchides, and thus on balance it seems most likely that common salt was the principal, if not the only, parting agent used at Sardis.

It is possible that some alum or other sulphatic material might have been used, since sulphates have been detected on some sherds,[19] albeit not in the same quantities as the chlorides. It is also possible that in some instances a mixed salt–alum or salt–ferric sulphate cement (that is, a chloride–sulphate mixture) might have been used, as mentioned by Pliny,[20] especially as this mixture seems to attack the silver at a lower temperature than common salt alone.[21] This would have been an important consideration in the Sardis process, which seems to have operated at relatively low temperatures (see p. 205). There is evidence that nitrates, in the form of saltpetre, were used for gold refining in ancient India and in the medieval Islamic world (see p. 34). However, there are no classical references to the use of saltpetre and no surviving physical evidence for nitrates, either singly or in combination. A mixed saltpetre–common salt cement (that is, a nitrate–chloride mixture) seems unlikely as it could have generated a gaseous version of *aqua regia*, capable of dissolving the gold.[22] The soluble nitrates could not be expected to survive, but the absence of precipitated gold in the parting-vessel walls and the low gold content of the silver in the sample from litharge cake 44393V (see Table 6.2, p. 160), strongly suggest that gold was not lost in this manner and, consequently, that nitrates were not used.

There is little or no record of a brick-dust or clay carrier in the ancient descriptions, and Agatharchides quite specifically states that at the end of the process nothing was left inside the parting vessel except the refined gold.[23] Notton established that the process worked perfectly well in a sillimanite crucible using a cement of common salt without a carrier at all, the active components in the clay being supplied by the walls of the vessel. At the end of the process, there were no residues in the vessel. However,

when brick dust was added a distinctive slag was formed on the inside of the scorifier, from which the gold had to be separated 'as far as was possible', directly contrary to Agatharchides' statement. However, all the medieval and later descriptions do specify a carrier of either clay or, more usually, brick dust. Significantly, many of the medieval pieces examined by Bayley (see p. 178) are slagged in a manner reminiscent of Notton's pieces.

The evidence from Sardis itself is generally negative, with little or no evidence for the slags encountered by Notton, but it is always difficult to make interpretations from negatives. Such evidence as there is comes from the two heavily slagged and vitrified ceramic fragments, 45686P and 45687Y, typical of smelting-furnace walls (see p. 167). The slag on the fragments was very rich in lead, and they were found in the heap of litharge cakes and ceramic fragments. Thus they can be reasonably interpreted as coming from furnaces where the silver-rich debris of the cementation process was smelted with lead to absorb the silver, prior to cupellation to recover the silver. The slag layers on the walls contained a large number of quartz fragments, which were identical to those in the coarseware Lydian cooking pots and in the bricks of the cementation furnaces, but not found in the furnace wall immediately adjacent to the slagging. The implication is that old parting vessels and furnace bricks were being treated here, but there was no evidence for much finer material such as clay or brick dust, which should have been preponderant in the smelting charge if it had been used in the process. Conversely, some of the gold foils, however, do have pinkish silicate deposits on the surface, which appear to be cemented together by the silver chloride. These deposits could be interpreted as the remains of adhering cement, especially as they are visually identical to deposits on the replication foils which had been treated with a cement containing one part common salt to one part brick dust (see p. 186).

The function of the carrier seems to have been partly to provide some of the chemicals and surfaces needed to form the reacting gases, and partly to act as an inert support for the gold. In this latter capacity, it would perform two functions: to absorb the volatile silver chloride, thereby inhibiting its evaporation from the parting vessel, and to keep the finely divided impure gold separated and evenly distributed throughout the vessel. The salt itself would do this whilst solid, but above 804 °C the salt would melt and all the gold would sink to the bottom of the parting vessel. Notton's experiments took place at about 800 °C, but even if the salt were molten the consequences of the pieces coming together would not have been serious as only three pieces were being treated. In the real ancient processes, where hundreds of very thin foils or thousands of granules would have been treated in each parting operation, it would have been vital to keep them apart. Thus, if there really was no carrier in the cement beyond the salt itself, then this would be quite strong evidence that the temperatue should have been below 800 °C (see below). From the medieval period on, the addition of an inert carrier enabled the processes to operate at significantly higher temperatures.

The other materials mentioned by Agatharchides, apparently as part of the cement, namely bran, lead and tin, seem on balance unlikely. As Notton established, they actively hindered the process.[24]

Evidence for graduation or quartation

The earliest detailed European descriptions mention the additional stage of refining which, by Ercker's time, had become known as graduation, or as quartation according to Libavius writing in the late sixteenth century.[25] It was also called quartation in eighteenth-century England.[26]

It is necessary to consider the possibility that graduation could already have been carried out at the time of the Sardis refinery. In the early accounts[27] of this process, the impure gold was refined by cementation in the usual way and then diluted with large quantities of copper and refined again.[28] This became standard practice in the modern volumetric assay of gold, except that the added metal is silver rather than copper, and it is mixed with the gold at the start of the operation.[29] The processes are discussed elsewhere, but briefly the main function of the excess of added metal was to be dissolved by the assay acids, thereby creating an open structure right through the gold, which brought the acid to the heart of the metal and greatly shortened the diffusion path to the nearest surface. As mentioned above, graduation could be advanced as an explanation for the high copper contents of many of the litharge cakes found at Sardis. However, there is no evidence of copper on the surface of any of the parting-vessel sherds, but it could be argued that these are all from a putative primary refining, where the primary flakes and granules of impure gold would have been treated. By this argument, the foils could be regarded as belonging to the secondary stage, after the gold had been melted and mixed with copper. A problem with this argument, however, is that many of the foils have abundant silver, including some with a composition suggesting that they are from the old electrum coinage before cementation treatment. The foils also have very little copper, especially when compared with the amounts required by the post-medieval sources.

Thus there is no real evidence for the graduation process having being carried out at Sardis. The work of Bayley on parting debris from Roman and Anglo-Saxon Britain likewise failed to detect anything which could be taken as evidence for graduation. Although graduation is mentioned in the traditional European accounts for use with salt cementation, it is always as a further refinement after the main treatment, and neither Agricola nor Ercker (see pp. 60 and 65) seem wholly convinced of its efficacy. In fact, it seems only to have become routine for the later liquid-phase parting treatments with mineral acids rather than with the vapour-phase salt cementation.[30]

Operating parameters of the process

Three principal parameters are considered here: the temperature, the environment within the parting vessel and the duration of the process.

Temperature

The evidence for the temperature of the cementation process at Sardis comes from the examination of the ceramics and furnace bricks from the operation, together with the contemporary electrum and gold coins. The combined evidence from these sources suggests the process operated at a relatively low temperature. Firing-temperature studies on the parting vessels suggest temperatures of the order of 800 °C, but these could well be the original firing temperatures and thus represent no more than the maximum temperature the process could have attained. The study on a mud brick from a furnace, which should have been substantially unfired apart from the operation of the furnace, also suggested a temperature maximum in the region of 800 °C (see p. 166).

If the cement was of common salt without a carrier, then the temperature should not have risen above the melting point of the salt (804 °C) to prevent the finely divided particles of precious metal from sinking and lying together at the bottom of the vessel.

The presence of the PGE inclusions in the Lydian coins of refined gold suggests a very low temperature for the parting process (see Appendix 4, p. 242). In particular, volatile osmium chloride is generated by exposure of osmium sponge to chlorine gas from temperatures of 650 °C. Thus if the temperature of the parting process was much in excess of 650 °C, the inclusions should at the very least be seriously depleted in osmium. Comparison of the osmium content of the inclusions in the Lydian electrum coins, where the metal was not refined, and of the Lydian refined gold coinage does suggest a slight but significant decline in the osmium content in the inclusions relative to the ruthenium and iridium, but the metal is still there in quantity (see p. 173). This analytical conclusion is supported by the actual appearance of the PGE inclusions in the gold coins that have been through the refining process. If depletion of the osmium from the PGE inclusions in the Lydian refined gold coins had taken place to a significant extent, then one would expect the surfaces of the inclusions to appear porous, but this is not in fact the case. This evidence from the inclusions in the coins suggests that the maximum temperature attained was not much in excess of 650 °C, especially as the parting process was likely to have been of long duration.

The presence of silver chloride permeating the furnace walls might be considered evidence of higher operating temperatures, as silver chloride is only completely volatile at temperatures in the region of 1000 °C. However, Notton found in his experiments that the silver had migrated through the walls of the parting vessel at 800 °C. In fact, silver chloride begins to volatilise at temperatures as low as 260 °C.[31] Salt and water vapour react together in the presence of silica to produce hydrogen chloride at temperatures as low as 600 °C, and this would react with the iron oxides in the clay of the parting vessel, producing volatile ferric chloride which in turn would generate chlorine. Thus at temperatures as low as 600 °C, three very reactive gases would be present in the parting vessel.

The low temperature of the process postulated above should find support in the absence of any trace of salt glazing either on the furnace walls or on the parting vessels. This strongly suggests that sodium vapour was not able to attack the silica in the ceramic walls and thus the temperature should have been below 800 °C. However, this absence of salt glazing is not restricted to Sardis but seems to be a general phenomena on salt parting vessels. Bayley[32] noted that the various Romano-British and Saxon parting vessels had a thin glaze, quite distinct from any ash glaze, on the outside but no glaze on the inside. She commented: 'It is not clear to me why the glaze should be only found on the outer surfaces, but Notton's 1974 experiments produced a similar effect.' In fact, Notton does not actually comment on the glaze or otherwise on his parting vessel, but from the excellent colour photographs which accompany his 1974 article it is clear that, before the experiment, the sillimanite crucible was perfectly clean and pale buff in colour both inside and out. After the experiment, the inside was still quite clean but the outside had a thin brown coat. This could have been a salt glaze, but, as Bayley notes, it is very difficult to see why it should only be on one side when the conditions should be similar on both sides. It was possibly

a thin coherent layer of iron oxides from the breakdown of ferric chloride, preferentially deposited on the outside of the vessel, because the concentration of chloride ions necessary to reconstitute ferric chloride is less than that on the inside of the vessel that contained the salt.

Thus the combined evidence from three separate materials suggest temperature limits for the cementation process of between 600 and 800 °C, with the evidence of the osmium in the PGE inclusions favouring a temperature that was towards the lower end of the range, but the evidence of the furnace brick suggests it may have been near the upper end.

Environment inside the parting vessels

The conditions inside the porous parting vessels are likely to have reflected their immediate environment: namely, a slow-burning wood fire, which would give hot, damp and quite oxidising conditions ideally suited to the parting process as determined by the experiments discussed on pp. 175–83. The generally oxidising conditions are confirmed by the distinctive colours on the surfaces of the parting vessels and furnace walls of the iron salts, which are now of hematite. There is no suggestion of the darker reduced forms, such as magnetite, Fe_3O_4, for example.

Duration of the process

As discussed on pp. 182 and 188, the time required for the removal of the silver is a function of temperature and of the thickness and grain size of the metal to be treated. For thin foils and flecks, the thickness is likely to be predominant, but for thicker foils and granules the grain size will be more important. It can be calculated that the purification of foil 60 µm thick takes about 20 minutes at 850 °C, about 4 hours at 700 °C and about 36 hours at 600 °C. Using the same formula, it can be shown that the purification of a foil whose thickness is 1270 µm (1.27 mm) will take about 6 days at 850 °C, confirming Hall's experimental results.

The surviving Sardis foils seem to have been predominantly in the 50–100 µm range, but at least one, sample 16A, was over 1000 µm thick, and this may be more typical of the majority of the original foils. The grains of freshly mined gold could be much thicker than this, as exemplified by sample 30A.

It was not possible to make an estimate of the duration of the process directly from the ceramics because the temperatures involved were too low for vitrification, upon which such estimates would have been made, to take place. Some indirect evidence is provided by the surviving foils.

The SEM study of gold foils, samples 19A, 26A and 33A, showed the heating had continued for long after the process had ceased. This was very likely due to the unintentional exhaustion of salt in the parting vessel, but it shows that the heating had continued for long enough afterwards for the complete homogenisation of the composition of the foils to take place (see pp. 145–51).

These observations are corroborated both by the description of Agatharchides, which specifies five days, and by the experiments of Hall and of Notton, which took similar lengths of time to achieve separation.

Many of the early descriptions suggest that the process was not usually carried out in one operation but rather in a stepwise fashion until the desired degree of purity was attained.[33] The condition of the foils from the Sardis refinery could be interpreted as providing some support for this. Almost all the foils showed signs of having been in the process but the composition was very variable, clearly showing that some at least had been removed before the parting process was complete. This could be taken as evidence for the stepwise procedure.

An alternative explanation could be that the foils were test pieces. It is reasonable to suppose that a little gold would have been removed periodically for touchstone assay to check on the progress of the process. These surviving foils could be seen as those which, for whatever reason, failed to make it back into the refining process. However, all three of the partially treated foils that were sectioned (samples 19A, 26A and 33A) tell a very different story. Grain-boundary porosity was fully established through the pieces, showing that parting was underway (Figs 5.53 to 5.55, 5.58 and 5.59, pp. 148–50 and 152–3). If the process had been interrupted in mid-operation, one would expect both the localised and general variations in the silver contents to be preserved, as was certainly the case with the equivalent pieces from the replication experiments (see Figs 9.33, p. 198, and 9.34, p. 143), being greater in the centre of the grains, etc. However, the silver content is absolutely constant throughout in each foil. The only explanation for this is that the salt had run out, thereby stopping the process, but the heating had continued for a considerable period, allowing the homogenisation of the content throughout the metal. This rather suggests that the stepwise processing was not deliberate, but came about because too little salt was added or the sealing of the parting vessel was inadequate. Thus the operatives allowed the process to continue for the full time unaware that there was no longer any salt present.

At the end of the process, when the gold was considered to be of sufficient fineness, it would have been removed from the parting vessel. The foils could have been individually picked out, but the granules and flakes must have been recovered by careful washing.

It would naturally be expected that the gold would have been melted in purpose-made crucibles. In fact, there is no evidence at all for purpose-made crucibles anywhere at the refinery (with the possible exception of a rim fragment: item 57 in Appendix 2, p. 231). The gold seems to have been melted exclusively on sherds of coarseware cooking vessels reused as melting trays, which are found in quantity at the site. These have been exposed to intense heat, causing vitrification, which in some instances has run over the breaks, showing they were already sherds, not complete vessels, when used. The small scale of these scraps of ceramic, where the gold was probably melted with the aid of blowpipes such as the one recovered from the PN refinery (see Appendix 2, item 58, Fig. A2.9, p. 231),[34] show that only small quantities of gold, amounting to no more than a few grams, could have been melted in each operation. Gold droplets have been found on the end of one of the square-section clay tuyeres (T97.6:10722, item 1 in Appendix 2, p. 223). It seems very improbable that all the gold from a parting operation could have been accommodated on one of these sherds, and it seems more likely that the operations performed on the sherds were to melt small samples of metal from the parting vessels, sufficient for touchstone assay at various stages in the process.

In addition, several other sherds have globules of silver in a lead-rich glaze on their concave surfaces[35] and it seems that these are the remains of the fire assay of silver.

The vitrified surfaces both of the sherds used for gold melting and of the lead-rich sherds used for silver assay usually contain lumps of calcium carbonate, which are either not present or not as common in the ceramic body beneath. This suggests that small quantities of lime were being added as a flux.

The absence of purpose-made crucibles is very noticeable, especially from an operation functioning on an industrial scale. In fact, there seems to be no evidence at all for the melting of the main body of the gold at the refinery itself after the parting was complete. It is even possible that the still-sealed parting vessels were taken straight to the mint for the gold to be removed under careful scrutiny and tight security, with the empty parting vessels returned for reuse or to recover the silver salts with which they were impregnated.

Efficiency of the separation and gold recovery

The effectiveness of the gold-refining process in separating other metals from the gold is demonstrated by the purity of the gold in the contemporary Lydian gold coinage, containing only traces of silver, typically of the order of 0.5% (Tables 7.1 to 7.4, pp. 170–1). Small amounts of gold would be lost during the process, either incorporated within the silver or left as small flakes of particulate metal on the surface of the parting vessels and as larger globules on the melting sherds, as evidenced by the surviving sherds. A small amount would have been lost by evaporation, although this loss should have been very small at the temperatures at which the refinery operated. There is some information on the overall gold losses during the refining processes, given by the analysis of litharge cake 44393V, which contained so much silver that the associated gold could be easily quantified (Table 6.2, p. 160). The sample contained 13.0% silver and 0.06% gold, showing the silver would contain about 0.5% gold. If it is assumed that all the gold lost in the cementation furnace became incorporated in the silver, and that a typical charge of alluvial gold would contain about seven parts gold to three parts silver, then the overall loss of gold would be about 0.2%. In Notton's experiments, 0.1% was lost using common salt alone in a crucible, but rather more was lost when brick dust and common salt were used together on an open scorifier. At 780 °C, the loss was 0.5%; at 820 °C, the loss had increased to 2.3%.[36]

Recovery of silver from the debris of the process

The principal evidence for the recovery of silver from the spent parting vessels and from the walls of the old furnaces are the numerous small hollows scooped in the dust and lined with clay in the immediate vicinity of the furnaces (see Figs 4.17 to 4.21, pp. 79–81, and Fig. 10.1). Although the assumption that these hollows were used for cupellation still seems eminently probable, as evidenced by the high concentration of lead found in samples of some of the clay linings, some others of the hollows are almost certainly the remains of the small hearths where the gold was melted on the sherds. It is significant that the vitrified sherds with the gold globules were concentrated in the vicinity of the hollows and that the tips from two of the square-section tuyeres have small droplets of gold contained within the vitrified surface (see p. 92 and Appendix 2, items 8 and 13, pp. 224–5). Gold droplets and a foil (sample 25, p. 217) were found within one hearth.

The cupellation operations would have been carried out to recover silver from the parting debris and also on the impure scrap gold prior to cementation. Silver would have been recovered from the silver salts permeating the used parting vessels and furnace bricks by smelting with lead followed by cupellation.[37]

To achieve this, the refractories would first have been crushed and smelted with lead in a furnace. Under the reducing conditions of the furnace, the silver salts would have been reduced to metal, which would then have been absorbed by the lead. An appropriate source for the lead would, of course, be the litharge cakes recycled from previous cupellation operations, and this is probably what was intended for the small heap of cakes found at Sardis together with two pieces of furnace lining, 45686P and 45687Y. These latter have a heavily slagged and vitrified inner face which is rich in lead. As noted above, the slagged layer also contains many quartz fragments, which are larger and more concentrated than in the adjacent furnace wall, strongly suggesting that the fragments came with the material charged into the furnace. The fragments are similar in size to those found in the refractories of the parting process that would have become enriched in silver. The most likely interpretation is that these two pieces of slagged lining are from the furnaces where the silver was extracted from the spent refractories into lead.

The argentiferous lead would then have been cupelled to recover the silver. In this process, the argentiferous lead was heated to temperatures in excess of 1000 °C and subjected to a strong oxidising blast of air. Under these conditions, the lead rapidly oxidised to lead oxide (litharge, PbO), which would be molten at these temperatures, together with any other base metals present. The silver,

Fig. 10.3 Hypothetical arrangement of a square-section tuyere placed to blow air into one of the small hollows that were used either to melt the gold or to recover the silver by cupellation. In this instance, a piece of litharge has been placed in the hollow.

Fig. 10.4 Hypothetical arrangement of a piece of vitrified bread tray to form a cupellation hearth around the small hollows in which one of the litharge fragments has been placed.

Fig. 10.5 Fragment of litharge cake (M97.9:10743) from the north-east dump with the impression of the molten silver in the centre top, sunk down through the litharge. (See Appendix 2, item 61, p. 231.)

together with any residual gold, would be in the form of molten metal floating on top of the litharge 'like oil on water', in Pliny's apt phrase (Fig. 4.36, p. 90).[38] This process typically took place at temperatures in excess of 1000 °C, at which temperature most ceramics show evidence of vitrification, and it seems likely that it is to this stage in the refining process that the partially vitrified and lead-glazed fragments of bread-tray tiles, tuyere and some of the bricks belong (Figs 10.3 and 10.4).[39] It must be stressed here that, although the vitrified ceramics were found in the vicinity of the hollows, none was found in direct association either with each other or with small hollows, and the reconstructions illustrated in Figs 10.3 and 10.4 are entirely conjectural.

On some of the litharge fragments from the Sardis refinery, the impressions left by the precious metal are clearly visible. Sometimes the precious metal floated on the surface (Figs 4.35 and 4.36, p. 90), but on others it does seem to have sunk down through the litharge (Fig. 10.5). The surviving impressions left on the three conjoining fragments suggest an original volume of about 20 ml when complete[40] and thus it is estimated that each cupellation operation would have recovered about 200 g of silver.

The fragments of the bread-tray tiles recovered from the refinery area all showed extensive burning at just one end (see p. 167). From the drips of vitrified and lead-glazed material, it is clear that the tiles were always in a horizontal position when strongly heated. The burnt bricks were burnt on just one face. Assuming that these refractories were associated with the small hollows, then their function could be to build up the sides of the otherwise very shallow installations to better contain the charcoal fuel, which could otherwise be blown away by the blast from the bellows. Biringuccio describes just such an arrangement.[41]

The tuyeres are described on pp. 91–3 and listed in Appendix 2, pp. 223–6. Two types of tuyere are represented, straight and angled. Both types have square external profiles, which suggests that they lay on the ground to serve surface installations[42] such as the cupellation and melting hearths (Fig. 10.3). The vitrification on the ends of most of the fragments shows that they had been subjected to high temperature, and are thus not associated with the parting furnaces.

Summary

The gold to be treated at the Sardis refinery was in a variety of forms, reflecting different origins. Some were the granules and flakes of the natural gold–silver alloy, washed from the nearby placer deposits of the Pactolus and other streams running from Mount Tmolus. More of the impure gold was in the form of thin foils, some of which had the same composition as the Lydian electrum coinage. Scrap gold from other sources is probably also represented here, together with possible residues from the putative gold washeries, which could have been extracted with lead. This latter gold would have been recovered by cupellation, and the other gold in the foils seems very likely also to have been cupelled as a preliminary treatment, as evidenced by the absence of PGE inclusions in the foils. The preliminary cupellation treatment would have been to remove base metals prior to the salt parting process, which would remove the silver from the gold.

The impure gold, either as grains or foils, was treated in parting vessels which had the form and ceramic body similar to that of the contemporary Lydian coarseware cooking vessels. The parting vessels almost certainly had some form of lid, even if this was only a seal of clay.

The round vessels would have sat singly on the central plinth of the cementation furnaces with a carefully controlled fire all around them to ensure even heating, burning wood which would have provided the necessary moisture as well as heat.

The active agent in the cement used in the parting vessel was probably just common salt, by itself, although the presence of alum or some other sulphatic material cannot be excluded. Evidence for an inert carrier, such as brick dust or clay, which was common in the later traditional salt parting processes, is less certain. Inert carriers are not mentioned in any of the early sources, but the surface deposits on some of the treated gold foils from Sardis do have the appearance of a cement with an inert carrier.

The process certainly operated at a low temperature, even relative to the medieval salt parting processes. Microscopic examination of the fabric of the mud brick of the furnace suggests temperatures no higher than around 800 °C. Pure salt melts at 804 °C (impure salt somewhat lower) and thus if it was also to act as the support for the dispersed gold then the temperature had to be below this. The strongest evidence for a low-temperature process is the survival of the osmium–indium–ruthenium inclusions in the refined-gold coinage, with the content of the easily oxidised osmium only slightly depleted compared with the inclusions in the coins of unrefined gold. At about 600 °C, salt and moisture begin to react in the presence of silica to produce hydrogen chloride from which chlorine is generated. Thus the most likely temperature range is between 600 and 800 °C. If ferric chloride vapour was the principal active agent then the temperature could have been even lower.

It is difficult to estimate the duration of the process, although the metallurgical structure of the gold foils suggests they were subjected to prolonged heat. Given the low temperatures employed at the Sardis refinery, and the thickness of at least some of the foils and grains, the process would necessarily have taken a long time, probably to be measured in days rather than hours, as suggested by both the ancient sources and the modern replication experiments of Hall and of Notton.

It would have been necessary to test the gold at frequent intervals in the process. This would have been performed by touchstone, and it is likely that the numerous sherds of coarseware, reused as melting trays, result from the melting of small quantities of gold in various stages of purity – before, during and after the parting process – ready for touchstone assay.

The purity of the gold was high, as evidenced by the contemporary gold coinage, and the loss of gold caused by the parting process quite low, estimated at about 0.2%. There is no evidence for graduation or any other secondary refining process.

At the conclusion of the parting process, the gold would have been removed from the parting vessels and melted ready for coining etc., although almost certainly these operations were not performed at the PN refinery itself. Broken and worn out parting vessels and furnaces were smelted with lead to extract the silver which they had absorbed. The silver was then recovered from the lead by cupellation in the hearths at the PN site.

Notes

1 Craddock (1989; 1995, pp. 12–17).
2 Bayley (1985). See also, for example, Bayley (1992a) and Ottaway (1992) for a detailed description of the finds from a more typical urban metalworking area, Coppergate in Anglo-Scandinavian York, which included a wide range of ferrous and non-ferrous metalworking as well as the refining of precious metals.

3 Fragments of furnace lining, 45686P and 45687Y, which are interpreted as belonging to the furnaces where the used parting vessels and furnace bricks were smelted with lead, prior to the recovery of the silver by cupellation (see pp. 167 and 204). (Note: the numbers with a letter suffix used here to identify the sherds were assigned by the British Museum. See also concordance, Appendix 6, p. 251.)

4 Topkaya (1984); Uzkut and Semerkant (1980).

5 Burstein (1989, p. 66). See also p. 34.

6 Morrisson et al. (1987, esp. pp. 208–9); Antweiller and Sutton (1970).

7 Waldbaum (1983, p. 186). See also p. 100.

8 See Appendix 3, p. 234. It is not impossible that the residues left after washing could have been treated with molten lead to extract more of the remaining gold. Such impure gold would have then been recovered from the lead by cupellation. This gold would then have been made into foils prior to cementation refining.

9 Three large dribbles were identified on sherd 45680Q (see p. 113).

10 There is support for this from some of the early descriptions of gold-refining operations which commence with a cupellation operation before the parting process (see pp. 33, 47 and 69).

11 It is unlikely that the copper in the litharge originates from graduation-type processes, in which copper was deliberately added to the metal to be refined (see p. 205 and Appendix 5, p. 250, note 15). This process is not thought to have taken place at Sardis, as much of the gold was clearly treated in its original form without melting, and none of the foils of gold has the appreciable quantities of copper necessary, even though the gold foils are in various stages of treatment. In addition, little copper was found on the parting-vessel surfaces.

12 Coarsewares account generally for only about 5% to 10% of Lydian ceramic assemblages at Sardis (Andrew Ramage, personal communication), but almost 100% of the sherds used in the melting and refining process are of coarseware.

13 The one sherd, 46849Y, of a fineware that does seem to have been used as a parting vessel shows signs of thermal cracking (Figs 5.44, p. 137, 5.45 and 5.46, p. 125).

14 Sherds 44401Z and 45671S (see Fig. 6.4, p. 161).

15 Some of the traditional Indian and Japanese processes did not use closed vessels (see pp. 41 and 49).

16 Bayley (1992c, pp. 54–5).

17 Craddock et al. (1998b, Plate 19).

18 It should be reiterated again here that there is no evidence, literary or otherwise, for the use in antiquity of sulphur or of sulphides as parting agents. (See p. 51, note 45, and p. 52, notes 72 and 76.) For the medieval and post-medieval usage, see p. 67.

19 Melting sherd 45671S and some others. Sulphates are not uncommon in the ceramics from the site, originating either from the clays used or later from contamination during burial. Thus the sulphates in the melting sherds may have no significance for the process.

20 *Pliny: Natural History*, 33.84 (Rackham 1952, p. 65). See also p. 35.

21 d'Elhuyar (1790, Vol. II, p. 200). See also p. 176.

22 Eighteenth-century authorities such as Lewis (1763, pp. 154 and 155) and Howard (n.d. [1788], p. 1069) specifically warn against this, although some of the cements recommended by the earlier authorities, such as Agricola and Ercker, do contain chlorides and nitrates together (see pp. 59 and 64).

23 Burstein (1989, p. 30).

24 See discussion on pp. 34–5 and 179. Note the presence of lead in the surface of parting sherd 44401Z, and in the sectioned globules from melting sherd 45680Q. Tin was found on the surface of melting sherd 45673Z and on the surface of the gold globules of melting sherds 45673Z, 45675V and 45680Q, and foil 29A.

25 *Alchemia*, see p. 70, note 36.

26 Sisco and Smith (1951, pp. 190–1); Howard (n.d. [1788], p. 1069). See also p. 66.

27 See pp. 37, 38 and 64.

28 Another function of the added copper would be to promote the in-situ formation of chlorine from the hydrogen chloride.

29 See Appendix 5, p. 250, note 15.

30 Percy (1880, p. 385).

31 Mellor (1923, pp. 190–1).

32 Bayley (1992c, p. 55).

33 For example, Abū al-Faẓl's description of parting in sixteenth-century India (Blochmann in Percy 1880, pp. 379–81) and Manṣūr ibn-Baʿra's description of the process at the mint of Cairo in the thirteenth century (Levey 1971). See also pp. 42 and 46 respectively.

34 Waldbaum (1983, pp. 143–4, Plate 55, Item 955).

35 Sherds 44397Y, 45678P, 45682X, 45688W and 45681Z (see pp. 117–22).

36 Notton (1971).

37 This process was widespread through the ancient world and thus will not be described in detail here. For the early history of the technique, see Gale and Stos-Gale (1981), Craddock (1995, pp. 221–31) and Gowland (1917–18). There is a detailed technical discussion by Conophagos (1980) of the cupellation processes likely to have been used by the Greeks at Laurion. Good contemporary descriptions of cupellation in post-medieval Europe were made by Agricola (Hoover and Hoover 1912), Biringuccio (Smith and Gnudi 1942) and Ercker (Sisco and Smith 1951). The best general descriptions of the traditional cupellation process are given by Percy (1870 and 1880).

38 *Pliny: Natural History*, 33.95 (Rackham 1952, p. 72). Note that the silver, being denser than the litharge, should sink (see Fig. 10.5). However, as Bayley (1992c) has shown, the litharge cakes were rarely pure liquid litharge, but rather a pasty mass incorporating the clay of the hearth lining beneath (see p. 167). From post-medieval times it was usual for the hearth lining to be made of bone ash, which efficiently absorbs the poisonous lead-oxide fumes. The origins of the use of bone ash in the cupellation process are discussed at length in Craddock (1995, pp. 228–9). It would seem that bone ash began to be used for cupels on urban metalworking sites in the Roman period, but probably no earlier than the Middle Ages for the hearths at the main primary production sites. The analyses of the litharge cakes from Sardis show that bone ash was not used in the cupellation process. At Kition (Cyprus), recent work in the levels relating to the Late Bronze Age has shown that the bone found there was not used in metal refining, as previously suggested (Karageorghis and Kassianidou 1999).

39 Note that none of the bricks still in situ in the cementation furnaces has any trace of vitrification.

40 A second, smaller litharge fragment, of which about half survives (weight 0.4 kg), has an impression of about 2.5 ml, suggesting an original volume of about 5 ml. (See Appendix 2, p. 231, item 60.)

41 *Pirotechnia*, Book VII, 3b (Smith and Gnudi 1942, p. 292). In a section describing melting processes on top of the forge with a bellows-blown charcoal fire, it states:

> Keep the crucible ... held upright and well covered in the charcoal. In order to do this, some keep a half circle of sheet iron on the bed of the forge and some do it with brick ends. These things are used only for holding the charcoal, so that it may be closer together.

42 Tylecote (1981).

The Significance of the Sardis Refinery in the Classical World

A. Ramage and P.T. Craddock

The use and accumulation of gold, as bullion, coin or jewellery, was widespread in the ancient world. Reference to it as a measure of importance or social standing in the international and domestic spheres was commonplace, even if the connection was not directly stated. Thus the lists we find of gold as tribute or as offerings to gods and kings were by implication not just neutral announcements, but comparisons with the contributions of others. The writers who offer the various figures may have had their own agendas and perhaps made a selective list, but their purpose of making some sort of comparison that was readily understood remains clear. If the comparison were to hold up, reference to a common standard of fine gold would have been required and the importance of the process for obtaining that standard of purity would have to be recognised. The value of gold, which really depended on its purity or fineness, was commonly an issue in the ancient Mediterranean world. We have frequent references to its testing,[1] but only rare mentions of the process that was able to bring uniform purity to the varied fineness of the products of many different deposits.

There is no evidence for the cementation process for the refining of gold during the Bronze Age, and there are no analyses of the finds from Lefkandi or the jeweller's hoard from Eretria to suggest its use during the Dark Ages.[2] Inferences about the understanding and use of the cementation process in Late Geometric Greece may be valid, but the finds at Sardis are the only certain evidence for its use so far known from the ancient world.

It is interesting to speculate on why the crucial developments in the refining of natural gold took place when and where they did. The evidence at present suggests that the refining of gold to remove base metals, notably copper and lead, which had been deliberately added, was regularly carried out from at least the third millennium BC. However, there seems to have been no attempt to remove the silver that the natural gold almost always contained, even though the necessary chemical technology certainly existed, and had been used almost since the inception of goldworking to enhance the surfaces of gold artefacts. Here, then, we have an apparent contradiction, at least to minds used to the concepts of modern European science.

It would seem that here was a technology awaiting a problem, a problem which was finally posed by the introduction of coinage, that is pieces of precious metal where the weight and the purity were guaranteed. Without a concept of elements with unique and unchanging compositions and properties it was quite logical to believe that the properties of gold, or of any other metal, would have varied depending on where it was found. A silver-rich, light-coloured gold would have been regarded as still being in a state of development. Equally, it must have been at least partially comprehended that most of the differences amongst golds from different locations were in fact due to variations in the silver content. It would have been clear from the surface enhancement treatments that something was being removed from the gold, and on treating the debris of the process, silver would have been recovered. Conversely, it could not have escaped notice that some of the light or white natural golds could have been imitated by the addition of silver to ordinary gold.

Before coinage came along, this would have caused no real problem: gold from different localities would have had different properties. All that refining would have achieved would have been a reduction in the weight of metal, sometimes quite appreciable. But once the concept of coinage was envisaged, then clearly the metal needed to be of consistent purity. Thus we believe that the Sardis refinery is likely to be amongst the earliest, if not the very first refinery in the world. It is surely suggestive that the first coins were of native alluvial gold, with the composition carefully adjusted by the addition of small quantities of silver to bring the gold content down to a fixed composition, just below the minimum gold content found in the metal from the Pactolus. The smiths lacked the technology to raise the gold content, and so were forced to reduce it to a constant composition. Clearly, such a situation where the smiths had no real ability to manipulate the composition of the gold was unacceptable and within a very short time the chemistry of the existing surface enhancement treatments was applied to the total removal of the silver throughout the gold. With that the true refining of gold and the control over the composition of precious metals was born. The salt cementation technology, once established, was to remain the usual method of gold refining for the next two millennia.

In hindsight it seems natural and fitting that the material evidence should come from Sardis, in view of its close association with the winning and distribution of great stores of gold. At the Pactolus North refinery, we have uncovered unique evidence, from which it has been possible to reconstruct the ancient processes otherwise only hinted at in many ancient sources. This has been achieved by examining the ceramic and metallic remains from the works with the best analytical tools we could muster.

Given that Herodotus makes the point about the Lydians' introduction of a bimetallic coinage with separate gold and silver coins, and claims that they are credited with the invention of struck coinage itself, we might suppose that cementation was seen as a new process. This idea must remain a hypothesis until many more pieces of gold have been analysed and the composition of native gold from additional sources is better known. It is not clear what impact the use of the cementation process had upon the contemporary world, in that silver rather than pure gold was used for coinage among the Greeks during the Archaic period. An exception must be made for the electrum coinages of East Greek cities like Cyzicus and Lampsacus, as well as Mytilene on the island of Lesbos, but some of those are agreed to be artificial mixtures.[3] Such coinage as the Persians used followed the late Lydian pattern of minting in pure gold and pure silver. Traditional scholarship has them continuing to use the mint at Sardis for their own purposes and eventually issuing what the Greeks called darics and sigloi.[4] The daric was named for King Darius the Great (521–486 BC), whose troops tried to invade Greece several times and were decisively beaten in 490 BC at the Battle of Marathon. Siglos was the Greek equivalent for the Phoenician 'shekel'; it was the standard silver coin in the western Persian Empire and 20 of them were equivalent to one gold daric.

Although the evidence for cementation at Sardis is so far unique, it is hard to imagine that subsequently there were not many such establishments scattered around the classical world. Although full purification might not have been necessary for the production of jewellery, it must surely have been required for the gold coins that were issued in great quantities in the fourth century BC and after, because of the varied sources of the raw material. We hope that the Sardis finds will help others in the field to recognise similar pieces from their own excavations and fill out the picture of an important economic activity. There is every reason to think that almost all of these installations would have been using the same process, even if some improvements in technique were made over the years. This conclusion stems from the fact that the parting process is recorded in a great many of the scraps of testimony from the ancient and medieval world and is extraordinarily similar to the important processes described in the technical literature of the Renaissance (reviewed in Chapter 3, pp. 54–65). These works allot considerable space to what we can consider the traditional process, even at a time when the more modern use of strong acids had become feasible. Nevertheless, the antique methods were still considered the more economical. Thus we are brought to an understanding of the remarkable continuity between the techniques of antiquity and those of early modern times.

The results of the intense analysis of the Sardis material have enabled us to make more sense of the spotty and apparently contradictory accounts of ancient processes that have been preserved for us from all over the ancient and medieval world, from China to Germany. We can now point with more confidence to the specific misunderstandings in the descriptions of the process by the early writers, be they historians or philosophers. They were probably not familiar with the day-to-day workings of an industrial workshop, and perhaps the explanations were not the model of clarity. The literature as a whole reflects this situation and our information is usually gained from asides or short explanations of technical terms that were probably obscure even to many ancient readers.

The finds at Sardis have shed light on numerous inconclusive or garbled attempts to set out several processes required for the purification of precious metals. At present, our discoveries offer the only archaeological evidence to confirm the ancient tradition.

Notes

1 An example is the well-known but frequently misunderstood story about Archimedes in the bath and the gold crowns related by Vitruvius (see Appendix 5, p. 245). The assay of gold in antiquity is dealt with in Appendix 5, pp. 245–50.

2 For gold objects from Lefkandi, see Jones (1980). For the jeweller's hoard from Eretria, see Themelis (1983).

3 Kraay (1976, p. 262).

4 Kraay (1976, p. 31).

Inventory and Descriptions of the Gold Samples

Compiled from Sardis records by A. Ramage

Abbreviations

* level in relation to a datum of 100 (when prefixing a decimal number)

E east

N north

S south

W west

BASOR *Bulletin of the American Schools of Oriental Research*

SPRT *Sardis from Prehistoric to Roman Times* (Hanfmann 1983)

Sample 1 (Fig 9.1, p. 139)

Findspot: Cupellation area A, W265.40/S344.00 *86.15

Weight: 10 mg

Previously published in *SPRT*, p. 39, no. 1

Piece of gold foil (9 mm long × 2 mm wide), highly pitted on both surfaces with several scratches, which may be intentional. One or two edges may have been cut and folded but others are very jagged and thin. Perhaps pieces of this size were used to increase the gold content in small dumps for early coinage. Colour designated as 'rich' gold.

Sample 2 (Fig 9.2a, p. 139)

Findspot: Cupellation area A, W265.40/S344.00 *86.15

Weight: 10 mg

Previously published in *SPRT*, p. 39, no. 2

Three pieces of dark gold foil, same as sample 1 (i.e. 'rich' gold).

Sample 3 (Fig 9.3, p. 140)

Findspot: Cupellation area A, W265.00/S344.00 (From sifting) *86.00

Weight: 7 mg

Previously published in *SPRT*, p. 39, no. 3

Folded bead of dark gold which is quite porous. The inside pores have a pink cast, which may be an indication of copper or iron impurity. The bead was subsequently opened out into a foil (see Fig. 9.3, p. 144).

Sample 4 (Fig 9.4, p. 140)

Findspot: Cupellation area A, test trench, W264.00–265.50/S343.20–345.00 *86.15–*86.00

Weight: 20 mg

Previously published in *SPRT*, p. 39, no. 40

Two medium-sized pieces of gold foil, dark in colour (similar to sample 1). The thin ridge running down one piece (the profile indicates thickening of the metal) may represent a crack in the working surface on which the foil was made. The piece is bent over so it is impossible to see the opposite side of the foil without heating it.

Sample 5 (Fig 9.5, p. 141)

Findspot: Cupellation area A, test trench, W264.00–265.50/S343.20–345.00 *86.15–*86.00

Weight: 10 mg

Four tiny pieces of dark gold foil, with the same surface as sample 1.

Sample 6

Findspot: Cupellation area A, test trench, W265.00–265.40/S342.50–344.00 *86.15–*86.00

Weight: 20 mg

Four pieces of dark gold foil.

Sample 7 (Fig. 4.45, p. 92)

Findspot: Cupellation area A, test trench, W265.00–265.40/S342.50–344.00 *86.15–*86.00

Weight: 30 mg

Large piece of gold foil folded several times upon itself.

Sample 8

Findspot: Cupellation area A, test trench, W265.00–265.40/S342.50–344.00 *86.15–*86.00

Weight: 10+ mg

Nine small pieces of dark gold foil.

Sample 9

Findspot: W260.00–263.00/ S341.00–345.00 *86.30–*86.20
Weight: 20 mg

Nine very small pieces of dark gold foil.

Sample 10

Findspot: Cupellation area A, test trench, W265.00–265.40/ S342.00–344.00 *86.00–*85.60

Weight: 10 mg

Four small pieces of dark gold foil.

Sample 11 (Fig. 4.46, p. 131)

Findspot: Cupellation area A, test trench, W265.00–265.40/ S342.00–344.00 *86.00–*85.60

Weight: 30 mg

One large piece and three small pieces of light gold foil. There are two lines on the large piece, one of which was purposely done with a wedge-shaped tool. The other line has an intentional dent.

Sample 12 (Fig. A.1.1)

Findspot: Cupellation area A, test trench, W265.00–265.40/ S343.20–345.00 *86.15 down

A1.1 Gold bead (sample 12).

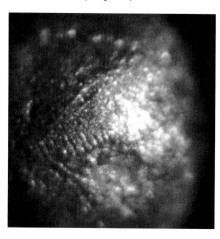

A1.2 Gold chunk, top view (sample 14).

Weight: Not taken

Large bead of gold which shows good dendritic formation.

Sample 13

Findspot: Cupellation area A, test trench, W265.00–265.40/ S343.20–345.00 *86.15 down

Weight: 40 mg

Six pieces of dark gold foil.

Sample 14 (Figs A1.2 and A1.3)

Findspot: Cupellation area A, test trench, W265.00–265.40/

A1.3 Gold chunk, cut surface (sample 14).

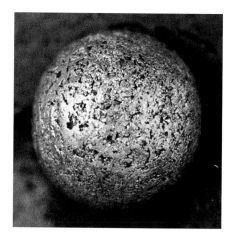

A1.4 The largest gold bead found (sample 17).

S343.20–345.00 *86.15 down

Weight: 52 mg

Chunk of light-coloured gold with a finer surface than sample 1. It has been melted and shows obvious traces of cutting by a sharp instrument, perhaps a chisel, in at least four places.

Sample 15

Findspot: Cupellation area A, test trench, W263.50–264.00/ S342.50–344.00 *85.90

Weight: 30 mg

Two gold droplets of light-coloured gold with beautiful dendritic formations.

Sample 16 (Sectioned, see Table 5.3, p. 145, and Fig. 5.51, p. 146)

Findspot: Cupellation area A, test trench, W263.50–264.00/ S342.50–344.00 *85.90

Weight: 50 mg

Large piece of gold foil, light in colour; less porosity than sample 15.

Sample 17 (Fig. A1.4)

Findspot: Cupellation area A, test trench, W263.50–264.00/ S342.50–344.00 *85.90

Weight: 20 mg

Single gold bead, the largest one found with a diameter of over 1 mm. Surface is very porous.

Sample 18

Findspot: Cupellation area A, test trench, W263.50–264.00/ S342.50–344.00 *85.90

Weight: 30 mg

Six pieces of light gold foil. All have smooth surfaces. One has a good dendritic layer, denoting an unhammered surface.

Sample 19 (Sectioned, see Table 5.3, p. 145, and p. 147; NAA analysis, see Table 7.1, p. 170)

Findspot: Cupellation area A, test trench, W265.30/S344.15 *85.65

Weight: Not taken

Large piece of light gold foil (25 × 20 mm, 300 µm thick) with red and black corrosion products adhering, and possible intentional scribed working lines. The surface is somewhat less porous than sample 1 but more porous than other lighter coloured samples.

Sample 20 NAA analysis, see Table 7.1, p. 170)

Findspot: Cupellation area A, W266.85/S342.90 *86.00

Weight: Not taken

Large piece of light gold foil (30 × 19 mm) with smoother surface. It also has a line down its centre, which is the trace of several pieces of metal pounded together. Planishing marks are not visible. There are several small dents but it is doubtful whether these are planishing marks. This is an interesting piece which, according to NAA analysis, contains 42.3% silver and 1.2% copper.

Sample 21 (See Table 5.3, p. 145; NAA analysis, see Table 7.1, p. 170)

Findspot: Cupellation area A,

W264.70/S343.30 *86.10

Weight: 180 mg

Previously published in Hanfmann 1983, p. 40, Fig. 59

This is the largest cut piece, 4 × 3 × 2 mm. Three sides are cut but the fourth is not. Its top is somewhat rippled and has crack marks where a chisel probably wrenched the metal. One corner shows definite striations of engraved lines. At the bottom, bits of metal that were not cut leave a ragged edge.

Sample 22

Findspot: Cupellation area A, W265.00–266.80/S342.00–343.50 *86.15–*85.90

Weight: 20 mg

Four pieces of dark gold foil.

Sample 23

Findspot: Cupellation area A, W265.00–266.80/S342.00-343.50 *86.15–*85.90

Weight: 20 mg

Eight small pieces of dark gold foil.

Sample 24 (See Table 5.3, p. 145; NAA analysis, see Table 7.1, p. 170)

Findspot: Cupellation area A, W262.50–265.00/S342.50–345.00 *86.15–*86.00

Weight: 80 mg

Five pieces in all: one large piece of light gold foil; one medium-large fragment showing some small dendritic formation; three small pieces of the same material as the largest piece in the group. One of the small pieces is very thin, almost like gold leaf, and is folded several times.

Sample 25

Findspot: Cupellation area A, W264.00–266.60/S344.00–347.60 *c.*86.00 (In dark ash layer of a cupel)

Weight: 40 mg

Three dark gold balls and three small pieces of dark foil.

Sample 26 (Sectioned, see Table 5.3, p. 145, and p. 147; NAA analysis, see Table 7.1, p. 170)

Findspot: Cupellation area A, W264.00–267.00/S344.00–346.00 *86.10–*85.80

Weight: Not taken

Three pieces of medium-sized dark gold foil. One small piece has a much smoother surface than the others, so it might have been in a different area.

Sample 27

Findspot: Cupellation area A, W264.00–266.60/S344.00–347.60 *85.70

Weight: 35 mg

Three pieces of dark gold foil.

Sample 28 (Figs 9.5, p. 141, and A1.5)

Findspot: Cupellation area A, W264.00–266.60/S344.00–347.60 *85.55

Weight: 30 mg

Two pieces of light gold foil and one dark gold ball. The foils show dendritic shapes as well as a smooth surface. There is a small crack in one of the pieces. The smaller piece of foil has several blotches on its back (see Fig. A1.5). In 1968, this piece

A1.5 Gold foil with blotches on one face (sample 28).

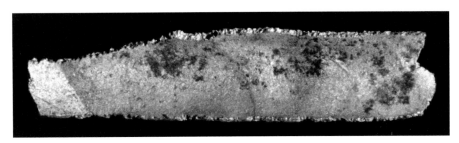

A1.6 Strip of gold, curved in its cross-section like part of a tube (sample 29). View of concave side.

exhibited a strange series of spots over its entire surface. There are now hardly any traces of discolouration and those that have remained have been photographed. They appear to be reddish-brown under high magnification and more yellow-brown under low magnification. The surface is covered with a series of parallel striations.

Sample 29 (Fig. A1.6)

Findspot: Cupellation area A, W264.00–266.60/S344.00–347.60 *85.55

Weight: 100 mg

This fragment is possibly the most important evidence for goldworking at Pactolus North. It is 13 mm long × 3 mm wide and *c.* 100 µm thick. It is curved like a section of a tube and looks as though it was cast, as there is evidence of dendritic formations on the interior and exterior surfaces. (However, it would be extremely difficult to cast something this thin.) Also, it is quite brittle, which further suggests that it could not have been bent into its shape.

The exterior surface is somewhat pitted, although it is not 'dark' gold. Adhering to this surface are patches of red-brown material and grey material, neither of which seem to be soil deposits (which are clearly seen under a microscope as sandy brown incrustations).

Sample 30 (Sectioned, see Table 5.3, p. 145, and p. 148; Figs 4.44, p.131,

and 5.57, p. 152)

Findspot: Cupellation area A, W264.00–266.60/S344.00–347.60 *85.55

Weight: 200 mg

A blob of gold, some of whose edges are quite bevelled. Perhaps at a later date the piece was subjected to considerable pitting and abrasion. It has very few distinct or dendritic formations on the surface. It is quite bumpy and unlike any other piece in the collection. Even close investigation of its crevices reveals little dendritic formation. Scientific examination (see p. 148) has confirmed that this is a piece of alluvial gold.

Sample 31

Findspot: Cupellation area A, W265.20/S344.35 *85.48

Weight: 200 mg

Three pieces of medium-sized gold foil.

Sample 32

Findspot: Cupellation area A, *c.* W266.00/S345.00 *85.30

Weight: Not taken

Small piece of light gold foil.

Sample 33 (Sectioned, see Table 5.3, p.145, and p. 148)

Findspot: Cupellation area A, *c.* W261.50/S345.00 *85.40

Weight: 30 mg

Medium-sized piece of light gold foil, 8 mm on its longest dimension, 6 mm wide, roughly triangular in shape. It has an even surface.

Sample 34

Findspot: Cupellation area A, W265.50/S342.50 *85.40 (Bottom of trench, fallen from scarp over winter)

Weight: 10 mg

Long (4 mm) piece of foil, folded. All its edges are rough. The surface is rather porous and somewhat granular.

Sample 35

Findspot: Cupellation area B, test trench, W273.00–277.30/ S343.00–343.50 *86.10

Weight: Less than 1 mg

This sample appears to be the size of a grain of sand (<1 mm). It is probably a small globule from the interior of a typical coarseware pot.

Sample 36

Findspot: Cupellation area B, test trench, W277.00/S342.90 *86.20

Weight: Less than 1 mg

This is a remarkably narrow ribbon of gold foil, 20 mm long × 2.5 mm wide. It has a porous surface, similar to sample 34.

A1.7 Smooth side of a triangular piece of gold (sample 38) at close range, showing striations from cutting. (See also Fig. A1.8.)

A1.8 Rough side of the triangular piece of gold (sample 38) shown in Fig. A1.7.

Sample 37

Findspot: Cupellation area B, test trench, W277.30/S343.20 *86.00 (From earth)

Weight: Less than 1 mg

This piece measures 5 × 1 mm. Its surface is the same as sample 36.

Sample 38 (Figs A1.7 and A1.8)

Findspot: From sieve

Weight: 4 mg

The sample is triangular in shape, 1.5 mm long × 1 mm wide. It is rough on one side and not particularly porous. The other side is smooth, as if it had been cut or sliced from a larger piece. This surface is marked by parallel striations.

Sample 39 (Fig. A1.9)

Findspot: Unit II, test trench, W279.40/S343.25 *85.90

Weight: 919 mg

Large dribble, 12 mm long × 6 mm wide. The surface is smooth with several lumps of various dimensions protruding from it. It was possibly caused by uneven cooling. One major appendage is deeply grooved, having the same characteristics of a fine smooth surface with small lumps protruding. Two smaller appendages seem to have tiny globules adhering.

Sample 40

Findspot: Unit II, test trench, W278.80/S343.25 *85.80

Weight: 4 mg

Dark, thin piece of gold foil, folded on itself like an envelope. It is 4 mm long × 2 mm wide. Its surface is quite porous. It has no cut edges. Possibly post-cementation.

Sample 41

Findspot: NW dump, *c.* W277.50/S332.50 *85.50 (In earth and charcoal)

Weight: 68 mg

Forty-five pieces of gold foil with various colours and textures. Several pieces have a red-brown accretion, some are deep gold in colour, while others have indistinct edges.

Sample 42

Findspot: NW dump, W277.50/S332.50 *85.50 (In earth and charcoal)

Weight: 1–2 mg

Light, thin gold foil, 4 mm long × 2 mm wide. Triangular in shape. There is one cut edge; all the others are rough. The surface is smooth, not yet having been made porous by the action of the cementation process.

Sample 43 (Fig. A1.10, p. 144)

Findspot: NW dump, W277.50/S332.50 *85.50 (In earth and charcoal)

Weight: 42 mg

Forty-five pieces of gold which are light in colour. As opposed to sample 41, this group is homogeneous in colour.

Sample 44

Findspot: NW dump, W275.00–279.00/S331.00–333.00 *85.40–*85.30

Weight: 25 mg

The largest piece of foil in this sample is 4 mm long × 2 mm wide. The sample consists of 17 pieces, of which two very small pieces are much more porous and darker than the rest, which shows distinctly in their photographic images. This sample may contain pre- and post-cementation foils.

Sample 45

Findspot: NW dump, W275.00–279.00/S331.00–333.00 *85.40–*85.30

Weight: 8 mg

Different textures of pre- and post-cementation foil.

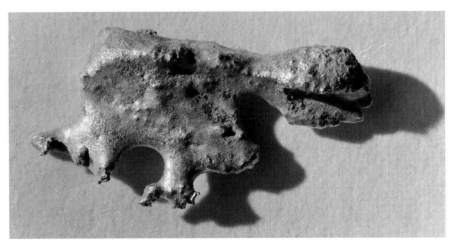

A1.9 Large gold dribble of irregular shape (sample 39).

Sample 46

Findspot: From sieve

Weight: 50 mg

Fifty-seven pieces, most of which are small and of thin, light gold foil. Three pieces are of dark cemented foil. Several pieces retain a red-brown surface.

Sample 47

Findspot: NW dump, W275.00–279.00/ S331.00–333.00 *85.50–*85.20

Weight: 16 mg

Ten pieces of light gold foil. One piece is 7 mm long × 4 mm wide, with a fine smooth surface. Another is reddish-brown on the edges and 4 mm long × 2 mm wide. The other eight pieces are light in colour and distinctly smaller. One of them is a shade darker in colour and possibly cemented.

Sample 48

Findspot: NW dump, W275.00–279.00/ S331.00–333.00 *85.50–*85.20 (From sieve)

Weight: 29 mg

Some 30 pieces of foil, the largest of which are *c.* 3 mm long; one thin strip is slightly longer. Mostly they are pieces of light gold foil. Six small pieces are of porous dark foil, probably post-cementation. Six

pieces have reddish-brown surfaces. All the rest are light gold. No cut edges were noticed.

Sample 49

Findspot: NW dump, W275.00–279.00/ S331.00–333.00 *85.50–*85.20 (From sieve)

Weight: 9 mg

Eleven pieces of gold, the largest of which is 2 mm square. Three pieces of light gold foil, four pieces of reddish-brown foil, and three pieces of dark gold foil, probably cemented. One piece is quite dark gold.

Sample 50

Findspot: W274.10/S328.00 *85.00

Weight: 3 mg

One piece of folded, medium-dark foil, 2 mm long with no cut edges.

Sample 51

Findspot: W279.00/S334.00 *85.10–*85.00

Weight: 2 mg

Lump of cemented dark gold, 1.5 mm long × 1 mm wide. Neither surface is very porous, but the gold itself is rather dark. Triangular in shape.

Sample 52

Findspot: Unit I floor, W278.50/

S338.80 *85.80–*85.70

Weight: 3 mg

Triangular-shaped piece of cemented dark foil, 3 × 2 mm, with a porous surface.

Sample 53

Findspot: From excavation dump (presumably from W275.00–279.00/ S331.00–333.00 as samples 44, 45, 47–49)

Weight: Less than 10 mg

Piece of medium-dark gold foil, roughly rectangular in shape. Not as porous as sample 52. Almost 4 mm long × 2.5 mm wide, after unfolding.

Sample 54

Findspot: Unit I, W279.10/S340.10 *86.80 (Just beside a cupel)

Weight: Less than 1 mg

Tiny speck of gold. This piece may be from a pot but is so small that it is difficult to tell.

Sample 55

Findspot: W297.00/S325.00 *84.80

Weight: Less than 10 mg

Three pieces of foil. Two pieces are of medium-dark gold. The larger is 2 mm long × 1 mm wide. The third piece, of light gold, is 5 mm long × 2 mm wide.

APPENDIX 2

Inventory and Descriptions of Finds of Equipment and Supplies

A. Ramage

Set out below are the specific details of the individual items mentioned in the descriptions of the dumps and working areas (see pp. 72–98).

Some 40 pieces are also referred to in the analytical sections (see pp. 99–156 and 157–68). The different categories of ceramics come first, beginning with the bellows nozzles; then follow the metal pieces, both artefacts and raw materials.

Also listed are varied quantities of different materials that have not been inventoried or described separately but are stored together in the laboratory at Sardis. In some cases, groups of items were noted but discarded. These are recorded at the end of the appropriate section.

Charcoal was found in concentrations sufficient to indicate that this was the primary fuel used in the hearths and furnaces. Ash was convenient and clean for lining the hearths but not, apparently, required for the chemistry of the process.

Brick remains from previous cementation furnaces, most of which had crumbled, gave some of the surrounding earth its distinctive red colour. In addition to the reddened fill of the disintegrated pieces, there are several whole bricks of different types that survive, as well as many pieces that belonged to other examples (Fig. A2.1). Two main categories make up the different processes. The bricks of the first category were made from the same fabric as that which was used for the furnaces, and may in fact be detached from dismantled furnaces (see p. 160). The bricks

Fig. A2.1 Broken pieces of brick, from the supports in the furnaces.

Fig. A2.2 Bread tray, at upper left, used like paving or a smooth working surface in the cupels.

of the second category were formed from well-levigated clay and well fired. These pieces were used in the cementation furnaces as supports, to judge from examples surviving in situ (Fig. 4.30, p. 88).

Broken pieces of bread tray were used for edging the cupellation hearths as the thick accretions of lead glaze and vitrification attest. Many were apparently used like paving blocks among the cupels to provide a smooth working surface (Fig. A2.2) and may have been exposed to heat or spills of litharge by accident. In his chapter on melting with a ladle,[1] Biringuccio describes how a 'crown of two or three brick ends is built around it on top of the rim so that the charcoal may be held better and in greater quantity'. A procedure like this would explain the vitrified edges of the bread tray and pottery pieces which were found. It would also explainwhy the bricks have an abrupt change from glazed to unfired surfaces, if the melting of the lead were done directly in the cupels, as is clear from the profile and the undersurface of the litharge cakes.

Abbreviations

Abbreviations for units are not included.

*	level in relation to a datum of 100 (when prefixing a decimal number)
Diam.	diameter
E	east
Est.	estimated
H.	height
Int.	interior
L.	length
N	north
S	south
Th.	thickness
W	west
W.	width
Wt.	weight
BASOR	Bulletin of the American Schools of Oriental Research
SPRT	Sardis from Prehistoric to Roman Times (Hanfmann 1983)

Dimensions

In the inventory entries, all dimensions are given in metres without stating the unit.

Bellows nozzles or tuyeres
See also pp. 92–4

1   T97.6:10722 (Figs 4.47 to 4.49, p. 93, and 4.50, p. 131)

Findspot: Cupellation area A, W264.00–266.00/
S345.00–347.00 *85.75–*85.45
H. 0.052; W. 0.059–0.068; L. 0.076; Diam. of air tube
0.03 × 0.022

Previously published in *SPRT,* p. 39, Fig. 57; *BASOR*
199 (1970), p.24, Fig.14

Four joining pieces forming complete squarish outer
and round inner section of a flat-faced tapering nozzle.
Intense heating has melted away some of the tip, which
is grossly misshapen with bits of vitrification adhering.
Extensive scoriation and vitrification are evident on
three sides, including blobs of glass containing minute
beads of gold at the narrower end. The base is not
vitrified and may have been set on the ground and not
exposed to intense heat. The body clay is micaceous
and the interior colour varies from grey to brown
to red.

2   T97.8:10724. Archaeological Museum, Manisa
(Fig. A2.3)

Findspot: Cupellation area A, W264.00–266.00/
S345.00–347.00 *85.75–*85.45
H. 0.055; W. 0.037; L. 0.075; Diam. of air tube 0.017

Two joining fragments; the exterior is heavily vitrified
with large holes, the underside is not vitrified. The
same shaped nozzle and equivalent body to item 1. The
air passage shows striations from manufacture, when a
reed was probably used to create the hole. The clay was
well levigated, very like standard Lydian pottery or
architectural terracottas (see p. 158). Although there
is extensive blobby vitrification on three sides, the
underside shows the only complete reddening. The
unvitrified clay that remains on the side is very thin.

3   T97.7:10723 (Figs 4.51 and 4.52, pp. 89-90)

Findspot: NW dump, W275.00–279.00/S329.80–330.70
*85.50–*85.20

Fig. A2.3 Item 2, tuyere: interior view showing the air passage and
the texture of the vitrified edge at top.

L. 0.058; W. 0.040; H. 038; Diam. of hole 0.025

Three joining fragments of an angled nozzle for
directing air downwards onto the lead for complete
oxidation. Grey and lightly vitrified on exterior
surface. Interior brick red.

4   T97.9:10725

Findspot: NW dump, W275.00–279.00/S331.00–333.00
*85.50–*85.00
L. 0.035; W. 0.032; Th. 0.01–0.013

Small fragment of an angled tuyere; smooth dark
vitrification with streaks of purplish red. Clay body
exactly as that of item 3, and this piece may have been
part of it since they came from the same general area.

5   T97.21:10737

Findspot: NW dump, W275.00–279.00/S331.00–333.00
*85.50–*85.20
L. 0.058; H. 0.031; W. 0.030

Corner of a tuyere, near the tapered end, as the inner hole is near the outer wall at one end. Coarse, micaceous fabric with small inclusions and reddish colour like item 1.

6   T97.10:10726

Findspot: Cupellation area B, W273.00–276.00/ S342.00–345.00 *86.75–*86.40
Max. W. 0.066; L. 0.078; Th. 0.018–0.034; Diam. of flared hole *c.* 0.033 at rear, int. 0.016

Nearly half of a well-made tuyere with one side and back edge preserved and the flare quite clear. The corners are rounded off and thumbprints are visible over the entire exterior surface. Interior is sienna red and exterior is mottled from brown to black. It has cracked, perhaps from initial firing.

7   T97.11:10727. Archaeological Museum, Manisa

Findspot: W275.30/S329.90 *85.30
L. 0.12; H. 0.059–0.0775; W. 0.0485

Almost complete side of a tuyere with a flared interior towards the rear edge, which is completely preserved and smoothed. Texture is red and gritty towards the centre, and grey and bubbled towards the edge. There are several large angular inclusions and some organic temper like narrow-stemmed grass. It was not so highly fired in the interior as towards the outside faces. The (assumed) underside is flatter than the right, vertical side or the remains of the top. The lower part of the piece is denser and less bubbled than the upper part. The inference is that this particular tuyere was set at ground level with just the tip, which is now lost, projecting over the edge of the cupel. The diameter of the air tube before the flare at the back was at least 25 mm (more probably a little over 30 mm).

8   T68.16:7825. Archaeological Museum, Manisa (Fig. A2.4)

Findspot: NE dump, W260.00–262.00/S336.00–338.50 *85.70–*85.50
L. 0.12; W. 0.052; Diam. of air hole over 0.02

Previously published in Waldbaum 1983, no. 956

Grossly misshapen side of a tuyere with an accumulation of glassy green vitrification adhering to two surfaces. Several very small beads of gold were observed with a hand lens. The piece is well vitrified along 95 mm of its length and the remainder was probably vitrified also, but there is a distinct lessening of scoriation and vitrified deposit as the wall thickens and the interior clay becomes browner. The curvature of the air hole that matches this progression indicates that this might have been the right-hand side in relation to the operator of the bellows. Many fragments of crushed quartz, several up to 2 mm square, were used as temper, and there was some organic temper as well. The latter was probably a narrow-stemmed grass, to judge from the impressions in the clay. Interior horizontal lines suggest the possibility of building around a stick or pushing one through. This feature can be observed on many other examples and might be considered standard practice even if the impressions do not survive. The reed *Arundo donax*, which seems to have been used in roofing,[2] is an excellent candidate for the pattern making.

9   T97.12:10728

Findspot: Cupellation area B, W273.70/S347.30 *86.30
L. 0.084; W. 0.039; H. 0.064

Part of a short tuyere with its back edge preserved. Vitrification begins abruptly 37 mm from it, suggesting that there may have been some sort of daubed protection for the intermediate nozzle and the skin of the bag (see p. 92). Some vitrification in front. Splaying at rear is apparent but not really measurable. The length of the air hole is 80 mm. The clay body is rather like the bread-tray mixture. Note: this sample was not examined petrographically.

Fig. A2.4 Item 8, tuyere: interior view, showing the curved air passage.

10 T97.13:10729

Findspot: W275.00/S329.00 *85.40
L. 0.058; W. 0.039; H. 0.033; Th. 0.014–0.021

Piece of tuyere with vitrified end and glassy material on the tip. It shows the square outer section and a fragment of the air passage. The reduced surface is assumed to be the bottom which would have lain near the charcoal. Inside and 'higher' is redder.

11 TS139a

Findspot: Fringe of NE dump, W262.00–265.00/ S335.00–337.00 *85.80–*85.60
H. 0.03; L. 0.045; Diam. of air tube 0.01

Outer surface is scoriated and vitrified; interior is brick red.

12 T97.16:10732

Findspot: W268.50–270.00/S329.00–330.00 *85.80–*85.65
L. 0.03; H. 0.045; Th. 0.019; Diam. of air tube 0.02

Small vitrified section.

13 T97.14:10730

Findspot: NW dump, W275.00–279.00/S331.00–333.00 *85.50–*85.00
L. 0.025; W. 0.016; Th. tapering from 0.015 to 0.005

Small vitrified fragment with tiny gold beads trapped in the vitrification. Part of the curve of the air tube is preserved and there is an abrupt change in colour within it from red to grey as the vitrification gets thicker on the exterior. Presumably, the oxygen in the air flowing over the tip of the tube was sufficient to change the colour back to red.

14 TS139i

Findspot: Cupellation area A, W264.00–266.00/ S345.00–347.00 *85.75–*85.45
L. 0.074; W. 0.03; H. 0.046

The tip is thickly covered with black, green and yellow vitrification. The vitreous surface is 6 cm along the length of the exterior. Large bubbles at tip. Green vitrification occurs only after the area showing previous violent bubbling. Vitrification continues into

tip of interior nozzle but the clay is only buff in colour for about 10 mm and then turns to typical brick red. Vitrification at the tip is as thick as 14 mm.

15 T97.15:10731

Findspot: Cupellation area A, W264.00–266.00/ S345.00–347.00 *85.75–*85.45
L. 0.041; W. 0.037; Diam. of air tube 0.013

This small piece shows heavy vitrification, about 3–5 mm thick, on the exterior. Beneath the vitrification is a layer of purplish clay, and a crack between this and the brick-red layer suggests the effect of the extreme temperature difference. Large chunks of quartz protruding from the vitreous matter.

16 T97.17:10733

Findspot: NE dump, W260.00–265.00/S336.00–338.50 *86.00–*85.60
L. 0.072; W. 0.049; H. 0.035

One corner of a tuyere; the air passage is not evident. Coarse fabric with some voids and a bit of grey pitted vitrification on one surface.

17 T97.18:10734

Findspot: NW dump, W275.00–276.00/S327.00–329.30 *85.50–*85.15
L. 0.052; W. 0.048; H. 0.029

Exterior corner of a tuyere; the air passage is not evident. Accretion of grey vitrification at one end.

18 T97.19:10735

Findspot: NW dump, W277.30–278.00/S329.80–330.70 *84.90–*84.70
L. 0.066; W. 0.035; H. 0.028

Corner of a tuyere; its outer edge and air passage are preserved. Coarse fabric with obvious inclusions. Thumbprint visible on exterior.

19 T97.20:10736

Findspot: NW dump, W276.00–278.00/S330.50–332.00 *85.40–*85.30
L. 0.058; W. 0.048; H. 0.019

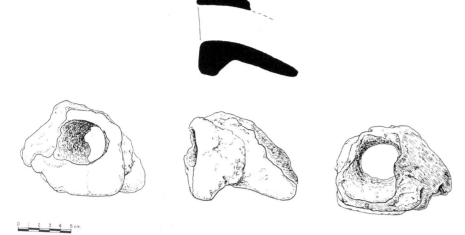

Fig. A2.5 Item 20, tuyere: possible air passage for a larger installation than the usual cupels.

Part of one side of a tuyere just preserving the curve of the air passage. Coarse fabric with large inclusions.

20   T97.25:10744 (Fig. A2.5)

Findspot: Unit I, W278.00–280.00/S337.00–342.00
c. *85.80
L. 0.099; W. 0.111; H. 0.078

Large lump of gritty, partly-fired mud with a substantial curving hole in it (*c.* 40 × 30 mm) and a flattish spot which has a much blacker reduced aspect, as if it had been nearer the heat. It appears to be part of an air inlet integral with the side or the bottom of a hearth or furnace, different from any that were found. The texture and change in colour of the mud look very much like the bottom of some of the cupels. It is quite clear that this piece could not have belonged to the cementation furnaces because they are reddened throughout. It is almost as difficult to associate it with the cupels that were found because of its scale. One end of the hole appears slightly more smoothed but the other end is very much broken as it merges with the flat section.

At the time of excavation, the present author described this piece as 'unfired or barely used bellows nozzle'. He still thinks that it is part of the built-in air supply for one of the hearths, although it clearly has not been anywhere near the hottest part of the fire. If this assessment is accurate (and it has been challenged from the beginning[3]), it requires one to contemplate the existence, somewhere, of a much bigger hearth or series of hearths with a much larger and less portable set of bellows than is required for the fired and partially vitrified tuyeres listed above.

Bread tray with vitrification

21   T97.26:10745 (Fig. 4.38, p. 91)

Findspot: NE dump, W260.00–265.00/S336.00–338.50
*86.00–*85.80
L. 0.080; W. 0.047; H. 0.073; Th. 0.017–0.028

Heavily vitrified bread tray and a fragment of a coarseware sherd on top, all fused on one face. It must have sat flat on the ground at the edge of the hearth with other pieces on top. One can see in the section that it was part of the raised edge, because one side is flat and the other is slightly curved.

22   T97.37:10779

Findspot: W265.00–268.00/S349.00–351.00
*86.30–*86.20
L. 0.087; W. 0.054; Th. 0.019

Small piece of bread tray with shiny drips of vitrification. No original edge.

23   T97.43:10787

Findspot: NW dump, W276.00–280.00/S329.00–331.00
*85.30–*85.20
L. 0.068; W. 0. 069; Max. Th. 0.039; Th. of tray 0.020

Bread-tray rim fragment with vitrification on the end (top of rim).

24 T97.39:10783

Findspot: NW dump, W277.80–280.00/S334.00–336.00 *86.00–*85.00
L. 0.077; W. 0.071; Th. of rim 0.03

Bread-tray rim fragment with black, glassy vitrification on end and underside.

25 T97.27:10746 (Fig. 4.37, p. 90)

Findspot: NW dump, W276.00–280.00/S329.00–331.00 *85.30–*85.20
L. 0.123; W. 0.067; Th. of tray 0.016, with vitrified lumps 0.021

Three joining pieces of bread tray vitrified on broken edge. There is quite a distinct difference between the vitrified and unchanged surface on the top (original) side of the bread tray.

26 T97.29:10748

Findspot: NE dump, W260.00–262.00/S336.00–337.00 *86.20
L. 0.077; W. 0.052; Max. Th. 0.028

Fig. A2.6 Item 29: raised edge of bread tray with vitrified edge at lower right, perhaps used as a pusher.

Bread-tray rim with vitrification. Exceedingly coarse pebbly texture.

27 T97.28:10747

Findspot: NE dump, W260.00–262.00/S336.00–337.00 *86.20
L. 0.061; W. 0.049; Th. of rim 0.014, with vitrification 0.020

Flat piece of bread tray with overhanging vitrification flow.

28 T97.34:10767

Findspot: NW dump, W275.00–280.00/S329.00–331.00 *85.50–*85.30
L. 0.099; W. 0.080; Th. 0.016

Vitrified and discoloured bread tray; no edge preserved. Vitrification on underside.

29 T97.24:10740 (Fig. A2.6)

Findspot: W269.00–272.00/S340.00–342.00 *86.15–*86.05
L. 0.152; W. 0.049; Max. H. 0.044

Side rail of bread tray with vitrified end. Perhaps used as a pusher or something that could be manipulated more easily than the smaller but wider potsherds at the edge of the hearth.

In addition to the nine pieces of bread tray described above, there are 13 other pieces with one or more of the characteristics of the individual pieces above, of which the largest is 80 × 60 mm. Some are vitrified right through and some just vitrified on the edge. One piece with a uniform yellowish crust on top has a dark upper surface.

Bread tray without vitrification

30 Many fragments of bread tray without vitrification were found (see Fig. A2.2, p. 222) and noted in the field books day by day, but most of it was discarded in the field or after consideration at the laboratory. Also in the bag of miscellaneous bread tray, but not drawn or photographed, was a small piece of vitrified brick, probably furnace lining rather than something to raise the wall of the cupel.

Coarseware with vitrification

31 P97.131:10758 (Fig. A2.7a, p. 144)

Findspot: Cupellation area B, W273.00–274.00/
S343.00–345.00 *86.40–*86.25
W. 0.09; H. 0.05; Est. diam. 0.16;
Th. 0.006–0.0085 because of the vitrification

Three joining rim sherds from a cooking pot with
heavy vitrification on the interior.

32 P97.132:10759 (Fig. A2.7b, p. 144)

Findspot: Cupellation area B, W273.00–274.00/
S343.00–345.00 *86.40–*86.25
0.07 × 0.07; Th. 0.008

Rim sherd from a tableware dinos. Heavy, glassy,
bubbled vitrification on exterior: the big bubble at
edge of rim retains the imprint of wood grain from the
charcoal, demonstrating more strongly that prepared
charcoal was the main fuel, as had already been
inferred from the quantity in the north-east dump
(see p. 86).

33 P97.129:10756
Findspot: Cupellation area A, W264.00–266.60/
S344.00–347.60 *86.00–*85.85
0.049 × 0.044; Th. 0.009

Grey coarseware vitrified with prominent gold globule.

34 P97.137:10764

Findspot: Edge of NE dump, W260.00–265.00/
S336.00–338.50 *86.10–*86.00
0.052 × 0.041; Th. 0.005

Grey coarseware sherd with interior vitrification with
gold globules adhering.

35 P97.133:10760

Findspot: Unit I, W280.60–281.30/S337.00–337.50
*86.30–*86.10
0.054 × 0.041; Th. 0.008 at neck, 0.01 for body with
vitrification

Gold dust cooking-pot sherd;[4] neck and body area.
Well smoothed on the exterior, vitrified on the interior

with gold globules adhering, higher up on the body
than usual.

36 P97.134:10761

Findspot: Cupellation area A, test trench,
W264.00–265.50/S343.20–345.00 *85.70–*85.30
0.065 × 0.045; Th. 0.01

Body sherd of gold dust ware with vitrification and
gold globules. Exterior well smoothed.

37 P97.135:10762

Findspot: Cupellation area B, test trench, western part,
W275.80–277.30/S343.00–343.50 *86.00–*85.25
0.042 × 0.033; Th. 0.011

Vitrified coarseware with gold; red exterior, reduced
interior. White quartz temper is particularly clear on
grey surface.

38 P97.145:10768

Findspot: NW dump, W277.30–278.00/S329.80–330.70
*84.90–*84.70
0.048 × 0.038; Th. 0.0097

Grey coarseware sherd with a thin layer of vitrification
on the interior, a considerable number of small gold
globules and an unusual irregular blob. Gritty white
temper very obvious on both sides.

39 P97.144:10778

Findspot: NW dump, W280.00–283.00/S329.00–332.00
*85.05–*84.80
L. 0.11; W. 0.10; Th. 0.017, with vitrification 0.022

Thick piece of flattish pithos reused as a crucible. One
third of the interior is vitrified and has a trail of smooth
glaze, which has dribbled over the broken edge to the
exterior.

40 P97.130:10757

Findspot: Unit I, W280.60–281.30/S337.00–337.50
*86.30–*86.10
L. 0.06; W. 0.039; Th. 0.011

Vitrified coarseware without gold globules.

Coarseware without vitrification

A great deal of coarseware without vitrification was found and noted in the field books day by day, but most of it was discarded in the field as featureless cooking pots.

41 P97.139:10766

Findspot: Cupellation area B, W273.00–274.00/S343.00–345.00 *86.40–*86.25
L. 0.175; W. 0.118; Th. 0.009; Est. rim diam. 0.21

Large red cooking-pot rim and upper body with handle. Curious circular stamp on exterior near rim.

Plainware

42 P97.136:10763

Findspot: NE dump, W260.00–265.00/S336.00–338.50 *86.00–*85.80
W. 0.12; H. 0.072; Th. 0.01

Three joining plainware sherds from a large water jar; locally vitrified on one edge with light vitrification like a brown glaze surviving in some areas. They must have cracked during the operation, because the vitrification has dribbled into the cracks and, although they join, they cannot be properly fitted together.

43 P98.199:11023

Findspot: Cupellation area B, test trench, W272.30–277.80/S346.00–346.70 *86.20–*85.70
L. 0.035; W. 0.032; Th. 0.006

Small plain potsherd with vitrified edge; probably used around the hearth.

44 P97.138:10765

Findspot: Cupellation area B, test trench, W276.20–277.40/S340.00–341.70 *85.90–*85.00
L. 0.056; H. 0.033; Th. 0.007

Greenish yellow vitrification spread all over a red painted dish with a fragment of coarseware attached as if with glue. The surface is crackled and flaking. The vitreous material seems to have been much less viscous than usual.

45 P98.200:11024

Findspot: NW dump, W275.00–279.00/S331.00–333.00 *85.50–*85.20
L. 0.046; W. 0.046; Th. 0.007

Small painted sherd with vitrification on interior and top.

46 P97.128:10755

Findspot: c. W297.00/S325.00 *84.8
0.041 × 0.027; Th. 0.0035

Grey body sherd from fine skyphos; most of the interior covered with thin skin of yellowing litharge. Three of the four edges are 'glazed'. Small spot of local vitrification on exterior.

Brick

47 T97.23:10739

Findspot: Cupellation area A, W265.00–266.00/S345.00–346.00 *85.00
0.215 × 0.145; Th. 0.077; Wt. 2550 g

Two joining edges and outer face preserved. The outer face is a dark grey with a whitish, peeling crust. There are troughs on the surface that look like finger marks. Behind this surface, the clay is much yellower than right in the middle, where it is a rich red. It appears to be part of a brick from a support of one of the cementing furnaces.

48 T97.22:10738

Findspot: Furnace area A, furnace 2, W281.00/S343.00
0.15 × 0.09; Original thickness of 0.071 preserved

Well-levigated clay with a few small pebbles. One surface is distinctly yellow shading to grey and has an obvious trough in it like item 47. The piece looks very much like one of the supporting bricks preserved in their original position in furnace 2 of Furnace area B (see Fig. 4.30, p. 88).

49 T97.33:10753

Findspot: W223.00–224.00/S339.00–340.00 *88.40
0.08 × 0.08
Found in 1963

Piece of highly vitrified brick from a furnace; fine clay with large inclusions of quartz.

Fig. A2.8 Item 53: fragment of brick with a glazed and scoriated surface.

50 TS150

Findspot: NE dump
0.13 × 0.085; Th. 0.048

Originally three pieces; two discarded. Well vitrified with a bubbled surface and considerable dripping, yet not too deep into the core the body is brick red.

Brick: pieces of glazed surface

51 TS119

Findspot: NE dump, W260.00–265.00/S336.00–338.50 *86.00–*85.60; and cupellation area A, W263.50–265.50/S342.00–345.00 *86.15–*85.60

Ten pieces comprising a variety of lumps, chips and slivers of vitrified brick covered in red and green vitrification.

52 T97.40:10784

Findspot: W294.00–295.00/S327.00–328.00 *85.10
L. 0.255; W. 0.180; H. 0.124

Substantial mass of vitrified brick (3.3 kg) with an irregular surface. It is as if several pieces had fused together after dropping from the walls of a furnace. No single piece is large enough to be an actual brick, even though the dimensions correspond to about half a

common-sized brick. The upper surface is glazed with a dull red-green colour. There are large inclusions of quartz that indicate brick rather than pottery. Below the surface are thousands of tiny bubbles.

53 T97.41:10785 (Fig. A2.8)

Findspot: Cupellation area A, W264.00–266.60/ S344.00–347.60 *86.00–*85.85
L. 0.195; W. 0.150; H. 0.095

Fragment of brick with a glazed and scoriated surface. It seems to have been subjected to intense heat on its top and two sides. The top is most affected, with runs of vitrification being quite obvious. The scoriation bubbles are usually less than 1 mm in diameter.

54 T97.42:10786

Findspot: Cupellation area A, W264.00–266.60/ S344.00–347.60 *86.20–*85.85
L. 0.134; W. 0.113; H. 0.089

Fragment of a furnace wall brick. One side is heavily covered with green and red vitrification, 15 mm thick. The vitrification starts around one corner and may indicate where the brick was set next to another and simply began to deform. It is unlikely that both edges were exposed. Both the upper and the lower surfaces are preserved. Large pebbles were used for temper; one is 30 mm long.

Vitrified materials

55 G98.1:11020

Findspot: Cupellation area A, test trench, W264.00–265.50/S343.00–345.00 *86.15–*85.70
Largest piece: L. 0.034; W. 0.023; Th. 0.011

Remains of much denser and compacted vitrified material compared with cupel edges or scoriated sherds. Thin trails not attached to brick or other supports suggest an extremely high temperature.

56 G98.2:11021

Findspot: Cupellation area A, test trench, W264.00–265.50/S343.20–345.00 *85.70–*85.30
Largest piece: L. 0.03; W. 0.017; Th. 0.007

Numerous fragments of dark green, glassy vitrification.

Crucible rim

57 P97.147:10788

Findspot: W297.00/S325.00 *84.80
H. 0.067; Max. W. 0.059; Max. Th. 0.018

Piece of crucible rim vitrified on the inside; made ad hoc; low-fired on the outside.

Metal

58 M68.19:7819 (Figs A2.9 and A2.10)

Findspot: Cupellation area B, test trench, western part, W272.30–277.80/S346.00–346.70 *86.20–*85.70
Ext. diam. 0.008–0.023; Diam. of air hole 0.003–0.004

Previously published in *BASOR* 199 (1970), p. 24, Fig.14; *SPRT,* p. 39, Fig. 57; Waldbaum 1983, Plate 55, no. 955

Iron blowpipe nozzle; heavily corroded and encrusted with sand.

Litharge cakes

See also Table 6.2, p. 160, and Figs 6.7 to 6.10, pp. 164–5.

All three inventoried litharge pieces are part of TS122, which consisted of 12.5 kg from the area of the north-east dump (W262.00–265.00/S336.00–338.50 *86.30–*86.15; see p. 90).

Fig. A2.9 Iron blowpipe nozzle (item 58): length 83 mm.

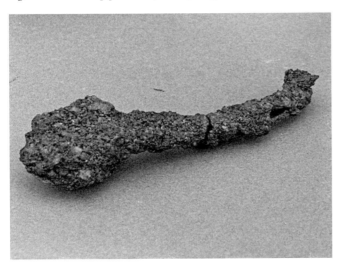

59 M97.7:10741 (Figs 4.35 and 4.36, p. 90)
Diam. 0.153; Diam. of central circular depression for silver button 0.062; Wt. 1130 g

Complete litharge cake composed of four joining pieces; lacking a small triangular piece in the centre and a few bits at the edge.

60 M97.8:10742

Diam. 0.136; Diam. of button *c.* 0.045; Wt. 365 g

About half of a litharge cake.

61 M97.9:10743

0.075 × 0.068; Max. Th. 0.017; Est. diam. of central circular depression for silver button *c.* 0.05

Separate small piece of a litharge cake.

The rest of the bag from the north-east dump and other pieces from different areas represent about 12 more cakes of roughly the same module.

Fig. A2.10 X-ray image of blowpipe nozzle in Fig. A2.9.

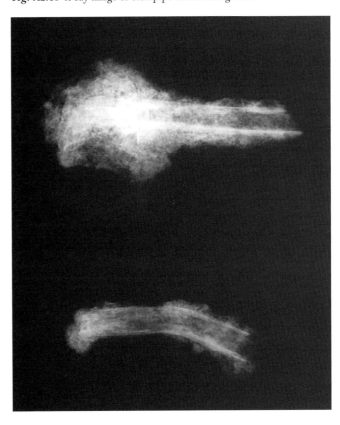

Lead dribble

62 M97.10:10752

Findspot: NE dump, W260.00–265.00/S335.00–336.50
*85.90–*85.70
L. 0.024; W. 0.011; Th. 0.0029; Wt. 4.34 g

Small congealed dribble of lead looking like crumpled wire with flecks of charcoal caught in the middle. Its dark colour makes it seem more metallic than the litharge.

Notes

1 *Pirotechnia,* Book VII, Ch. 3-1 (Smith and Gnudi 1942, p. 290).
2 See Ramage (1976, p. 7, footnote 5) for use of *Arundo donax* in roofing.
3 'It is felt that this is probably a lump of fired wall clay or mud brick into which an animal may have burrowed.' (Sidney Goldstein, *Final Laboratory Report*, 1969, Appendix VI.)
4 Gold dust ware refers to a yellowish monochrome pottery decorated with a micaceous wash that gives it a metallic sparkle.

Early History of the Amalgamation Process

P.T. Craddock

Through the Middle Ages both gold and silver metals were often separated from their finely crushed ores or other metals by treatment with mercury, which forms amalgams with precious metals. The gold and silver could then be recovered from the mercury by squeezing through a medium such as chamois leather, followed by distillation. There is little evidence for the use of amalgamation in earlier periods, but it is possible that lead might have been used to perform a similar function (see p. 235).

Woolley recovered over a kilogram of liquid mercury dispersed through sandy layers dating from the fifth century BC, at the Greek trading colony of Al Mina on the Syrian coast, and also at Seleucia. He claimed it would have been used to extract gold.[1] However, this discovery well predates any recorded use of mercury in gold recovery; possibly the mercury was intended for use in the amalgam gilding of copper, which is first attested in the Greek world during this period.[2] It must, however, be noted that mercury, being a very dense liquid, has a propensity for sinking through the ground and thus the possibility exists that the Al Mina mercury could have penetrated from higher levels.

The histories of the use of mercury to extract gold and silver are very different, and will be considered separately.

Gold extraction

Vitruvius and Pliny both describe the use of mercury to recover or to refine gold. Vitruvius[3] states that gold could be recovered from worn-out golden embroideries by burning them and treating the wet ashes with mercury, which amalgamated with the gold. The resulting amalgam was then pressed through cloth which allowed the mercury to pass but retained the precious metal.

In his *Natural History,* Pliny[4] stated that gold was the only thing which mercury attracted and that consequently this property gave rise to an excellent method of refining gold. He then went on to describe how the impure gold was to be shaken up in an earthenware pot with mercury. The gold amalgamated, leaving the impurities behind. The gold could be recovered from the mercury by squeezing the amalgam through well-dressed hides (chamois leather?), which left the gold behind.

It is not immediately obvious whether the gold was being recovered from ores or from worn-out gilded materials, as described by Vitruvius. The passage occurs in Book 33 on silver, in sections dealing with mining, and mercury is mentioned in the context that it occurs in silver mines. It is significant that there is no mention of the use of mercury in the sections in which the recovery of gold from its ores is described in some detail, Pliny obviously writing from direct experience of the mining and collection of gold. Pliny is in error when he states that only gold amalgamates with mercury. Several other metals also form amalgams, including silver, although admittedly somewhat less readily than gold, but amalgamation does not part gold and silver. The method could have been used to recover gold from gilded metal, but that is not really refining. One suspects that Pliny had in mind treatments such as Vitruvius describes, namely, the separation and recovery of metallic gold, dispersed in artefacts of other materials. Davies[5] also considered it very unlikely that mercury was used in classical antiquity to recover gold from its ores.

None of the analyses of material from Sardis has detected mercury, which is a very invasive and persistent material. If it had been used in any quantity in the Pactolus North area, one would have expected to find small

Fig. A3.1 Amalgamation of gold concentrates and the recovery of mercury by distillation of the amalgam, as practised in sixteenth-century Germany. **Key:** *A* lazy Heinz tower, *B* side chambers, *C* earthenware pot used as receiver, *D* earthenware still head, *E* blind head with spout through which water can be poured, *F* performing amalgamation, *G* squeezing mercury through a leather bag, *H* iron pot with top removed, *K* top of iron pot, *L* leather bag through which mercury is squeezed, *M* remelting gold in airstream produced by bellows. (From *Lazarus Ercker's Treatise on Ores and Assaying*, A.G. Sisco and C.S. Smith, 1951, p. 113)

amounts absorbed by the foils and the gold droplets on the melting-sherd fragments, for example. In fact, none was detected (see Tables 5.1, pp. 103–5, 5.2, p. 116, and 5.3, p. 145; and p. 185). The earliest written evidence of the use of mercury in the extraction of gold from its ores is given by al-Hamdānī[6] in the treatise he compiled in Yemen in the mid tenth century. However, as this is the earliest detailed technical work on the subject to survive in the West, it is quite possible that the use of mercury could well predate al-Hamdānī; certainly from his descriptions this and other amalgamation processes seem already to have been well established.

The earliest extant European account of the extraction of freshly mined gold with mercury is found in the treatise of Theophilus, which was probably compiled in the early twelfth century.[7] There it states that gold was extracted from the sands of the Rhine using mercury. More detailed descriptions of the European amalgamation processes in use in the sixteenth century are to be found in technical books such as the *Probierbüchlein*[8] and the works of Biringuccio,[9] Agricola[10] and Ercker[11] (Fig. A3.1).

The thirteenth-century writer al-Birūnī describes the process as used in the Islamic world as follows:[12]

After pounding the gold ore or milling it, it is washed out of its stones and the gold and mercury are combined and then squeezed in a piece of leather until the mercury exudes from the pores of the leather. The rest of the mercury is driven off by fire.

He also describes an apparently very basic process carried on to recover gold directly from the Sind river:[13]

At its sources there are places in which they dig small pits under the water, which flows over them. They fill the pits with mercury and leave it for a while. Then they come back after the mercury has become gold. This is because at its start the water is rapid and carries with it particles over the surface of the mercury which picks up the gold leaving the sand to pass away.

Treatments using lead

It is possible that, in antiquity, lead was used in an analogous way to mercury to extract gold.[14] That is, molten lead would have been mixed with the finely ground ore to absorb and concentrate the very finely dispersed metal. It is also possible that this process is contained in the well-known description of gold parting by Agatharchides (see p. 34), where apparently both lead and common salt were added to the gold. The concept of using lead to concentrate ore or metal was certainly known in antiquity for the smelting of silver. Thus, in the smelting of the so-called 'dry' silver ores (i.e. those without lead), lead was deliberately added to the furnace charge specifically to absorb the small amounts of dispersed silver that would otherwise have been lost.[15] The process would have been akin to the liquation of argentiferous copper[16] with molten lead, but treating powdered ore instead of molten copper.

Lead was used in a similar manner for the treatment of pyritic gold ores. In the post-medieval descriptions, the ores were smelted and the gold was released from the resultant slag by treatment with lead.[17] There are no clear contemporary accounts of the smelting of pyritic gold ores in antiquity but they were certainly processed by the Romans, and the gold recovered from the slags with lead, as the investigation of smelting sites in Portugal has revealed.[18] There are no ancient reports of the use of lead to concentrate small quantities of gold from the crushed ores, although certain passages in the *Arthaśāstra*,[19] for example, seem to suggest that the gold from the mine was expected to contain lead. This could have arisen either from the treatment of the crushed ores, or from the auriferous slags if pyrites ores were being treated.

Either way, it offers another potential explanation for the presence of lead on the parting-vessel sherd 44401Z (British Museum code) from the Sardis refinery.

Silver extraction

There is no direct reference to the recovery of silver by amalgamation in the classical literature. The earliest extant reference is again in al-Hamdānī.[20] However, as noted above for gold, his descriptions give the impression of a well-

Fig. A3.2 *Tortas* of ground silver ore, common salt, *magistral* and mercury being kneaded by teams of horses on the flat surfaces, *patios*, that gave their name to the process. (From *Handbook of Metallurgy*, C. Schnabel and H. Louis, 1905, Vol. 1, p. 766)

established process, suggesting that amalgamation of silver had been practised for some time previously. Mercury was used to recover finely dispersed metallic silver from the ores or from the slags and dust in the workshops. The ores, slags and dust were finely ground and treated with mercury to form the amalgam. The silver was recovered by a crude distillation process, in which the amalgam was placed inside a pot which was inverted over another pot sunk into the ground and containing water. The upper vessel was heated, causing the mercury to volatilise and recondense in the lower vessel, leaving the silver in the upper vessel.

There are other, quite detailed descriptions given in other Islamic sources.[21] That given by Manṣūr ibn-Baʿra of Cairo in his thirteenth-century treatise on the production of gold and silver, specifically for the production of coinage, will suffice as an exemplar:[22]

> … a strong, concave stone mortar and a strong pestle are taken as large as the open hand. Ore of the refining mixture of the first night is put in by itself. Then, a little water is admixed, and it is all well pulverized with a pestle until it is very fine. Then, mercury is added in the amount desired. The pulverization is continued until all the silver in that body is mixed well with the mercury. It is then softened; water is poured over it washing it so that it overflows but care must be exerted that no mercury escapes with the water. The mercury is then put in a thin goatskin which is twisted well to squeeze and to filter it. The mercury comes out of the walls of the goatskin and the silver [amalgam] remains as a walnut. This is placed in a clay pot over the fire so that the mercury which remained with the silver will be separated. The silver is melted and the weight determined. The same is done to all the ore on a continuous basis.

Note that there is no reference here to any other chemicals beyond mercury and water, and thus these processes were for the recovery of finely divided silver *metal* by amalgamation. The processes for extracting silver from its *mineral ores* are not attested before the sixteenth century. The first references to the treatment of silver *salts* are contained in some of the editions of the *Probierbüchlein*,[23] compiled in the early sixteenth century, in which the recovery of silver salts from the spent cement of the salt parting process with mercury is specified. The better-known and certainly more detailed account, which is usually claimed as the earliest description, is given in Biringuccio's *Pirotechnia*,[24] published in 1540, where, together with familiar accounts of the recovery of gold and silver metals with mercury, there is a description of how certain silver *ores* could be successfully amalgamated. The finely crushed ores were to be agitated with common salt, vinegar, corrosive sublimate (mercuric chloride), verdigris (copper acetate) and mercury for an hour or two in a mill, during which time the silver would pass into the mercury, from which it could be recovered by distillation.

Biringuccio was at pains to explain that this was a new process, for the secrets of which he had paid a German handsomely. In essence, it is a description of the famous Patio process (Fig. A3.2),[25] which is generally believed to have been developed in Mexico by Bartolome de Medina[26] at least ten years later. He apparently succeeded only after long and well-recorded experimentation, but made no mention of an earlier European process. Thus the process seems to have been developed in Germany for the recovery of silver from the spent cements of the parting processes, and was successfully adapted for treating ores in the Americas. It was only adopted from Mexico by the European silver mines as a means of treating ores in the seventeenth century. De Medina first tried common salt and mercury alone, but not until vitriol (copper and iron sulphates, known locally as *magistral*) was included, did he achieve success. The process used little or no fuel and was widely used in the Americas to treat silver ores that contained little or no lead, known as dry ores,[27] until supplanted by the cyanidation processes in the late nineteenth century.[28]

Notes

1 Woolley (1937–8).
2 Although mercury gilding in the West is rare before the Roman Imperial period, there does seem to be some evidence for its limited use: see Craddock (1977, 1985 and 1995, p. 302) and Williams and Ogden (1994, p. 253). See Oddy (1993a) and Oddy et al. (1990) for accounts of gilding in classical antiquity, and La Niece (1990 and 1993) for a general history of silvering.
3 *Vitruvius on Architecture*, 7.8.4 (Granger 1934, Vol. II, p. 117).
4 *Pliny: Natural History*, 33.99 (Rackham 1952, p. 77).
5 Davies (1935, p. 57).
6 al-Hamdānī, *Kitāb jawharatain al-ʿatīqatain* (Allan 1979, pp. 7, 8; Dunlop 1957, p. 29). See Toll (1968) for a complete translation.
7 *On Divers Arts: The Treatise of Theophilus*, 49 (Hawthorne and Smith 1963, p. 123).

8 *Probierbüchlein* (Sisco and Smith 1949, esp. pp. 123, 132–4, 136–7, 152 and 187).

9 *The Pirotechnia of Vannoccio Biringuccio* (Smith and Gnudi 1942, p. 30).

10 *Georgius Agricola: De Re Metallica* (Hoover and Hoover 1912, pp. 294–9).

11 *Lazarus Ercker's Treatise on Ores and Assaying* (Sisco and Smith 1951, pp. 106–8).

12 al-Hassan and Hill (1986, p. 246).

13 al-Hassan and Hill (1986, p. 247).

14 Aitchison (1960, Vol. 1, p. 173) also suggests this.

15 See Craddock et al. (1985) and Craddock (1995, pp. 216–21) for a description of such a process as carried out in antiquity at the great silver mines of Rio Tinto, Spain, contra Halleux (1985, p. 54), where it is stated erroneously that the lead was added to the copper matte.

16 See Percy (1880, pp. 304–499), Agricola (Hoover and Hoover 1912, pp. 491–544) or Ercker (Sisco and Smith 1951, pp. 224–53) for detailed descriptions and histories of the liquation process. The process was certainly used by the Romans in Britain (Gowland 1900, 1915 and 1917–18).

17 Agricola (Hoover and Hoover 1912, pp. 378–90), Ercker (Sisco and Smith 1951, pp. 108–9).

18 See Bachmann (1993) for a description of the pyritic smelting process carried out by the Romans at Três Minas and Campo de Jales in northern Portugal.

19 *Arthaśāstra*, 2.13.5–7 and 9 (Kangle 1972, p. 111). See also pp. 33 and 50, note 4.

20 *Al-Hamdānī: Die bieden Edelmetalle Gold und Silber*, XX–XXI (Toll 1968, pp. 28–30, 248–74).

21 Allan (1979) and Levey (1971), for example.

22 Manṣūr ibn-Baʿra, *Kashf al-āsrar al-ʿilmīya bidār al-ḍarb al-Miṣriya* (Levey 1971, pp. 59–60).

23 *Probierbüchlein*, recipe 158 (Sisco and Smith 1949, p. 152). See also this volume, p. 57, as noted by Hoover and Hoover (1912, p. 279).

24 Smith and Gnudi (1942, pp. 384–5).

25 For descriptions of the Patio process, see Muspratt (1860, pp. 1856–60), Percy (1880, pp. 576–656) and Schnabel and Louis (1905, pp. 760–82). See Bargalló (1969) for a more general discussion of the treatment of silver ores with mercury.

26 Probert (1969).

27 Crozier (1993).

28 See Dorr (1936), for example, for a description of the cyanidation process. The history of the process is described by Habashi (1989).

The Platinum Group Element Inclusions

P.T. Craddock

Gold from placer deposits often contains small silvery inclusions of the platinum group elements. Although routinely removed from modern goldwork, they are quite common in ancient gold artefacts, including coins.[1] They are especially prevalent in both Lydian coins of unrefined electrum and those of refined gold, (see Fig. 7.1, p. 138) but strangely the gold foils recovered from the Pactolus North refinery site, in various stages of refining, have no trace of them at all. Clearly, this situation demands an explanation, and the solution contains vital clues to the whole refining process as practised at the PN refinery and elsewhere in antiquity.

The inclusions are tiny granules made up of varying combinations of some of the platinum group metals, which comprise iridium, osmium, palladium, platinum, rhodium and ruthenium. The majority of PGE inclusions found in the gold deposits, and thus in ancient goldwork, are of variable osmium–ruthenium–iridium composition.[2]

The granules are not associated with gold in the primary deposits, but occur in basic and ultra-basic rocks, typically olavines in association with magmatic base-metal sulphides and chromite. Where gold flakes, eroded from the primary deposits, wash over such material they can pick up the PGE granules.

The general physical and chemical properties of the naturally occurring platinum group alloys are well known,[3] but there is much less familiarity with them as inclusions in gold, quite simply because, for the last century, they have been routinely separated in the gold-refining processes such as electrolysis.

The PGE inclusions are noted for their resistance to most forms of chemical and physical attack, but this aspect has possibly been over stressed by recent authors when discussing their presence in ancient metalwork.[4] Also, although classified as a group, the physical and chemical properties of the six metals (platinum, palladium, rhodium, osmium, iridium and ruthenium) are very different. In particular, platinum, palladium and rhodium are described as being soluble in gold,[5] thus they would not be expected to be found regularly as inclusions in gold.[6] In fact, no inclusions containing these elements have ever been found in the Lydian coins, whose inclusions are alloys of iridium–osmium–ruthenium, all of which are insoluble in gold.[7] The metals are noted for their very high specific gravity, such that the inclusions rich in osmium (22.57) and iridium (22.42) usually sink to the bottom in molten gold, itself one of the densest substances known (19.32), but palladium, rhodium and ruthenium have specific gravities of only 12.02, 12.41 and 12.41 respectively. Thus an inclusion rich in ruthenium could well float.

Finally, and most significantly, the platinum group metals are noted for their resistance to chemical attack, but in fact ruthenium and to a lesser extent iridium and osmium, which are the most common constituents of the PGE inclusions, are chemically quite active.[8] These latter elements are all more or less susceptible to oxidation at elevated temperatures, and this poses a problem in their determination by fire assay.[9] Thus Bowditch[10] noted that during the traditional cupellation process, performed at between 900 and 1300 °C , almost all of the osmium will be lost as the volatile tetroxide, OsO_4, whilst about 50% of the ruthenium and 30% of the iridium will be oxidised and absorbed by the cupel. If this level of loss were repeated in the PGE inclusions, then it would be sufficient to break them up and any remaining metal would disperse as much smaller, almost invisible, particles in the gold.

Only relatively recently has it been fully appreciated that the *adamantes* of classical antiquity referred in part unambiguously to the PGE inclusions.[11] Not only were their physical properties described in some detail, but also methods by which they could be removed from gold.

Given the background of the great difficulties encountered in their removal from gold in the nineteenth century, it could seem little short of incredible to suggest that the ancients knew how to remove the inclusions.[12]

In reality they can be reacted and removed from gold without too much difficulty by traditional processes such as cupellation, and some ancient sources suggest as much (see p. 241). However, it must be stressed that there is no evidence that the PGE inclusions were routinely removed from gold in antiquity. It seems to be the case that the technology was there if required, but in general the presence of inclusions was tolerated. Thus there is little reference to difficulties in their removal from gold prior to the nineteenth century, when quite serious problems were caused by their presence in the gold from some of the new placer deposits discovered in the nineteenth century, notably in California and Australia.[13] It seems that this was due to the changes in gold-refining technology taking place at just the time that the new placer deposits containing the PGE inclusions were being discovered. Thus some of the traditional methods of precious-metal refining, notably cupellation, removed the PGE inclusions, but the processes of gold refining which were used through much of the nineteenth century had no effect on them. They are immune to dissolution in all the mineral acids which were used in the acid parting of gold. The Miller process of purifying gold by bubbling chlorine gas through the molten impure gold was introduced in the 1860s and rapidly became one of the standard methods of gold refining.[14] Although the platinum group metals are attacked by chlorine,[15] the Miller process removed only very slowly, if at all, the solid PGE inclusions suspended in the molten gold. Only with the introduction of electrolytic refining at the end of the century was there a standard method which would routinely separate the PGE inclusions from the gold.

Some of the difficulties in accepting the ability of earlier goldsmiths to remove the PGE inclusions stem from misconceptions and mistakes concerning the properties of the inclusions and thus their fate when subjected to the traditional gold-refining processes of cementation and cupellation.

Ogden states that the PGE inclusions of iridium–osmium–ruthenium in gold and silver are unaffected by cupellation at 1063 °C. This seems unlikely in view both of the well-established propensity of osmium to oxidise and volatilise at temperatures well below this, and of the combined practical experience of the analysts. Ogden also claimed that at low temperature (below 650 °C) the inclusions would be little affected, but that above that temperature they were liable to attack by chlorine.[16] This is

a reasonable surmise given that the usual method of making the volatile osmium tetrachloride, $OsCl_4$, is to pass chlorine gas over osmium sponge at between 650 and 700 °C.[17] However, the survival of the inclusions in the Lydian coins of purified gold suggests that when the osmium is in combination with the other components of the inclusions, mainly iridium, it is much less susceptible to halogen attack.[18] It seems likely that there is a general increased tendency for the osmium to oxidise and volatilise as the temperature rises, thereby bringing about the break-up of the inclusions.

Zwicker has carried out some experiments to investigate the effects of cupellation and salt cementation on PGE inclusions in gold and its alloys.[19] For the cupellation experiment, an artificial alloy containing equal quantities of iridium, osmium and ruthenium was made. Granules of this alloy were melted with a base gold containing silver and copper.[20] The alloy was then wrapped in lead and cupelled at 1100 °C for 15 minutes in air on a bone-ash cupel. After this treatment, the PGE inclusions were still present although blackened by oxidation. This is initially surprising as the experiment was meant to replicate the conditions and times of the usual cupellation assay process, which routinely disperses the inclusions (see p. 238). However, the concentration of the platinum group metals in the artificial alloys used in the experiment was vastly greater than that encountered naturally and the cupellation times were very short compared with those for real cupellation carried out as a production process.[21] Thus the results of these experiments do not invalidate our suggestion that the absence of inclusions in the foils was due to prior cupellation.

For Zwicker's salt cementation experiments, natural PGE granules predominantly of iridium–osmium–ruthenium were used. These had been obtained from sources in the Ural Mountains, but they are similar in composition to those found in the Lydian coins. One group of the inclusions was added to pure gold and the other to base gold containing about 50% gold and equal parts of copper and silver. Both of these test alloys were then melted with borax and remelted with salt in an open crucible at 1100 °C. In both cases, the PGE inclusions were totally dispersed. On analysis, the pure gold was found to have 200 ppm each of iridium, osmium and ruthenium, but none could be detected in the base gold. Zwicker suggested that the difference was due to the different solubilities of the PGEs in gold and in gold alloys, but in fact the solubilities of all the constituent metals in both gold and its alloys are negligible.[22] It seems more likely that at the elevated temperature used in these experiments the inclusions had partially oxidised and broken down. The difference between the two experiments could then be explained because the inclusions had remained dispersed in the pure gold, which was not far above its melting point and thus fairly viscous, but had evaporated from the molten base gold, which would have been much more liquid at 1100 °C. Zwicker concluded, on the basis of his experiments, that the PGE inclusions were removed by the salt cementation process. However, he was operating with *molten gold*, several hundred degrees above the usual temperatures of salt cementation, which relies on solid-state diffusion, and it seems more likely that it was just high-temperature oxidation that was responsible for the removal of the inclusions in Zwicker's experiments.

Early references to PGE inclusions and their removal from gold

The Lydian coins of electrum and of gold often contain quite conspicuous steely-grey inclusions of osmium–iridium–ruthenium (see Fig. 7.1, p. 138), although, as noted above, they are not in the gold foils found at the PN refinery. They also seem to be less common in the contemporary gold jewellery from Anatolia.

Inclusions must have been a familiar nuisance to gold workers, as evidenced, for example, by one of the *opus interassile* gold bracelets from the recently discovered Roman treasure at Hoxne in Suffolk, where the goldsmith had come up against an inclusion and been forced to cut around it. He would no doubt have agreed with Eissler,[23] some 1500 years later, that inclusions were 'of vexatious annoyance to the manufacturer of jewellery'.

Thus it was likely that inclusions would be mentioned in the ancient literature, and there are indeed references to very hard, resistant materials found in gold, known variously as *adamans* or *adamas*. Unfortunately, the words were somewhat indiscriminantly applied to gemstones as well, specifically diamond, but where the *adamans* and *adamas* were said to occur in gold, it has been suggested that PGE inclusions were meant.[24]

The earliest references to *adamas* in gold are to be found in the works of Plato. Thus in the *Timaeus*[25] he states:

> Of all the kinds of water which we have termed 'fusible', the densest is produced from the finest and most uniform particles: this is a kind of unique form, tinged with a glittering and yellow hue, even that most precious of possessions, 'gold', which has been strained through stones and solidified (i.e. placer gold). And the knot of gold, which is very hard because of its density and black in colour, is called 'adamant'.

The significance of the specification of placer gold is, of course, that the PGE inclusions are only found in gold from secondary deposits. Halleux and Barb[26] have both suggested that hematite was referred to here but this seems inherently unlikely. The description of the *adamas* as being black is interesting as normally the inclusions are silvery, only going black after the gold has been subjected to a refining process (due to the oxidation of the osmium).

In *The Statesman (Politicus)*,[27] Plato, describing how unworthy men shall be separated from the true ruling class, alludes to the purification of gold:

> ... the refiners (of gold) first remove earth and stones and all that sort of thing: and after that there remain the precious substances which are mixed with the gold and akin to it and can be removed only by fire – copper and silver and sometimes *adamant*. These are removed by the difficult processes of smelting and tests, leaving before our eyes what is called unalloyed gold in all its purity.

The especial significance of this section is not only the recognition of the PGE inclusions, but the easy assurance that they could be separated from the gold. This seems inherently unlikely, but some ancient recipes (see below) apparently do describe methods for the removal of the *adamas* from gold.

Ogden and Nicolini have independently recognised that the *adamas* described by Pliny in his *Natural History*, although in a section on gemstones, must in part refer to the PGE inclusions.[28] *Adamantes* are normally understood to mean gemstones, specifically diamond, but in this instance it does seem that descriptions of both diamond *and* the PGE inclusions are run together. To us this is a surprising conflation, but Pliny really did believe the inclusions were a form of diamond. At 37.55, Pliny noted that:

> *Adamas* was the name given to the *auri nodus* (literally knot of gold), found very occasionally in mines in association with gold and, so it seemed, formed only in gold.

He noted that specimens were no larger than a cucumber seed and not unlike one in colour. Pliny then stated that as many as six kinds of *adamas* had been recognised, including the Indian, which was not formed in gold, and had a certain affinity with rock crystal, which it resembled in respect of its transparency and its smooth faces meeting at six corners. This surely must refer to diamond, but he then went on to relate at 37.57 that:

> The rest have a silvery pallor and are liable to be formed only in the midst of the finest gold. All these

stones can be tested upon an anvil, and they are so recalcitrant to blows that an iron hammer head may split in two and even the anvil itself be unseated. Indeed, the hardness of 'adamas' is indescribable, and so too that property whereby it conquers fire and never becomes (affected by ?) heated.[29] Hence it derives its name, because, according to the meaning of the term in Greek, it is the 'unconquerable force'. One of these stones is called in Greek 'cenchros' or millet seed, and is like a millet seed in size. A second is known as the Macedonian and is found in the gold mines of Philippi. This is equal in size to a cucumber seed.

This must surely be a description of the PGE inclusions: they are resistant to both deformation and melting, their two most obvious properties, and are found in gold mines.

At 37.58, Pliny describes yet other varieties of *adamas* that have different properties, the *siderites* that shine like iron and can be pierced by other *adamas*. At 37.59, Pliny relates that the *adamas* can be broken up by being hammered only after having been steeped in goat's blood that is still warm, a detail that appears in later descriptions of the *adamas* (see p. 242). Finally, at 37.60, he states that:

When an 'adamas' is successfully broken it disintegrates into splinters so small as to be scarcely visible. These are much sought after by engravers of gems and are inserted by them into iron tools because they make hollows in the hardest materials without difficulty. The 'adamas' has so strong an aversion to the magnet that when it is placed close to the iron it prevents the iron from being attracted away from itself. Or again, if the magnet is moved towards the iron and seizes it, the 'adamas' snatches the iron and takes it away.

This is rather more problematic; the first part could refer to mounting engraving tools with either diamond splinters or PGE inclusions. No surviving tools have been found mounted with either, but inclusions of osmium–iridium from Californian placer gold were mounted on the nibs of the eponymous Osmiroid pens from the nineteenth century until quite recently.[30] The reference to magnetic properties would seem to rule out diamond as, in common with other gemstones, it is not magnetic. It is true that the pure PGE inclusions found in gold are also non-magnetic, but the unassociated granules, known as perodites, often contain iron minerals and *are* magnetic. Ogden, for example, quotes the description by E.A. Smith of platinoid granules from the Urals that 'attracted iron filings more powerfully than an ordinary magnet',[31] noting the

similarity to Pliny's remarks on the *adamas*. This could be taken as evidence that people in the Old World, as well as in South America, had access to platinum group metals that were not associated with gold.[32] However, recent analysis has shown that this is not so.

There is a description of *adamans* in the *Mappae Clavicula*[33] which would also seem to refer to PGE inclusions rather than gemstones. The *Mappae Clavicula* is a strange compilation of metallurgical and other recipes that date from late antiquity, with distinct echoes of Pliny's *Natural History*, and other even more distant works. In Chapter 126, it states that:

The stone *adamans* is engendered from cadmia (usually zinc carbonate or oxide, but possibly lead oxide is meant in this context) and in the cooking of gold, in the first cooking of the lump. After the first cooking, while you are breaking up the lump {for the whole lump is easily broken up lightly} yet the [*adamans*] stones remain, some small some large, which neither iron nor any other stone can overcome. This itself is stronger than all; it is vanquished by lead alone, and this is the power of lead.

Take female (metallic) lead, soft and malleable, and melt it and throw in there the piece of *adamans* that you want to disintegrate; and heat up the lead over a slow fire; and, as it [the *adamans*] begins to disintegrate, immediately pick it up with tongs, and cover it with olive-oil soap, smoothly and very cleanly because it may be weak. For it is more fragile than glass, and softer than lead, since it may be melted in lead. Take it out of the soap to disintegrate and the water will fall out and free it from the soap.

Then put as much as you wish carefully in a big fire and let it heat for 2 or 3 hours, until it is completely white-hot throughout. Afterwards take it out and wash it, and there will emerge *adamans* which fire does not overcome, nor does it shatter when struck, and it does not wear down when you work with it. By this means you can copy everything you want to work on.

This passage left Smith and Hawthorne completely mystified, perhaps understandably since they understood *adamans* to be either diamond or corundum. However, if PGE inclusions were being referred to here, then a very different interpretation becomes possible. The first paragraph may well refer to the use of lead or litharge to collect gold from the ore. In the first stage of refining, cupellation is carried out, where, as the *Mappae Clavicula* states, *adamas*, which neither iron nor any stone can overcome, is vanquished by lead.

What follows is a description of a quite remarkable treatment, apparently to recover the *adamantes* from the gold. The long, slow heat could well generate the hard, brittle gold–lead intermetallic compound, Pb_3Au, which is molten at about 300 °C. If the *adamantes* were of osmium, ruthenium and iridium, the usual PGE inclusions found in ancient gold, then they would be unaffected by a low-temperature process and could conceivably be concentrated by allowing the melt to stand. They would tend to sink to the bottom of the melt, as the osmium and iridium are much denser than the molten intermetallic, and it is conceivable that they could be separated from the gold–lead melt. The subsequent treatments and strong heating could be to remove the remaining lead and possibly the gold. The *adamantes* so liberated were apparently then used to copy (work?) anything.

It must be stressed that this is a speculative interpretation, but one which does seem to fit with the information given and the general context of gold treatment. Also, the treatment of *adamantes,* where gemstones were more likely to have been intended, is described quite separately elsewhere in the *Mappae Clavicula,* at recipe 290 (see below).

As well as similarities with the passage on *adamas* in Pliny's *Natural History,* parts of the description seem to echo a somewhat similar passage in the *Arthaśāstra,*[34] written nearly a millennium before in India, where white gold was to be purified by treatment with lead.[35]

Adamantes are also mentioned at several places in the *Book of Minerals* compiled by Albertus Magnus in the thirteenth century.[36] He was apparently following classical sources, especially Pliny, and thus also conflates diamond and the PGE inclusions. He also notes[37] that the *adamas* were extremely hard, but could be destroyed by the blood and flesh of goat, once again following Pliny.[38] He states that the *adamas* could be destroyed by lead, and then describes the magnetic properties of the *adamas,* which implies that it is the PGE inclusions that are to be understood here, and their destruction by lead should then refer to the treatment of gold by cupellation. However, this specific reference to lead treatment was not taken from Pliny but seems to come from the *Opera* of Constantine the African,[39] or from the pseudo-Aristotelian work *De Lapidibus.*[40] The latter work Albertus quotes again in a later section dealing with magnetic materials:[41] 'Aristotle asserts that neither iron nor any stone can overcome *adamas,* but lead can since [lead] is the softest of metals.' This is usually interpreted as no more than a reference to mounting gemstones in lead prior to working, since another surviving medieval manuscript of the *De Lapidibus* describes in detail how the *adamantes* were to be

physically *mounted* in lead and beaten with a lead mallet.[42] The polishing of quartz and *adamans* mounted in lead after their treatment with goat's blood, is also described in the *Mappae Clavicula* at Chapter 290,[43] quite separately from the very different chemical process described in Chapter 126, already discussed on p. 241. This would seem to support the view that Albertus *was* referring to the polishing of gemstones but for the specific reference to their magnetic properties, which suggests that the PGE granules may have been meant.

Implications of the survival of PGE inclusions for the interpretation of the Sardis process

From all the aforementioned considerations, it would seem that the osmium- and iridium-rich PGE inclusions in gold, which are immune to attack by concentrated mineral acids, are liable to high-temperature oxidation, which creates volatile oxides and thereby breaks down the inclusions. Thus they are likely to be removed from gold by the traditional cupellation process of refining metals, which operates at about 1100 °C in an oxidising atmosphere. Conversely, the PGE inclusions will be relatively unaffected by the salt cementation process provided that the process operates at a low temperature. The temperatures at which the experimental replications of Notton and of Hall were run lay in the region from about 800 to 850 °C, which represent the highest temperatures permissible without risk of melting the impure gold. The petrographical examination of the relevant ceramic sherds from the Sardis refinery (pp. 159–60) also suggests that cementation temperatures in this range were used, although the survival of PGE inclusions in the Lydian coins of refined gold at the same abundance as that found in the coins of unrefined electrum, suggests a somewhat lower temperature may have been used.

However, the gold foils from the refinery site, which represent an intermediate stage in the recovery process, contain no inclusions at all. A possible explanation is that the gold in the foils originated from a putative primary gold source that would have been free from them. However, there is little direct evidence for the mining of primary gold at Sardis, and the composition of the foils suggests that they are of scrap gold, including some of the old electrum coins, not freshly mined metal at all. If this is so then the absence of the inclusions strongly suggest that they had been removed, probably during a pre-cementation cupellation to remove base metals such as copper.

The absence of the PGE inclusions from the foils suggests that this gold may not have been destined for use in coinage. It is possible that the removal of the PGE inclusions from the metal was at least noted, if not actually one of the deliberate reasons why the gold was cupelled. The gold from the purified foils could have been reserved for fine metalworking in jewellery, as noted already, where the presence of inclusions would have been a considerable nuisance. A technical survey, undertaken by members of the Sardis team, has found some evidence for this in the local contemporary gold jewellery, now in the Archaeological Museum, Manisa, and the Museum of Anatolian Civilisation, Ankara , where the inclusions seemed much less common in the more heavily worked pieces.[44]

Thus the PGE inclusions which caused such problems to the nineteenth-century gold refiners before the advent of electrolytic refining (being resistant to mineral acids and to the Miller process) could be successfully dispersed by the ancient process of cupellation. The reference in the ancient literature to the dissolution of the PGE inclusions by the agency of lead, discussed earlier, can now be seen most probably as reference to the cupellation process.

The presence of the PGE inclusions in the electrum coins and in the coins of refined gold show that neither the melting nor the salt cementation processes removed the inclusions, although the more reactive osmium was somewhat depleted in the inclusions in the refined gold. In total contrast, the foils from the PN site had no inclusions at all. The most likely reason for this absence is that the gold came from scrap jewellery and recycled coins, and had been cupelled prior to being hammered into foils.

Notes

1 Meeks and Tite (1980); Ogden (1976 and 1977); Nicolini (1990, p. 41).
2 Although they were recognised in Egyptian goldwork by Petrie in the late nineteenth century and presciently identified by him as containing osmium and iridium (Petrie and Quibell 1896, p. 66; Petrie 1915, p. 23), the revised edition of Lucas (1962, p. 224, note 7) continued to misidentify them as being of platinum.
3 Cabri (1972); Harris and Cabri (1973); Cabri and Harris (1975).
4 Notably Ogden (1977). See also Nicolini (1990, p. 41).
5 Ogden (1977).
6 However, Meeks and Tite (1980) found platinum metal in 22 of the 308 inclusions that they analysed in concentrations of up to 12%. Note also that PGE inclusions which are predominantly of platinum itself have occasionally been found in gold artefacts which have been cast and thus the gold had been molten; the well-known Celtic Warrior Fibula, previously known as the Flannery broach (Stead and Meeks 1996), provides a good example. Clearly, inclusions of platinum could survive intact through the melting and casting processes.
7 Meeks and Tite (1980), but contra Young (1972), and Whitmore and Young (1973).
8 Griffith (1967).
9 Potts (1987, p. 487), for example.
10 Bowditch (1973), repeated in Potts (1987, p. 489). See also Haffty et al. (1977).
11 Ogden (1977); Nicolini (1990, p. 41). See also p. 240.
12 See p. 241 and Craddock (1995, p. 113).
13 Thus, for example, Dana (1857, p. 310) noted that California gold caused 'injury' to jewellery because of its osmiridium (an alloy of osmium and iridium) content. The 1870 report of the London Mint (pp. 65 and 84–5) noted that some gold from the USA contained iridium. Most inclusions in the gold from the Australian and Californian goldfields were of osmiridium. In the mid nineteenth century, before the application of electrolytic refining, these were removed by decantation. The gold was melted in tall crucibles, allowing the dense inclusions to sink, and then the gold was decanted. Silver was then added to the remaining gold containing the inclusions and the process of decantation repeated several times, each time replacing the decanted metal with more silver until the osmiridium inclusions were in a gold–silver alloy. This was then granulated and parted with acid to remove the silver and the gold dissolved in *aqua regia*, leaving the osmiridium as a black powder (Rose 1915, p. 435). Ogden (1977) suggested that, in the nineteenth century, after decantation, the inclusions were separated from the gold by amalgamation, but it is difficult to see how this would work.
14 See Percy (1870, pp. 417–35); Rose (1915, pp. 450–64). See also p. 69.
15 Griffith (1967).
16 Ogden (1977).
17 Thorpe and Whiteley (1949, p. 135).
18 The osmium and iridium which make up the bulk of the PGE inclusions in the Lydian coins are in the form of solid solutions. The overall osmium content of the inclusions in the refined Lydian gold coins is lower than in the PGE inclusions in the Lydian electrum coins (Meeks and Tite 1980, Fig. 4, and this volume p. 173), showing that osmium had been lost preferentially during the putative salt cementation, as one would expect.
19 Zwicker (1998).
20 14 mg of the platinum group metal alloy was added to 54 mg of the gold alloy. These proportions, of approximately 25%, vastly exceed those found in natural placer-gold deposits, where the overall contribution of the PGE inclusions to the total mass of the gold does not exceed a few hundred parts per million at most.
21 Percy (1870, pp. 177–211) suggests that durations of several hours were common in the nineteenth century. Small-scale primitive processes may well have been shorter but were often repeated

several times. The evidence from the Sardis refractories (see p. 161) suggests times measured in hours rather than minutes.

22 Hansen and Anderko (1958); Elliot (1965).

23 Eissler (1900, p. 623).

24 Ogden (1976 and 1977); Nicolini (1990, p. 41), although doubts on this interpretation were expressed in McDonald and Hunt (1982, p. 3). Healy (1978, p. 35) remarks that 'platinum, iridium and osmium play no part in Greek and Roman mining and metallurgy'.

25 *Plato: Timaeus,* 59b (Bury 1929, pp. 146–7).

26 Halleux (1974, pp. 83–4); Barb (1969).

27 *Plato: The Statesman*, 303e (Fowler 1925, pp. 166–7).

28 *Pliny: Natural History,* 37.55–61 (Eichholz 1962, pp. 207–11).

29 The word used, *incalesco,* gives the idea of going through several stages up to a terminal point, which could perhaps be the melting point that cannot be reached. (Giumlia Mair, personal communication.)

30 McDonald (1960, p. 239).

31 Ogden (1977), quoting Smith (1947). See also Crangle (1959).

32 The only apparent material evidence from the Old World was the strange inlay reported on an Egyptian casket, which Berthelot (1900, p. 132, and 1901b) suggested was of platinum, on the rather shaky evidence that the inlay could not be made to dissolve in acids. Even so, this identification was generally accepted and published (Petrie 1915, p. 23; Partington 1935, p. 85; Lucas 1962, pp. 244–5; and McDonald and Hunt 1982, p. 2). However, an unpublished analysis by Ch. Éluère, brought to the present author's attention by A. Giumlia Mair (to both of whom he is grateful), has shown that there are no platinum group metals in the piece. In fact, the only known ancient use of platinum comes from certain localities in Ecuador and Colombia, where the gravels are rich in perodites, and from the Choco region of Columbia, where the gold deposits are also rich in platinum granules (Bergsøe 1937; Chaston 1980; Craddock 1995, pp. 119–21; McDonald and Hunt 1982, pp. 13–27; Scott and Bray 1980 and 1994). Artefacts of platinum were made in these regions

over 2000 years ago. Studies by Bergsøe (1937) and by Scott and Bray (1980 and 1994) showed they had been made by sintering the platinum granules in a gold matrix, and additionally there is some evidence for platinum plating on gold. These were extremely challenging techniques, as the experimental replications of Bergsøe and of Scott and Bray demonstrated. Similarly, European attempts to sinter platinum in the eighteenth and nineteenth centuries were only achieved with great difficulty (Chaston 1980). The only attempt at the production of platinum artefacts by sintering on a large scale – Russian roubles minted in the mid nineteenth century – was a failure (Bachmann and Renner 1984).

33 *Mappae Clavicula*, recipe 126 (Smith and Hawthorne 1974, pp. 45–6).

34 See p. 50, note 4.

35 *Arthaśāstra,* 2.13.5–6 (Kangle 1972, p. 111).

36 Wyckoff (1967, pp. 39–40, 56, 61, 70–1, 94, 103–4, 133 and 148).

37 *Albertus Magnus*: *Book of Minerals*, Book II, Tract. ii (Wyckoff 1967, p. 70).

38 *Pliny: Natural History,* 37.59 (Eichholz 1962, p. 211).

39 Constantinus Africanus (1536, p. 320).

40 *De Lapidibus* is possibly of Syriac or Persian origin but ascribed to Aristotle, where it was also stated that the *adamans* could be broken with lead. See Rose (1875), Ruska (1912), Thorndike (1960) and Wyckoff (1967, p. 263).

41 *Albertus Magnus: Book of Minerals,* Book II, Tract. iii (Wyckoff 1967, p. 148).

42 Rose (1875, pp. 389–90).

43 Smith and Hawthorne (1974, p. 76). Nassau (1984, p. 10) makes the interesting observation that the oft-repeated reference to softening gems with goat's blood actually relates to the controlled heat treatment to produce small cracks prior to dyeing, as part of gem-enhancement procedures.

44 Meeks et al. (1996).

Assaying in Antiquity

P.T. Craddock

The ability to determine the composition of gold was a prerequisite to the use of precious metals in coinage and to the control and regulation of the refining processes. The consistent composition of the earliest Lydian electrum series of coins, as reported in Table 7.1, p. 170, shows beyond doubt that the Lydians already had the ability to determine the composition of each batch of native electrum in order to be able to add the correct quantity of silver to produce a uniform electrum alloy for the coins.

The subject of the assaying of precious metals in antiquity has been discussed at length elsewhere,[1,2] and therefore will only be summarised here. There are quite detailed accounts of the traditional processes as practised in early post-medieval Europe,[3] although these were already more sophisticated than those available earlier, especially with the advent of mineral acids and precision balances. In medieval Europe there were seven recognised methods of testing gold, although not all were quantitative.[4] They were by: solution, the touchstone, density, taste, the action of fire, fusion and sublimation.[5] Al-Hassan and Hill[6] record an additional medieval Islamic method whereby the purity of the gold was assessed by noting the speed of solidification after removal from the furnace.

The three principal quantitative methods at least potentially available to the ancients were specific gravity determination, fire assay and the touchstone.

Specific gravity

The evidence for the use of specific gravity methods in antiquity is tenuous, and centres around the well-known anecdote of Archimedes devising a non-destructive method for determining the purity of some golden crowns, as recounted some centuries later by Vitruvius.[7] In brief, the method, as described there and by subsequent late antique and early Islamic accounts, determines the volume of the metal to be tested by filling a beaker to the very brim with water and weighing it, then immersing the metal in the water such that some is spilt over the side. The metal is removed and the beaker weighed again, determining the mass of water lost and thereby the volume displaced. The metal itself is weighed, and thus with volume and mass known the specific gravity can be calculated.

In common with many scientific and technical processes conceived in classical antiquity, it was a nice idea but impracticable with the apparatus available. The method may have worked with Archimedes' crowns or other bulky objects, but problems with the surface tension of water at the edges of the beaker would have rendered it hopelessly inaccurate for smaller masses of metal. The more familiar method of weighing the metal in air and then in water to determine the volume was first mentioned in about AD 500 in a poem, *Carmen de ponderibus*, by the Roman author Priscian. The application of the principle to a practical method seems to have been an early medieval, possibly Islamic, development with the introduction of the hydrostatic balance.[8] The earliest detailed description of a practical method is given in the *Mappae Clavicula*,[9] compiled in the ninth century where it is stated that:

> ... if you find any shaped [goldsmith's] work which you believe to be alloyed with silver and you want to know how much gold or how much silver is contained in it, take some silver or gold and, checking and examining the weight, make a lump of whichever metal you wish equal to the weight of the suspected work, and place both of these, namely the work and the lump on the [two] pans of the balance and immerse them in water. If the lump that you made was of silver, the work will

outweigh it; if it was of gold, the object will rise and the gold sink.

The description then went on to show how the percentage of silver could be calculated.

However, few other works on the refining and testing of precious metals mention specific gravity as a method, even in the post-medieval period. Ercker[10] mentions and even illustrates (Fig. A5.1) a variant of the method, but more as a potential method that should be feasible rather than as a regular working procedure, although he does state that it had a long history. Presumably the balances

available were not of sufficient accuracy.[11] Instead, the practical manuals describe fire assay and the touchstone. This situation is reflected in India as exemplified by the broadly contemporary *Ā-īn-i Akbarī*.

Fire assay

Fire assay removes the base metals from weighed quantities of silver or gold by cupellation with lead (Fig. A5.1), and silver from gold by cementation or with acids. The

Fig. A5.1 Cupelling, parting and performing a density assay on precious metals. **Key:** *A* assay furnace with assayer undertaking an assay, *B* iron tray on to which assay pieces are poured, *C* wooden shield with viewing slit to protect eyes and face when looking into the furnace, *D* parting flask for assaying gold, on its stand, *E* weighing auriferous silver in water. (From *Lazarus Ercker's Treatise on Ores and Assaying*, A.G. Sisco and C.S. Smith, 1951, p. 138)

remaining gold is then weighed. This was, and remains, a powerful method: in the words of the *Probierbüchlein,* 'The Most Reliable Assay and the Best is by Fire', but the accuracy was once again limited by the available balances. Almost certainly, the very early accounts of the fire refining of gold in Egypt and Mesopotamia[12] refer to cupellation processes, as discussed on p. 32. There is little or no evidence for the removal of silver from gold prior to the introduction of coinage, specifically the Lydian bimetallic coins coeval with the PN refinery. After that time, full fire refining was possible and there are several early Islamic[13] and Indian[14] treatises on gold purification with descriptions of practical fire assaying. A favourite method was to fire assay the samples under test together with gold of known purity to ensure that the actual firing was reliable and to enable strict comparability. To ensure the metal was completely pure, the fire assay was often repeated to constant weight. Fire assay is reliable, and indeed is still used by institutions such as the Goldsmiths' Company of London and the Swiss Federal Bureau for the Control of Precious Metals.[15] However, it is destructive, time consuming and requires both specialist skills and apparatus, and as such it was probably not in general use in antiquity. To quote Agricola:[16]

> For although the assay made by fire is more certain (than by touchstone), still, since we often have no furnace, nor muffle, nor crucibles, or some delay must be occasioned in using them, we can always rub gold or silver on the touchstone, which we have in readiness. Further, when gold coins are assayed in the fire, of what use are they afterwards?

Some assaying methods[17] merely melted the suspect gold. If, after cooling, the surface was white then it contained silver, but if rough and black then it contained either copper or lead.[18] This method would reveal whether the gold was base and had been extensively treated to enhance the gold content at the surface by depletion of the other metals. On melting the gold, its true composition would be revealed at the surface: silver would lighten the metal, but copper or lead would oxidise and appear dark. Thus, although this was not in itself a true assaying method, it could have been a sensible precursor to the application of the touchstone.

Touchstone

The most widespread method of assaying, from classical antiquity until well into the post-medieval period and beyond,[19] was with the touchstone.[20] Put very briefly, the metal to be tested was rubbed on a stone with a uniform, dense, dark and matt surface. This left a streak which was then compared with a streak made with a needle of precious-metal alloy of broadly similar composition, selected from a set of such needles of known composition (Fig. A5.2), until an exact match was obtained. This worked very well for binary combinations (gold with either copper or silver[21]), but failed when the gold contained both silver and copper in unknown combinations. This problem was only resolved in the medieval period when the introduction of mineral acids allowed the copper or other base metals to be selectively

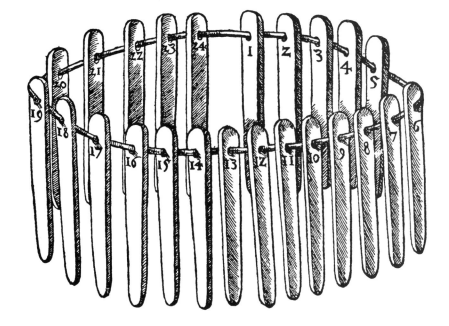

Fig. A5.2 A set of 24 touch needles containing from 1 part Ag : 23 parts Au to 23 parts Ag : 1 part Au. Agricola noted that needles of just gold and silver were not much used because most gold coins also contained some copper, and thus a more complex series of touch needles of alloys of all three metals were required in practice. (From *Georgius Agricola: De Re Metallica,* H.C. Hoover and L.H. Hoover, 1912, p. 255)

removed from the streak.[22] The problem would not have been too serious for natural gold, which tends to be a binary alloy with silver, but must have been much more serious with the complex artificial alloys of the artificers and alchemists (see p. 249).

Although the touchstone was used in Bronze Age Europe,[23] there is no evidence as yet that it was used by the ancient civilisations of the Middle East. The earliest references are from Greek sources of the sixth century BC, mentioning the *basanos* (touchstone).[24] There are also a number of descriptions of rubbing gold against the touchstone to provide the requisite colour comparison.[25] The earliest detailed descriptions are to be found in the *Arthaśāstra*[26] and in the work of Theophrastus on mineralogy,[27] both dating from the fourth century BC.

The *Arthaśāstra* states that:

> After first rubbing the gold [to be tested] on the touchstone, he (the Superintendent of gold) should afterwards rub the standard gold [on]. That with a streak of the same colour [as the standard] on places [on the stone] that are neither depressed nor elevated, he should know as properly tested; what is over-rubbed or lightly rubbed or powdered over with red chalk from underneath the finger-nail, he should know as deception. Gold, touched with the forepart of the hand smeared with *jāti*-vermilion or sulphate of iron dissolved in cow's urine, becomes white. The streak [of the gold] on the touchstone, that has filaments, is smooth, soft and lustrous is best.

The account continues with a description of the varieties of touchstone available.

Theophrastus' work contains one of the most detailed early accounts, and not only gives the accuracy of the method but specifically describes the stones from the Sardis area as being the most suitable for touchstones:

> **45** The nature of the stone which tests gold is remarkable, for it seems to have the same power as fire, which can test gold too (i.e. fire assay). On that account some people are puzzled about this, but without good reason, for the stone does not test in the same way. Fire works by changing and altering the colors, and the stone works by friction, for it seems to have the power of picking out the essential nature of each metal.

> **46** They say that a much better stone has now been found than the one used before; for this not only detects purified gold, but also gold and silver that are alloyed with copper, and it shows how much is mixed in each stater. And indications are obtained from the

smallest possible weight. The smallest is the *krithē*, and after that there is the *kollybos*, and then the quarter-obol, or the half-obol; and from these weights the precise proportion is determined.

> **47** All such stones are found in the river Tmolos. They are smooth in nature and like pebbles, flat and not round, and in size they are twice as big as the largest pebble. The top part, which has faced the sun, differs from the lower surface in its testing power and tests better than the other. This is because the upper surface is drier, for moisture prevents it from picking out the metal. Even in hot weather the stone does not test so well, for then it gives out moisture which causes slipping. This happens also to other stones, including those from which statues are made, and this is supposed to be a peculiarity of the statue.

Section 46 has usually been interpreted as the weights of impurities which could be detected in a one stater gold coin. Taken at face value, it suggests that quantities of one grain (*krithē*, about 0.06 g) could be determined in a stater of 8.72 g: that is, 1 part in 144, which, as Caley remarks, seems very doubtful.[28]

The comment on the new stone which could detect copper in the alloy is interesting. Note that Theophrastus distances himself from the claims with the cliché 'they say that'. It was not something that he was personally familiar with, and in practice no such stone was ever discovered. The statement does, however, emphasise the real problem of trying to deal with alloys of gold with silver and copper together in the absence of mineral acids.

Touchstones from the 'Tmolos', which is almost certainly to be identified with the streams running from Mount Tmolus, including the Pactolus, continued to be highly prized. Thus Pliny,[29] though quoting from Theophrastus, states that:

> With the mention of gold and silver goes a description of the stone called the touch stone, formerly according to Theophrastus not usually found anywhere but in the river Tmolus, but now found in various places. Some people call it Heraclian stone and others Lydian. The pieces are of a moderate size, not exceeding four inches in length and two in breadth. The part of these pieces that has been exposed to the sun is better than the part on the ground. When experts using the touch stone, like a file, have taken with it a scraping from an ore, they can say at once how much gold it contains and how much silver or copper, to a difference of a scruple, their marvellous calculation not leading them astray.

Summary

At the time of the very first electrum coinage, the specific gravity method almost certainly was not known, and fire-assay methods that could remove the silver by cementation were also not used. (That, after all, was to be the crucial development of the Lydians at Sardis in the following generation.) Thus the only method would have been the touchstone, which the Lydians developed to such a degree of perfection that thereafter the best touchstones were said to come from the Sardis region.

The Renaissance and later works dealing with gold all refer to the touchstone as the routine method of gold assay, usually used in conjunction with acid. Thus, for example, in Howard's *Dictionary of the Arts and Sciences*,[30] published in 1788, it states that:

> It is said that a mixture of *gold* with one third, or even with half it's weight of silver, acquires on cementation with verdigris, the colour of pure *gold*. The silver however is easily discovered on the touchstone by means of *aqua-fortis*: if a line be drawn with the compound metal, a little *aqua-fortis* applied upon the part, will eat out the silver; and the remaining *gold* will appear specked and disunited.

Thus an alloy with the right combination of copper and silver to preserve the gold colour would deceive the touchstone test by itself until the application of nitric acid to remove the base metals and silver.

Some of the various descriptions of goldworking and gold assaying quoted elsewhere in this work show that the touchstone was, and in some instances still is, widely used. For example, some of the official records made in India at the end of the nineteenth century[31] and other records made more recently in the twentieth century,[32] make it clear that the touchstone continues to be widely used. The touchstone was used until very recently at assay offices in the Netherlands and is still in use in Switzerland as a quick preliminary method capable of giving reliable and acceptable results even if not with the same precision as the modern analytical methods.

Thus it would seem that throughout antiquity, into the Middle Ages and beyond, the touchstone was the usual method of assay. This could explain the prevalence through these periods of recipes 'increasing' or creating false gold. They can be found in artisans' recipe books such as the *Mappae Clavicula* or the *Leyden Papyrus* and the works of the alchemists. They are myriad, but many describe the adulteration of real gold with combinations of silver (or arsenic) and copper to preserve or even enhance the colour of the gold. Such a process would typically be called 'increasing' or 'doubling' the gold.

In some ancient coin issues, the overall quantity of copper and silver added varies enormously, but the silver to copper ratio remains quite constant at an alloy which would preserve the colour of the metal.[33]

The recipes of the alchemists concentrate on the colour of 'their' gold, which was in reality debased, without any reference to specific gravity, or to fire assay. Clearly, the alchemists' gold could not pass these tests as we understand them. Historians of science have pondered this question[34] at length and sought both to explain the problem and to exonerate the alchemists from fraudulent intent by suggesting that the gold so produced, often disingenuously called by the alchemists 'their' gold, was consciously different from ordinary gold and not intended to deceive. At a more fundamental level, Hitchcock,[35] followed by many other eminent scholars, including no less a scientist than Jung,[36] have argued that the whole subject of the creation and refining of gold was no more than a symbolic metaphor for the redemption and regeneration of the soul. However, this has always seemed to be special pleading and the contemporary artisans' recipes give the same procedures but with much more prosaic motives.

Surveying the available methods of testing gold puts the alchemists in a harsher but more realistic light than that created by their modern apologists. It seems that the specific gravity method was hardly ever used until the precision balance was developed in the seventeenth and eighteenth centuries, and fire assay, involving both cupellation and cementation, was not really applicable as a routine method. Only the touchstone was in regular use and this, without the refinement of mineral acids to detect the presence of copper or other base methods in ternary alloys, could not be used with any accuracy at all in determining the fineness of gold debased with combinations of silver and other metals that preserved the golden colour. Thus there was no easy method of detecting these debased or 'increased' golds, and the primary motive of the alchemists to the very end was likely to have been simple gain. Writers such as Agricola were in no doubt at all about the fraudulent motives of the contemporary alchemists.[37] He remarks caustically on their singular lack of success in sixteenth-century Europe, stating:

> We do not read of any of them becoming rich by this art, nor do we see them growing rich; and that given the number of producers, if they were successful they would by today have filled whole towns with gold and silver.

Notes

1 One of the best historical surveys of the subject is that of Smith (1923–4), who also wrote one of the standard texts on traditional assaying methods (Smith 1947).

2 Éluère 1986; Halleux (1985); Hradecky (1930); Lord (1937); Moore and Oddy (1985); Oddy (1983, 1986 and 1993b); Ridgeway (1895); Smith (1923–4); Wälchi (1981); Wälchi and Vuilleumier (1985 and 1991a, b).

3 *Probierbüchlein* (Sisco and Smith 1949); Biringuccio (Smith and Gnudi 1942, pp. 136–41, 159–61, 202–5); Agricola, Book 7 (Hoover and Hoover 1912, pp. 219–65); Ercker (Sisco and Smith 1951, esp. pp. 65–6, 116–24); and Badcock (1677).

4 Quoted in Stillman (1924, p. 218).

5 As listed by Vincent de Beauvais in the late fifteenth century, and repeated in Stillman (1924, p. 218).

6 al-Hassan and Hill (1986, p. 247).

7 *Vitruvius on Architecture*, 9.9–12 (Granger 1935, pp. 203–7).

8 Halleux (1985, pp. 42–4); Smith and Hawthorne (1974, p. 56, footnote 132). Some of the medieval Islamic scholars ascribed the hydrostatic balance to Menelaos of Alexandria, *c.* AD 100 (Partington 1970, p. 206).

9 Recipe 194 (Smith and Hawthorne 1974, p. 56).

10 Sisco and Smith (1951, pp. 134–5).

11 See Stock (1969) for a succinct account of the development of the precision balance in post-medieval Europe.

12 Levey (1959a, b).

13 See Levey (1971) for the accounts of several medieval Islamic authors who were engaged in the production and refining of precious metal.

14 See Percy (1880, pp. 379–81). See also Blochmann's translation (Phillott 1927, p. 24), in which the use of the touchstone is also described (pp. 19–20). It is very noticeable that specific gravity determinations are discussed, not in the section on the mint with the rest of the refining and assaying methods, but in the section describing the properties of the elements (p. 42). Clearly, in sixteenth-century India specific gravity measurements were not regarded as a practicable assay method.

15 See the succinct but detailed recent accounts of the current assay procedures in Bachmann (1995), Forbes (1978), Evans (1991 and 1993), Wälchi and Vuilleumer (1991b) and Ammen (1997, pp. 302–24). The modern process, differing only in detail from the traditional fire-assay method, can be briefly outlined here (see Berzelius 1836 or Bugbee 1933, for example). First, the approximate composition of the sample is determined (the Swiss office uses the touchstone). The appropriate amount of additional silver is added to bring the ratio of silver to gold to an overall value of 3 to 1. This is necessary to ensure that the parting acid can penetrate the alloy easily and remove all the silver, both original and added. This was the rationale behind the graduation process described by Ercker and others (Sisco and Smith 1951, p. 190, and this volume, pp. 64–5) for the production of high-purity gold.

The sample and parting silver are then wrapped in lead foil and cupelled at temperatures of about 1150 °C for about 15 to 20 minutes, during which time all of the lead and any base metal will have oxidised and sunk into the cupels, which are now made of magnesia (MgO) rather than the traditional bone ash. The remaining button of precious metal (gold and silver) is mechanically cleaned of any surface debris from the cupel and then rolled out to give a large surface area. The acid parting then follows, with the strip placed in boiling nitric acid for about 20 minutes, during which time all the silver is dissolved out. The strip is then washed, annealed and weighed. It is usual to analyse reference samples from gold of known composition at the same time as a check on the accuracy of the assay. This was the practice in the past, very often using gold coins of known composition as the standard (see note 13).

16 Hoover and Hoover (1912, p. 252).

17 For example, Pliny's *Natural History* 33.59 and 33.127, and the *Leyden Papyrus*, recipe 43 for silver (Halleux 1981, p. 95).

18 Raub (1995) has shown that as little as 1% copper in the gold will perceptibly blacken the alloy on heating with a flame.

19 Thus, for example, it was the method universally used by the goldsmiths in northern India in the late nineteenth century (Maclagan 1890, p. 19; Baden-Powell 1868–72; see also this volume, pp. 43–5).

20 See Hradecky (1930), Wälchi (1981), Wälchi and Vuilleumier (1985) and Ammen (1997, pp. 128–9) for descriptions of modern touching.

21 Gowland (1910) observed that, in the nineteenth century, the Japanese could get to within a per cent of the true composition of a gold–silver alloy. But without acid they could not test ternary mixtures of gold, silver and copper.

22 Healy (1980a) states that it is possible to distinguish the composition of 'gold with silver and/or copper' by touchstone alone, citing Oddy and Schweizer (1972). However, reference to this paper, which compares various methods of analysing precious metals, shows (p. 178) that dilute nitric acid was used for base gold and a mixture of nitric and hydrochloric acids for fine gold in conjunction with the touchstone.

23 Éluère (1986) has published a touchstone from Choisy-au-Bac (Oise) dated to the Late Bronze Age, which had streaks of gold. It was found near a small gold ingot.

24 *Basanos* formed the root of the word used in Classical Athens to describe the torture of slaves specifically to extract reliable evidence from them, on the somewhat questionable grounds that it was illogical to expect a slave to give true testimony except under duress!

25 Especially Herodotus VII.10: '... if we rub alongside it with some other gold we will discern the finer [of the two].'

26 *Arthaśāstra*, 2.13.15–25 (Kangle 1972, p. 112).

27 *Theophrastus on Stones*, 45–7 (Caley and Richards 1956, pp. 54–5, 150–9).

28 Caley and Richards (1956, p. 155); Halleux (1985, p. 42).

29 *Pliny: Natural History*, 33.126 (Rackham 1952, p. 95).

30 Howard (n.d., [1788], p. 1069).

31 See pp. 43–5.

32 See p. 45.

33 Note, for example, the analyses of Celtic gold coins from Britain and north-west Europe (Cowell et al. 1987 and Cowell 1992), where although the range of debasement is very wide, the alloy used was a very constant two parts silver to two parts copper that would preserve the colour of the metal.

34 See Needham (1983, pp. 1–28), who neatly encapsulated the problem in the phrase 'aurifaction or aurifiction' (Needham 1974, p. 9, and 1980, p. 490).

35 Hitchcock (1857). See also Needham (1983, pp. 15–16).

36 Jung (1944).

37 *Georgius Agricola: De Re Metallica* (Hoover and Hoover 1912, pp. xxvii–xxix).

Concordance between Sardis Excavation Codes/Descriptions and British Museum Codes

Hardly any of the excavated material taken for analysis at the British Museum was catalogued for the excavation's inventory. After field-sorting of a much larger sample, most of the pieces were taken to the camp laboratory for closer examination and storage. Many were there given TS (technical study) numbers in general categories, e.g. TS 122, litharge, but without further specification. Similarly, the coarseware was separated into groups, which we have maintained in our previous list, defined by whether it was 'glazed' or not and whether this 'glazing' contained gold globules or not. The same procedure was followed for other items like bread tray, brick or litharge. These pieces were marked with the date of finding and their findspots, and they were eventually listed in Appendix VI of the *Final Laboratory Report* of 1969. The list was compiled by Sidney M. Goldstein, the assistant conservator, hence the use of SMG to refer to the list.

The following items are mostly ceramic materials, with the exception of some fragments of litharge. The samples for scientific examination were taken before the retrospective inventory was compiled, and such details as existed were taken from the bags in which the samples were contained. Where no number existed, an identification is given. For the gold material, the original sample numbers have been used.

The British Museum identification codes were assigned in the Museum's Department of Scientific Research.

Sardis excavation code/description	BM code
Parting and melting sherds	
July 7, 1969 SMG p. 9 #11	44396P
July 1969 SMG p. 9	44397Y
July 3, 1969 SMG p. 9 #4	44398W
TS154	44399U
TS123 sample 34 (three pieces)	44400Q
TS123	44401Z
TS130 sample 31	44402X
July 8, 1969 SMG p. 10 #4	45670U
July 19, 1969 SMG p. 10 #5	45671S
TS131	45672Q
TS132	45673Z
July 3, 1969 SMG p. 9 #4	45674X

Sardis excavation code/description	BM code
Unmarked melting sherd with stump of handle	45675V
TS125 sample 7 (several pieces)	45676T
July 7, 1969 SMG p. 9 #11	45677R
July 8, 1969 SMG p. 10 #4	45678P
July 10, 1969 SMG p. 9 #10	45679Y
July 10, 1969 SMG p. 9 #7	45680Q
July 11, 1969 SMG p. 11 #3	45681Z
July 12, 1969 SMG p. 9 #8	45682X
July 25, 1969 SMG p. 9 #1	45683V
June 29, 1969 SMG p. 11 #1	45684T
TS123.8	45685R
Rim from heavy gray coarse bowl	45688W
TS125 sample 7 (several pieces)	46849Y

Sardis excavation code/description	BM code	Sardis excavation code/description	BM code
Large sherd	46851Z	No number	51504Y
Tuyere fragments		*Other ceramics (coarsewares)*	
TS139i	47910X	1A (1–3)	47905R
TS139u	47232V	1A (1–3)	47906P
Lead smelting furnace lining fragments		Refired pot sherd	47907Y
TS149	45686P	*Litharge fragments*	
TS150, NE dump	45687Y	July 22, 1969 SMG p. 8 #2	44389Y
Bread tray		Aug 17, 1968 SMG p. 8 #4	44390Q
Heavily vitrified	47908W	Cupellation area B (with Cu stained bone)	44391Z
Heavily vitrified	47909U	1968 NE dump	44392X
Unburnt	45690X	1968 NE dump	44393V
Unburnt	45691V	TS122	44394T
Parting-furnace bricks		1968 NE dump	44395R
Sample 20	47668S	*Metal*	
Refired furnace brick, from sample 20	47911V	Iron blowpipe (Appendix 2, item 58)	45689U

Technical Glossary

Alluvial gold *see* **Gold**

Amalgamation This is the process of extracting gold and silver from their ores or from plating etc on base metals by forming an amalgam with mercury.

An amalgam is an alloy of mercury with other metals. Note that, by and large, the metals are not in solution in the amalgams; most metals, including gold and silver, are only sparingly soluble in mercury. The amalgam can be liquid, pasty or quite solid, depending on the mercury content.

Mercury forms amalgams with many metals, but those of gold and silver are most familiar in early metallurgy, having a long history as outlined in Appendix 3, p. 233. An amalgam with gold or silver can be formed by dipping the metal into hot mercury. When the mercury predominates, a gold amalgam is quite liquid, but becomes more pasty with gold contents in the range from 25% to 50% by weight. This was colloquially known as 'butter of gold'. Gold amalgams of this consistency were used in the fire gilding process (also known as mercury or parcel gilding). The silvery 'butter' was spread and brushed out over the cleaned surface of the metal to be gilt, and then heated to evaporate the mercury. Amalgams can be broken up by heating at about 400 °C, at which temperature the majority of the mercury is rapidly driven off, but it is difficult to completely remove the mercury.

Atomic absorption spectrometry (AAS) This is an analytical technique which is capable of measuring the concentrations of a wide range of elements present in inorganic materials such as metals and ceramics. A sample is taken (typically weighing 10–100 mg), dissolved and then sprayed into the flame of the instrument, where it is atomised. Light of a suitable wavelength for a particular element is shone through the flame and is absorbed by the atoms of the sample. The amount of light absorbed is proportional to the concentration of the element in the solution, and hence in the original sample. Measurements are made separately for each element of interest in turn to achieve a complete analysis. The technique is therefore relatively slow to use. However, it is very sensitive and can be used to measure trace elements as well as those present in major and minor amounts.

Bloating On heating to high temperatures, some clay-based ceramics exhibit a permanent expansion due to the formation of gas bubbles, which become trapped in the vitrifying clay (see also **Vitrification**). The resulting bubbly texture can be observed directly by eye or with the aid of magnification. The degree of bloating can provide an indication of the maximum temperature achieved during firing/use.

Cementation In a cementation reaction, the principal reactants are present as solids, at least at the beginning and end of the reaction, although the reaction usually takes place at an elevated temperature, during which a liquid or vapour is generated which attacks the solid. Direct reactions between solids are in reality not feasible because no matter how finely divided and intimately mixed the reactants, the contact at atomic level is minimal. In the traditional salt cementation **parting** (see below) process for separating silver from gold, an astringent vapour containing chlorine, ferric chloride and hydrogen chloride was generated, which attacked the silver in the impure gold, forming a vapour of silver chloride which evaporated from the gold.

Cupellation and scorification This is a process used for the extraction of silver from its smelted ores, or for the separation of base metals from precious metals, or for establishing the purity of silver by fire assay (see Appendix 5, p. 246).

In this process, the metal is mixed with lead, heated to about 1100 °C and subjected to a continuous blast of air. Under these intensely oxidising conditions, the lead oxidises to litharge (lead oxide, PbO), absorbing all the base metals, which have also oxidised. The precious metals, gold and silver, do not oxidise, but form a discrete coherent mass of molten metal, quite separate from the litharge.

A related process is known as **scorification,** in which the bulk of the lead was oxidised and removed under less stringent conditions prior to the main cupellation process.

Electrum *see* **Gold**

Emission spectrography (ES) In ES, a small sample is removed from the object to be analysed, mounted in a conducting graphite electrode and excited either by a spark source or by a continuous d.c. arc. The spectrum emitted by the elements present is in the visible or ultraviolet wavelength regions and can be recorded either on a photographic plate or electronically.

The method is extremely sensitive for most elements and can give a simultaneous analysis for up to 40 elements. However, due to inherent instability in the arc excitation, the method lacks sufficient precision for the quantitative measurement of major elements. These problems have now been largely resolved by **inductively coupled plasma spectrometry (**see below). ES was the method employed for most of the major analytical studies on antiquities up to the 1970s.

Energy-dispersive x-ray analysis *see* **Scanning electron microscope**

Gold In the present book the term is used fairly generally to mean a metal which is principally or all of gold.

The freshly mined and untreated gold from the sands and gravels of rivers, such as the Pactolus, is here termed **alluvial gold**. This will normally have contained silver, typically in the range from 5% to 40% by weight, plus a little copper, which rarely exceeds 1%. Alternative names for gold from sands and gravels are **placer** or **secondary** gold. This is because the water-deposited sands and gravels are sometimes known as placers, and they are constituted of redeposited or secondary materials from the erosion of the original primary geological context.

For artificial alloys of gold and silver, either deliberately or accidentally produced, the term **electrum** is employed here, following the original Greek usage. In current usage, electrum is used to describe both artificial and naturally occurring alloys. Fortunately, with these definitions it is still possible to use the usual numismatic term of electrum to describe the very earliest Lydian coins, because it has been shown that the natural alluvial gold–silver alloy was supplemented by additions of silver to achieve a constant composition in the coins.

Gravimetric analysis This method of quantitative analysis is capable of high accuracy but requires relatively large samples (in the region of 1 g) and is time consuming. Therefore, it has been largely superseded by instrumental methods. The sample is dissolved and then individual elements are isolated as precipitated compounds of known formulae. The precipitates are accurately weighed and, from the weight and the compound formula, the amount of each element in the original sample can be calculated.

Inductively coupled plasma spectrometry (ICP) ICP (or more fully ICP–AES, atomic emission spectrometry, or ICP–OES, optical emission spectrometry) is an analytical technique which is capable of measuring most elements found in materials such as metals, ceramics, slags and ores. In normal operation, the archaeological sample (typically weighing 10–50 mg) has to be dissolved before being introduced into a high-temperature plasma, where the light emitted by the sample is measured. It is a sensitive technique with a wide dynamic range, so that it can measure both trace and major elements in the same solution. It also rapid – 30 or 40 elements can be measured in a few minutes.

Liquation In general, liquation is a process by which metals with different melting points, and immiscible in the liquid phase, can be separated by carefully controlled melting. More specifically, it was employed to remove precious metals, mainly silver, from copper using lead. The two metals were melted together and stirred and worked to ensure good contact. The precious metals tended to go into the lead, which on cooling separated from the copper. The precious metals could then be recovered from the lead by **cupellation** (see p. 253).

Neutron activation analysis (NAA) NAA is an extremely sensitive technique of chemical analysis which may be used on samples of a few milligrams and can detect concentrations in the parts per million (ppm) range. Samples, removed by drilling or by abrasion onto a pure substrate material such as quartz (streak method), are exposed to a source of neutrons, usually in a nuclear reactor. Some of the elements in the material are thus converted to radioisotopes by neutron capture. The radioisotopes emit radiation in the form of a spectrum of gamma rays that are characteristic of the elements present and with intensities proportional to their concentrations. A major drawback is that lead cannot be determined.

Parting The traditional techniques for the separation of gold and silver are known as parting processes. Before the development of mineral acids, this was achieved by **cementation** (see p. 253) with acidic salts or sulphur. From the Middle Ages, acid parting became increasingly common. Nitric acid, HNO_3, was the acid usually employed at first, but the use of sulphuric acid, H_2SO_4, became increasingly common in the nineteenth century.

The hot aqueous acid solutions attacked the gold surfaces, proceeding down the grain boundaries of the metal and dissolving out the silver, thereby creating porosity and exposing the interior of the metal to the acid. In order to facilitate the formation of porosity, and thus the parting process, silver was added to the impure gold such that the overall silver content was about one part in four, hence the old English name of quartation for the process (see p. 65). The process was also known as graduation in the post-medieval period (see p. 64), but its origins lie back in the salt **cementation** process (see p. 38). Addition of silver is still a standard part of the volumetric acid-assay of gold (see Appendix 5, p. 250, note 15).

These traditional methods of acid parting were joined in the nineteenth century by electrolytic parting and the Miller process of bubbling chlorine through molten gold (see p. 69).

Petrography *see* **Thin-section petrography**

Placer *see* **Gold**

Refractory This term is used to describe materials capable of withstanding high temperatures, permitting their use in furnace linings and other situations exposed to strong heat. In current usage, the term is sometimes restricted to materials able to withstand temperatures in excess of 1500 °C. However, ancient processes rarely involved exposure to temperatures greater than 1200–1300 °C, so that, by modern standards, many of the ceramics used in metallurgical processes were not particularly refractory.

Scanning electron microscope (SEM) This a powerful imaging microscope which provides a wide magnification range (typically useful between ×10 and ×50 000) with a large depth of field. It is complementary to optical microscopy, but by using electrons instead of light it overcomes the optical limitations, particularly as the magnification increases. The SEM allows the topography of rough surface specimens, such as the Sardis sherds, to be seen in detail and recorded photographically. The SEM is often combined with an energy-dispersive x-ray spectrometer, which allows the sample composition to be determined by quantitative, qualitative or element-distribution methods (see below).

In the SEM operation, a narrow beam of electrons is accelerated by a high voltage down the 'electron column', where electromagnetic lenses focus and scan the beam in a raster pattern on the surface of the sample. The focused beam is extremely narrow at the sample and the working distance is relatively long, giving rise to the characteristic high magnification and depth of field. As the high-energy electrons hit the sample, some are scattered back and others knock low-energy electrons from the sample. These backscattered and secondary electrons are detected and give rise to images, on a TV or computer monitor, corresponding to the area scanned on the sample. Higher magnification is simply achieved by scanning smaller areas on the sample, while the image is displayed full-size on the monitor. Secondary electrons produce familiar 3D-type images of the surface, while the backscattered electrons produce images containing compositional information derived from the differences in the relative densities (mean atomic number) of the different regions of the sample.

Samples bombarded with the focused beam of high-energy electrons from the SEM emit x-rays which are characteristic of the elements present in the sample. These x-rays are detected by the solid-state detector (either of silicon or germanium) positioned close to the sample and are displayed on a monitor as a spectrum of intensity against x-ray energy. This technique is known as **energy-dispersive x-ray analysis** (EDX), and is analogous to standard **x-ray fluorescence** (see p. 256). The energies of the x-rays are characteristic of the elements present in the sample and their intensities depend upon how much of those elements is present.

The depth of penetration of the electron beam into a sample is around 1 micrometre (1 μm) in metals and up to a few micrometres in non-metals, so EDX analysis is essentially a surface analytical technique.

By scanning a large area of the sample, an average composition can be obtained, or the high resolution of the SEM can be used to analyse very small specific areas or tiny particles, such as the particles of gold on some of the Sardis sherds. This is referred to as microanalysis.

Scorification *see* **Cupellation and scorification**

Secondary gold *see* **Gold**

Slag This term is generally used to describe all the non-metallic durable debris of most pyrometallurgical operations. In the present book, it is used much more specifically for the vitreous debris of metal-smelting operations.

Specific gravity (SG) The specific gravity of a material is its density relative to water at the same temperature. The specific gravity can be used to estimate the gold content of uncorroded gold–silver alloys provided that there are no significant amounts of other metals present, e.g. copper. It is usually determined using Archimedes' principle, which involves comparing the weight of the object in air with that in a liquid. In practice, a dense, non-toxic organic liquid is used in preference to water to improve the precision of the measurement.

Thin-section petrography This is a standard geological technique, used in the examination of rocks and minerals by optical microscopy, but it can also be applied in exactly the same way to archaeological ceramic materials. A small fragment is mounted on a glass slide and then ground until it is *c.* 0.03 mm thick. At this thickness, most silicate minerals are transparent, allowing them to be identified from their diagnostic optical properties. Textural features, such as the size, shape and abundance of particular inclusions in the clay paste, can also be observed. In this way, the fabrics of ceramic materials can be described, characterised and compared.

Valency Valency is defined as the combining power of an atom, expressed in terms of the number of hydrogen atoms (or equivalent) with which it will combine. Thus, for example, one atom of chlorine, Cl, combines with one atom of hydrogen, H, to form HCl, hydrogen chloride, and so chlorine is said to be univalent. Sodium and silver also combine with one atom of chlorine, forming NaCl, sodium chloride, and AgCl, silver chloride, respectively, and are thus also univalent. Iron atoms can combine with two or three chlorine atoms, depending on the molecule formed, and these are di- or trivalent.

Vitrification At elevated temperatures, the clay body of a ceramic may start to melt, leading to the formation of a glassy phase on cooling. This is a progressive change and the degree of vitrification may provide some indication of the maximum temperature to which the ceramic was exposed during firing or use (see also **Bloating**).

X-ray diffraction analysis (XRD) This technique is used in the identification of a wide range of crystalline materials. It requires only a very small sample (less than the size of a pin-head for some techniques). The diffraction of x-rays by a crystalline material produces a pattern of lines or peaks which is characteristic of the structure of the material. By comparison with standard data, these patterns can be used rather like fingerprints to identify the material being analysed.

X-ray fluorescence analysis (XRF) This technique of chemical analysis (which, in principle, is non-destructive) analyses only the surface to a maximum depth of about 100 micrometres (0.1 mm). The material to be analysed is targeted by a beam of x-rays, which causes x-rays of characteristic energy to be emitted by the elements present in the material. The spectrum of x-rays emitted is either detected simultaneously by a solid-state system (energy dispersive) or scanned by dispersion through a crystal (wavelength dispersive). The spectrum indicates the elements present as well as their concentrations. The technique can give erroneous results on ancient artefacts, particularly those of metal, because the surface composition will be altered by corrosion effects. Accurate analysis of the bulk requires the area targeted by the x-ray beam to be abraded to expose the core metal. This is easier with energy-dispersive XRF than with wavelength-dispersive XRF, as the x-ray beam is usually smaller, often only 1–2 mm across.

Bibliography

Aitchison, L. 1960. *A History of Metals,* 2 vols. (London)

Allaire, B. 1996. The gold and silver refiners of Paris in the XVIth and XVIIth centuries, *Can. Inst. Min. Bull.* 89(1004), 76–82

Allan, J.W. 1979. *Persian Metal Technology, 700–1300 AD.* (London)

Allmand, A.J. and Ellingham, H.J.T. 1924. *Principles of Applied Electrochemistry,* 2nd edn. (New York)

Amandry, P. 1939. Rapport sur les statues chryséléphantines de Delphes, *Bull. corresp. hellénique* 63, 86–119

Amandry, P. 1977. Statue d'un taureau d'argent, *Bull. corresp. hellénique, Suppl.* 4, 273–93

Ammen, C.W. 1997. *Recovery and Refining of Precious Metals.* (New York)

Antweiller, J.C. and Sutton, A.L. 1970. *Spectrochemical Analyses of Native Gold Samples,* US Geol. Surv. Rep. (USGS–GD–70–003). (Washington)

Arias, P.E. and Hirmer, M.A. 1962. *A History of Greek Vase Painting,* trans. and rev. B.B. Shefton. (London)

Avaldi, L., Confaloneiri, L., Milazzo, M., Paltrinieri, E., Testi, R. and Winsemann-Falghera, E. 1984. Quantitative results of XRF analysis of ancient coins by mono-chromatic X-ray excitation, *Archaeometry* 26(1), 82–95

Bachmann, H-G. 1993. Zur Metallurgie der römischen Goldgewinnung in Três Minas und Campo de Jales in Nordportugal, in *Montanarchäologie in Europa,* ed. H. Steuer and U. Zimmermann,153–60. (Sigmaringen)

Bachmann, H-G. 1995. Gold analysis: from fire assay to spectroscopy – a review, in *Prehistoric Gold in Europe: Mines, Metallurgy, and Manufacture,* ed. G. Morteani and J.P. Northover, 303–16. (Dordrecht)

Bachmann, H-G. and Renner, H. 1984. Nineteenth century platinum coins, *Platin. Metals Rev.* 28(3), 126–31

Badcock, W. 1677. *A Touch-stone for Gold and Silver Wares ....* (London)

Baden-Powell, B.H. 1868–72. *Handbook of Economic Products of the Punjab.* (Lahore)

Bailey, K.C. 1929. *The Elder Pliny's Chapters on Chemical Subjects,* Vol. I. (London)

Barb, A. 1969. Lapis Adamas – Der Blutstein, in *Hommages à Marcel Renard,* ed. J. Bibauw, Vol. 1, 66–82. (Brussels)

Bargalló, M. 1969. *La amalgamación de los minerales de plata en Hispanoamérica Colonial.* (Balderas, Mexico)

Bayley, J. 1985. What's what in ancient technology, in *The Archaeologist and the Laboratory*, ed. P. Phillips, 41–4 + microfiche. (London)

Bayley, J. 1991a. Archaeological evidence for parting, in *Archaeometry '90,* ed. E. Pernicka and G.A. Wagner, 19–28. (Basel)

Bayley, J. 1991b. Processes in precious metal working, in *Archaeological Sciences 1989,* ed. P. Budd et al., 125–31. (Oxford)

Bayley, J. 1992a. *Anglo-Scandinavian Non-Ferrous Metalworking from 16–22 Coppergate.* Archaeology of York Fascicule 17/7: The Small Finds. (York)

Bayley, J. 1992b. Metalworking ceramics, *Medieval Ceramics* 16, 3–10

Bayley, J. 1992c. *Non-Ferrous Metalworking in England: Late Iron Age to Early Medieval.* PhD thesis, Institute of Archaeology. (London)

Bayley, J. and Barclay, K. 1990. The crucibles, heating trays, parting sherds, and related material, in *Object and Economy in Medieval Winchester,* ed. M. Biddle,

Winchester Studies 7 (ii): *Artefacts from Medieval Winchester*, Vol. i, 175–97. (Oxford)

Bayley, J. and Eckstein, K. 1997. Silver refining – production, recycling and assaying, in *Archaeological Sciences 1995*, ed. A.G.M. Sinclair, A.E. Slater and J.A.J. Gowlett, 107–11. (Oxford)

Beck, L. 1884. *Geschichte des Eisens in seiner technischen und kulturgeschichtlichen Beziehung*, Vol. 1. (Braunschweig)

Bellinger, A.R. 1968. Electrum coins from Gordion, in *Essays in Greek Coinage Presented to Stanley Robinson*, ed. G.K. Jenkins and C.M. Kraay, 10–15. (Oxford)

Bergsøe, P. 1937. The metallurgy and technology of gold and platinum among the pre-Columbian Indians, trans. C.F. Reynolds, *Ingeniørvidenskabelige Skrifter* A–44

Bergsøe, P. 1938. The gilding process and metallurgy of copper and lead among the pre-Columbian Indians, trans. C.F. Reynolds, *Ingeniørvidenskabelige Skrifter* A–46

Berthelot, M. 1888. *Collections des anciens alchimistes grecs*, Vol. 2. (Paris)

Berthelot, M. 1893. *Histoire des sciences; la chimie au moyen âge*, Vol. 2. (Paris)

Berthelot, M. 1900. Sur les métaux égyptiens; étude sur un étui metallique et ses inscriptions, in *Monuments Piot*, Vol.7, 121–41. (Paris)

Berthelot, M. 1901a. Sur les métaux égyptiens, *Annales du service des antiquités de l'Égypte* II, 160–1

Berthelot, M. 1901b. Sur les métaux égyptiens: présence du platine parmi les caractères d'une inscription hiéroglyphique, *Comptes rendus Séanc. Acad. Sci.* 132, 729–34

Berthelot, M. 1906. *Archéologie et histoire des sciences avec publication nouvelle du papyrus grec chimique de Leyde et impression original du Liber du Septaugenta de Geber.* (Paris)

Berzelius, J.J. 1836. *Lehrbuch der Chemie*, trans. F. Wöhler, 4th edn, Vol. 1. (Dresden and Leipzig)

Birgi, S. 1944. *Gold Deposits in the Region of Sardis (Salihli)*. Internal Report of Turkish Minerals Research Agency. (In Turkish). (Ankara)

Boardman, J. and Hayes, J. 1966. *Excavations at Tocra 1963–1965: The Archaic Deposits I*, British School of Archaeology at Athens, Suppl. Vol. 4. (London)

Bodenstedt, F. 1976. *Phokäisches Elektron-Geld von 600–326 v. Chr.: Studien zur Bedeutung und zu den Wandlungen einer Antiken Goldwahrung.* (Mainz)

Bolin, S. 1958. *State and Currency in the Roman Empire to 300 AD.* (Stockholm)

Borchers, W. 1904. *Electric Smelting and Refining: The Extraction and Treatment of Metals by Means of the Electric Current*, 2nd English edn. (London)

Bothmer, D. von. 1981. Les trésors de l'orfèvrerie de la Grèce orientale au Metropolitan Museum de New York, *Comptes rendus Académie des inscriptions et belles-lettres*, 194–207

Boussingault, J.B.J.D. 1833. Mémoire sur l'action du gaz acide hydrochlorique sur l'argent à une haute température; observation sur le départ sec, *Annales de Chimie et de Physique* 54, 253–63

Bowditch, D.C. 1973. A comparative study of three analytical methods for the collection and determination of gold and platinoids in precious metal bearing ores, *Bull. Aust. Min. Dev. Lab.* 15, 71–87

Bray, W. 1978. *The Gold of El Dorado*. Catalogue of exhibition at Royal Academy, London, 21 Nov 1978 to 18 Mar 1979

Bray, W. 1993. Techniques of gilding and surface-enrichment in pre-Hispanic American metallurgy, in La Niece and Craddock 1993, 182–92

Breglia, L. 1974. Interrogativi sulle 'creseidi', *Annali Scuola norm. sup. Pisa* IV (3), 659–85

Brunn, J.H., Dumont, J.F., de Graciansky, P.Ch., Gutnic, M., Juteau, Th., Marcoux, J., Monod, O. and Poisson, A. 1971. Outline of the geology of the Western Taurids, in *Geology and History of Turkey*, ed. A.S. Campbell, 225–55. (Tripoli)

Bugbee, E.E. 1933. *A Textbook of Fire Assaying*, 2nd edn. (New York)

Burstein, S.M. (trans. and ed.). 1989. *Agatharchides of Cnidus: On the Erythraean Sea.* (London)

Bury, R.G. (trans.). 1929. *Plato: The Timaeus.* (London: Loeb Classical Library)

Bury, S. 1991. *Jewellery 1789–1910: The International Era*, Vol. 1. (London)

Butler, H.C. 1922. *Sardis* I. *The Excavations*, Part I: *1910–1914.* (Leiden)

Buttrey, T.V., Johnston, A., MacKenzie, K.M. and Bates, M.L. 1981. *Greek, Roman, and Islamic Coins from Sardis*. Archaeological Exploration of Sardis, Monograph 7. (Cambridge, Mass)

Cabri, L.J. 1972. The mineralogy of the platinum-group elements, *Minerals Sci. Engng* 4, 3–29

Cabri, L.J. and Harris, D.C. 1975. Zoning in Os–Ir alloys and the relation of the geological and tectonic environment of the source rocks to the bulk Pt : Pt + Ir + Os ratio for placers, *Can. Minerologist* 13, 266–74

Cahill, N.D. 1990. Domestic buildings east of colossal Lydian structure, in C.H. Greenewalt Jr, N.D. Cahill, H. Dedeoğlu and P. Herrmann, The Sardis Campaign of 1986, *Bull. Am. Sch. Orien. Res. Suppl.* 26, 143–55

Caley, E.R. 1926. The Leyden Papyrus X, *Chem. Educ.* 3, 1149–66

Caley, E.R. and Richards, J.F.C. (trans.). 1956. *Theophrastus on Stones.* (Columbus, Ohio)

Chaston, J.C. 1980. The powder metallurgy of platinum, *Platin. Metals Rev.* 24(2), 70–9

Clews, F.H. 1925. The interaction of sodium chloride and alumina, *J. Chem. Soc.* 127, 735–9

Clews, F.H. and Thompson, H.V. 1922. The interaction of sodium chloride and silica, *J. Chem. Soc.* 121, 1442–8

Conophagos, C.E. 1980. *Le Laurium antique et la technique grecque de la production de l'argent.* (Athens)

Constantinus Africanus. 1536. *Opera ....* (Basel)

Cowell, M.R. 1992. An analytical survey of the British Celtic gold coinage, in *Celtic Coinage: Britain and Beyond,* ed. M. Mays, BAR Brit. Ser. 222, 207–33. (Oxford)

Cowell, M.R., Hyne, K., Meeks, N.D. and Craddock, P.T. 1998. Analyses of the Lydian electrum, gold and silver coinages, in *Metallurgy in Numismatics,* ed. W.A. Oddy and M.R. Cowell, Vol. 4, 526–38. (London)

Cowell, M.R., Oddy, W.A. and Burnett, A.M. 1987. Celtic coinage in Britain: new hoards and recent analyses, *Br. Numismatic J.* 57, 1–23

Craddock, P.T. 1977. The composition of the copper alloys used by the Greek, Etruscan and Roman civilisations. 2: The Archaic, Classical and Hellenistic Greeks, *J. Archaeol. Sci.* 4(2), 103–23

Craddock, P.T. 1980. The composition of copper produced at the ancient smelting camps in the Wadi Timna, in *Scientific Studies in Early Mining and Extractive Metallurgy,* ed. P.T. Craddock, Br. Mus. Occas. Pap. 20, 165–73

Craddock, P.T. 1985. Three thousand years of copper alloys: from the Bronze Age to the Industrial Revolution, in *Application of Science in Examination of Works of Art,* ed. P.A. England and L. van Zelst, 59–67. (Boston, Mass)

Craddock, P.T. 1989. The scientific examination of early mining and metallurgy, in *Scientific Analysis in Archaeology,* ed. J. Henderson, 178–212. (Oxford)

Craddock, P.T. 1995. *Early Metal Mining and Production.* (Edinburgh)

Craddock, P.T., Freestone, I.C., Gale, N.H., Meeks, N.D., Rothenberg, B. and Tite, M.S. 1985. The investigation of a small heap of silver-smelting debris from Rio Tinto, Spain, in *Furnaces and Smelting Technology in Antiquity,* ed. P.T. Craddock and M.J. Hughes, Br. Mus. Occas. Pap. 48, 199–218

Craddock, P.T., Freestone, I.C., Gurjar, L.K., Middleton, A.P. and Willies, L. 1998b. Zinc in India, in *2,000 Years of Zinc and Brass,* ed. P.T. Craddock, Br. Mus. Occas. Pap. 50, 2nd edn, 27–72

Craddock, P.T. and Giumlia Mair, A. 1993. *Ḥśmn-Km,* Corinthian bronze, *Shakudo:* black patinated bronze in the ancient world, in La Niece and Craddock 1993, 101–27

Craddock, P.T., Meeks, N.D., Cowell, M.R., Geçklini, A.E., Hook, D.R., Middleton, A.P and Ramage, A. 1998a. The refining of gold in the classical world, in *The Art of the Greek Goldsmith,* ed. D. Williams, 111–21. (London)

Cramer, J.A. 1741. *Elements of the Art of Assaying Metals,* trans. J. Mortimer. (London)

Crangle, J. 1959. Some magnetic properties of platinum-rich Pt–Fe alloys, *J. Phys. Rad.* 20, 435–7

Crindle, J.W. (trans.). 1882. Ctesias, Ancient India, *Indian Antiquary*

Crozier, R.D. 1993. Silver processing in Spanish America: the Patio and the Buytron processes, *Can. Inst. Min. Bull.* 86, 86–91

Curtis, C.D. 1925. *Sardis XIII, The Jewelry and Gold Work,* Part 1: *1910–1914.* (Rome)

Dana, J.D. 1857. *Manual of Mineralogy ...,* 2nd edn. (New Haven, Conn)

Darling, A.S. and Healy, J.F. 1971. Microprobe analysis and the study of Greek gold–silver–copper alloys, *Nature* 231(5303), 443–4

Davey, C.J. 1988. Some ancient Near-Eastern pot-bellows, *Levant* 11, 101–11

Davies, O. 1935. *Roman Mines in Europe.* (Oxford)

d'Elhuyar, F. 1790. *Bergbaukunde,* 2 vols. (Leipzig)

de Jesus, P.S. 1980. *The Development of Prehistoric Mining and Metallurgy in Anatolia,* BAR Int. Ser. 74. (Oxford)

Demakopoulou, K., Mangou, E., Jones, R.E. and Photos-Jones, E. 1995. Mycenaean black inlaid metalware in the National Archaeological Museum, Athens: a technical examination, *Annual Br. Sch. Athens* 90, 137–53

de Sélincourt, A. 1972. *Herodotus: The Histories,* 2nd edn., new intro. A.R.Burn. (Harmondsworth)

Dibble, C.E. and Anderson, A.J.O. (trans. and ed.). 1959. *Fray Bernardino de Sahagún: Florentine Codex: General History of the Things of New Spain,* Monographs of the School of American Research and the Museum of New Mexico, 14. (Santa Fe)

Dorr, J.V.N. 1936. *Cyanidation and Concentration of Gold and Silver Ores.* (New York)

Dubertret, L. and Kalafatçıoğlu, A. 1973. *Explanatory Text of the Geological Map of Turkey, Izmir.* (Ankara)

Dunlop, D.M. 1957. Sources of gold and silver in Islam according to al-Hamdānī, *Studia Islamica* 8, 29–49

Eichholz, D.E. (trans.). 1962. *Pliny: Natural History*, Vol. 10. (London: Loeb Classical Library)

Eissler, M. 1900. *The Metallurgy of Gold: A Practical Treatise on the Metallurgical Treatment of Gold-bearing Ores including the Assaying, Melting and Refining of Gold*, 5th edn. (London)

Elliot, R.P. 1965. *Constitution of Binary Alloys,* Supplement, Vol.1. (New York)

Éluère, Ch. 1986. A prehistoric touchstone from France, *Gold Bull.* 19(2), 58–61

Emmerich, A. 1965. *Sweat of the Sun and Tears of the Moon: Gold and Silver in Pre-Columbian Art.* (Seattle)

Evans, D. 1993. The Assay Office Report, *Goldsmith's Rev.* 52

Evans, D.W. 1991. Assaying and hallmarking in London, *Gold Technol.* 3(Jan), 2–8

Farag, M.M. 1981. Metallurgy in Ancient Egypt, *Bull. Met. Mus. Jap.* 6, 15–30

Fishlock, D. 1962. *Metal Colouring.* (Teddington)

Forbes, J.S. 1978. Assay methods 1478–1978, in *Touching Gold and Silver,* Catalogue of exhibition to celebrate 500 years of hallmarking, Goldsmiths' Hall, London, 7 to 30 Nov 1978, ed. S.M. Hare, 10–12. (London)

Forbes, J.S. and Dalladay, D.B. 1958–9. Metallic impurities in the silver coinage trial plates (1279–1900), *J. Inst. Metals* 87, 55–8

Forbes, R.J. 1948. *A Short History of the Art of Distillation from the Beginnings up to the Death of Cellier [Cellérier] Blumenthal.* (Leiden)

Forbes, R.J. 1971. *Studies in Ancient Technology,* 2nd rev. edn, Vol. 8. (Leiden)

Fossey, Ch. 1935. L'essai et affinage de l'or chez les Babyloniens, *Rev. études sémitiques* 4, 2–6

Fowler, H.N. (trans.). 1925. *Plato: The Statesman.* (London: Loeb Classical Library)

Frantz, J.H. and Schorsch, D. 1990. Egyptian red gold, *Archeomaterials,* 4(2), 133–52

Freestone, I.C., Craddock, P.T., Hegde, K.T.M., Hughes, M.J.

and Paliwal, H.V. 1985. Zinc production at Zawar, Rajasthan, in *Furnaces and Smelting Technology in Antiquity,* ed. P.T. Craddock and M.J. Hughes, Br. Mus. Occas. Pap. 48, 229–44

Gabelmann, H. 1965. *Studien zum frühgriechischen Löwenbild.* (Berlin)

Gale, N.H. and Stos-Gale, Z.A. 1981. Cycladic lead and silver metallurgy, *Annual Br. Sch. Athens* 76, 169–224

Gee, F.E. 1892. *The Jeweller's Assistant in the Art of Working in Gold: A Practical Treatise for Masters and Workmen, compiled from the Experience of Thirty Years' Workshop Practice.* (London)

Gill, T. 1822. Various processes employed in jewellery, in *Technical Repository*, Vol. 1. (London)

Giumlia Mair, A. and La Niece, S. 1998. Early niello decoration on the silver rhyton in the Museo Civico, Trieste, in *The Art of the Greek Goldsmith,* ed. D. Williams, 139–45. (London)

Godley, A.D. (trans.). 1920. *Herodotus,* Vol. 1. (London: Loeb Classical Library)

Goldstein, S.M. 1977. Reconstruction of the Lydian gold industry. Appendix to C.H. Greenewalt Jr, The Eighteenth Campaign at Sardis (1975), *Bull. Am. Sch. Orient. Res.* 228, 47–60.

Gordus, A.A., 1971. Rapid nondestructive activation analysis of silver in coins, in *Science and Archaeology,* ed. R.H. Brill, 145–55. (Cambridge, Mass)

Gowland, W. 1900. Remains of a Roman silver refinery at Silchester, *Archaeologia* 57, 113–24

Gowland, W. 1910. Arts of working metals in Japan, *J. Inst. Metals* 4, 4–41

Gowland, W. 1915. Metals and metal-working in Old Japan, *Trans. Proc. Jap. Soc.* 13, 19–100

Gowland, W. 1917–18. Silver in Roman and earlier times, *Archaeologia* 69, 121–60

Granger, F. (trans.). 1934–5. *Vitruvius on Architecture,* Vol. 2. (London: Loeb Classical Library)

Griffith, W.P. 1967. *The Chemistry of the Rarer Platinum Metals.* (London)

Gunther, R.T. (ed.). 1934. *The Greek Herbal of Dioscorides,* trans. J. Goodyer. (Oxford)

Gusmani, R. 1969. Der lydische Name der Kybele, *Kadmos* 8(2), 158–61

Gusmani, R. 1975. *Neue epichorische Schriftzeugnisse aus Sardis (1958–1971).* Archaeological Exploration of Sardis, Monograph 3. (Cambridge, Mass)

Habashi, F. 1989. One hundred years of cyanidation, in *All That Glitters: Readings in Historical Metallurgy,* ed. M.L. Wayman, 78–85. (Montreal)

Haffty, J., Riley, L.B. and Goss, W.D. 1977. A manual on fire assaying and the determination of the noble metals in geological materials, *US Geol. Surv. Bull.* 1445

Hall, E.T. 1953. *Quantitative Analysis by Secondary X-Rays for Use Particularly in Archaeology.* PhD thesis, University of Oxford

Hall, E.T. 1958. Some uses of physics in archaeology, *Year Book Phys. Soc.,* pp. 22–34

Hall, H. (ed.). 1896. *The Red Book of the Exchequer.* (London)

Hallbauer, D.K. and Barton Jr, J.M. 1987. The fossil gold placers of the Witwatersrand, *Gold Bull.* 20(3), 68–79

Halleux, R. 1974. L'affinage de l'or de Crésus aux premiers alchimistes, *Janus* 78, 79–102

Halleux, R. (trans. and ed.). 1981. *Les alchimistes grecs.* Vol. 1: *Papyrus de Leyde, Papyrus de Stockholm ....* (Paris)

Halleux, R. 1985. Méthodes d'essai et d'affinage des alliages aurifères dans l'antiquité et au moyen âge, in Morrisson et al. 1985, 39–77

Hammer, J. 1908. Die Feingehalt der griechischen und römischen Münzen, *Z. Numismatik* 26, 22–30

Hanfmann, G.M.A. 1972. *Letters fom Sardis.* (Cambridge, Mass)

Hanfmann, G.M.A. 1983. *Sardis from Prehistoric to Roman Times: Results of the Archaeological Exploration of Sardis, 1958–1975.* (Cambridge, Mass)

Hanfmann, G.M.A. and Ramage, N.H. 1978. *Sculpture from Sardis: The Finds through 1975.* Archaeological Exploration of Sardis, Report 2. (Cambridge, Mass)

Hansen, M. and Anderko, K. 1958. *Constitution of Binary Alloys.* (New York)

Harris, D.C. and Cabri, L.J. 1973. The nomenclature of the natural alloys of osmium, iridium and ruthenium based on the new compositional data of alloys from world-wide occurrences, *Can. Mineralogist* 12, 104–12

Hartmann, A. 1982. *Prähistorische Goldfunde aus Europa; spektralanalytische Untersuchungen und deren Auswertung,* Vol. II. (Berlin)

al-Hassan, A.Y. and Hill, D.R. 1986. *Islamic Technology: An Illustrated History.* (Cambridge, New York and Paris)

Hatchfield, P. and Newman, R. 1991. Ancient Egyptian gilding methods, in *Gilded Wood: Conservation and History,* ed. D. Bigelow, E. Cornu, G.J. Landrey and C. van Horne, 27–47. (Madison, Conn)

Hawthorne, J.G. and Smith, C.S. (trans. and ed.). 1963. *On Divers Arts: The Treatise of Theophilus.* (Chicago)

Head, B.V. 1892. *Catalogue of the Greek Coins of Ionia in the British Museum,* ed. R.S. Poole. (London)

Head, B.V. 1901. *Catalogue of the Greek Coins of Lydia in the British Museum.* (London)

Healy, J.F. 1974. Greek refining techniques and the composition of gold–silver alloys, *Rev. belge Numismat.* 120, 19–33

Healy, J.F. 1978. *Mining and Metallurgy in the Greek and Roman World.* (London)

Healy, J.F. 1979. Mining and processing of gold ores in the ancient world, *J. Metals (AIME)* 31(8), 11–16

Healy, J.F. 1980a. Greek white gold and electrum coin series, in *Metallurgy in Numismatics,* ed. D.M. Metcalf and W.A. Oddy, Vol. 1, 194–215. (London)

Healy, J.F. 1980b. Problems in mineralogy and metallurgy in Pliny the Elder's Natural History, in *Tecnologia, economia e società nel mondo romano: atti del convegno di Como, 27/28/29 Settembre 1979,* p. 191, note 122

Healy, J.F. 1981. Pliny the Elder and ancient mineralogy, *Interdisciplinary Sci. Rev.* 6(2), 166–80

Healy, J.F. 1989. Greek and Roman gold sources: the literary and scientific evidence, in *Minería y metalurgía en las antiguas civilizaciónes Mediterráneas y Europeas: Coloquio Internacional Asociado,* ed. C. Domergue, Vol. 2, 9–20. (Madrid)

Higgins, R. 1980. *Greek and Roman Jewellery,* 2nd edn. (London)

Hiorns, A.H. 1912. *Mixed Metals or Metallic Alloys,* 3rd edn. (London)

Hitchcock, E.A. 1857. *Remarks upon Alchemy and the Alchemists, Indicating a Method of Discovering the True Nature of Hermetic Philosophy ....* (Boston, Mass)

Hook, D.R. 1998. Inductively-coupled plasma atomic emission spectrometry and its role in numismatic studies, in *Metallurgy in Numismatics,* ed. W.A. Oddy and M.R. Cowell, Vol. 4, 237–52. (London)

Hoover, H.C. and Hoover, L.H. (trans. and ed.). 1912. *Georgius Agricola: De Re Metallica.* (London 1912)

Howard, G.S. n.d. [1788]. *The New Royal Cyclopaedia and Encyclopaedia; or Complete Modern and Universal Dictionary of the Arts and Sciences,* 3 vols. (London)

Hradecky, K. 1930. *Die Strichprobe der Edelmetalle.* (Vienna)

Hughes, M.J., Cowell, M.R. and Craddock, P.T. 1976. Atomic absorption techniques in archaeology, *Archaeometry* 18, 19–37

Hulme, E.W. 1940–41. Prehistoric and primitive iron smelting. Part II: The crucible processes of the East. *Trans. Newcomen Soc.* 18, 23–30

Hunt, L.B. 1976. The oldest metallurgical handbook, *Gold Bull.* 9(1), 24–31

Ivotchkina, N.V. 1993. The early Chinese Chu gold plates (5th–3rd century BC), in *Proc XIth Internat. Numismatic Congress,* ed. M. Hoc, 329–32. (Louvain-la Neuve)

Jacoby, F. 1923. *Fragmente der griechischen Historiker.* (Berlin)

Joannès, F. 1993. Metalle und Metallurgie, A.1: In Mesopotamien, in *Reallexikon der Assyriologie,* ed. E. Ebeling and B. Meissner, 96–112. (Berlin)

Jones, H.L. (trans.). 1923. *The Geography of Strabo,* Vol. 2. (London: Loeb Classical Library)

Jones, H.L. (trans.). 1929. *The Geography of Strabo,* Vol. 6. (London: Loeb Classical Library)

Jones, R.E. 1980. Analyses of gold objects from the cemeteries, in *Lefkandi I: The Iron Age,* ed. M.R. Popham and L.H. Sackett, 461–4. (London)

Jung, C.G. 1944. *Psychologie and Alchemie.* (Zurich)

Kangle, R.P. (ed.). 1972. *The Kauṭilīya Arthaśāstra,* 2nd edn, Vol. II. (Bombay)

Karageorghis, V. and Kassianidou, V. 1999. Metalworking and recycling in Late Bronze Age Cyprus: the evidence from Kition, *Oxford J. Archaeol.* 18(2), 177–88

al-Kāshānī, Abū al-Qāsim ʿAbd Allah. n.d. *Arāyis al-jāwahir wa nafāyis al-atāyib.* Reprinted 1966. (Teheran)

Kelly, L.G. 1990. *Basil Valentine: His Triumphant Chariot of Antimony, with Annotations of Theodore Kirkringius (1678).* (New York)

Klein-Franke, F. 1970. The knowledge of Aristotle's lapidary in the Latin Middle Ages, *Ambix* 17(3), 137–42

Kohler, E.L. 1995. *The Lesser Phrygian Tumuli.* The Gordion Excavations, 1950–1973: Final Reports, Vol. II. (Philadelphia)

Koucky, F.L. and Steinberg, A. 1982. Ancient mining and mineral dressing on Cyprus, in *Early Pyrotechnology: The Evolution of the First Fire-using Industries,* ed. T.A. Wertime and S.F. Wertime, 149–80. (Washington)

Kraay, C.M. 1958. The compositions of electrum coinage, *Archaeometry* 1, 21–3

Kraay, C.M. 1976. *Archaic and Classical Coins.* (Berkeley, Calif)

Kubaschewski, O. and Evans, E. 1958. *Metallurgical Thermochemistry,* 3rd edn. (Oxford)

Laffineur, R. 1974. L'incrustation à l'époque mycénienne, *Antiquité classique* 43, 5–37

Laist, J.W. 1954. *Copper, Silver and Gold. Comprehensive Inorganic Chemistry,* ed. M.C. Sneed, J.L. Maynard and R.C. Brasted, Vol. 2. (Princeton, NJ)

La Niece, S. 1983. Niello: an historical and technical survey, *Antiquaries J.* 63(2), 279–97

La Niece, S. 1990. Silver plating on copper, bronze and brass, *Antiquaries J.* 70(1), 102–14

La Niece, S. 1993. Silvering, in La Niece and Craddock 1993, 201–10

La Niece, S. 1995. Depletion gilding from third millennium BC Ur, *Iraq* 57, 41–7

La Niece, S. and Craddock, P.T. (ed.). 1993. *Metal Plating and Patination: Cultural, Technical, and Historical Developments.* (Oxford)

Leake, R.C., Bland, D.J. and Cooper, C. 1993. Source characterisation of alluvial gold from mineral inclusions and internal compositional variation, *Trans. Inst. Min. Metall.* 102 (May–Aug), B65–B82

Leather, J.W. and Mukerji, J.N. 1911. *Bull. Agric. Inst., Pusa,* 24

Lechtmann, H.N. 1971. Ancient methods of gilding silver: examples from the old and new worlds, in *Science and Archaeology,* ed. R.H. Brill, 2–31. (Cambridge, Mass)

Lechtmann, H.N. 1973. The gilding of metals in Pre-Columbian Peru, in *Application of Science in Examination of Works of Art,* ed. W.J. Young, 38–52. (Boston, Mass)

Lefond, S.J. (ed.) 1975. *Industrial Minerals and Rocks: Nonmetallics other than Fuels,* 4th edn. (New York)

Levey, M. 1959a. The refining of gold in ancient Mesopotamia, *Chymia* 5, 31–6

Levey, M. 1959b. *Chemistry and Chemical Technology in Ancient Mesopotamia.* (Amsterdam)

Levey, M. 1971. *Chemical Aspects of Medieval Arabic Minting in a Treatise by Manṣūr ibn-Baʿra.* Japanese Studies in the History of Science, Suppl. 1. (Tokyo)

Lewis, W. 1763. *Commercium Philosophico-Technicum: or, the Philosophical Commerce of Arts.* (London)

Liddell, D.M. 1926. *Handbook of Non-Ferrous Metallurgy,* Vol. II. (New York)

Lins, P.A. 1991. Basic properties of gold leaf, in *Gilded Wood: Conservation and History,* ed. D. Bigelow, E. Cornu, G.J. Landrey and C. van Horne, 17–23. (Madison, Conn)

Lord, L. 1937. The touchstone, *Classical J.* 32(Apr), 428–31

Lucas, A. 1962. *Ancient Egyptian Materials and Industries,* 4th edn, rev. J.R. Harris. (London)

McCombe, C. n.d. *The Development of the Cupola.* Reprinted from *J. Foundry Coke Tech. Serv. (FaCTS).* (London)

McDonald, D. 1960. *A History of Platinum, from the Earliest Times to the Eighteen-Eighties.* (London)

McDonald, D. and Hunt, L.B. 1982. *A History of Platinum and Its Allied Metals,* rev. edn. (London)

Maclagan, E.D. 1890. *Monograph on the Gold and Silver Works of the Punjab 1888–9.* (Lahore)

McLaughlin, D.H. and Wise, E.M. 1964. Sources and recovery of gold, in *Gold: Recovery, Properties, and Applications,* ed. E.M. Wise, 1–24. (Princeton, NJ)

Maniatis, Y. and Tite, M.S. 1978–9. Examination of Roman and medieval pottery using the scanning electron microscope, *Acta praehist. archaeol.* 9/10, 125–30

Maniatis, Y. and Tite, M.S. 1981. Technological examination of Neolithic–Bronze Age pottery from Central and Southeast Europe and from the Near East, *J. Archaeol. Sci.* 8, 59–76

Marshall, A. 1915. *Explosives, their Manufacture, Properties, Tests and History.* (London)

Matson, F.R. 1971. A study of temperatures used in firing ancient Mesopotamian pottery, in *Science and Archaeology,* ed. R.H. Brill, 65–79. (Cambridge, Mass)

Mattusch, C. 1977. Bronze and ironworking in the area of the Athenian Agora, *Hesperia* 46(4), 340–79

Meanley, P. and Byers, W. 1996. Something in the air, *Ceramic Rev.* 157(Jan–Feb), 16–20

Meeks, N.D. 1988. Backscattered electron imaging of archaeological material, in *Scanning Electron Microscopy in Archaeology,* ed. L. Olsen, BAR Int. Ser. 452, 23–44. (Oxford)

Meeks, N.D., Craddock, P.T., Geçklini, A.E., Hook, D.R., Middleton, A.P. and Ramage, A. 1996. The scientific study of the refractory remains and gold particles from the Lydian gold refinery at Sardis, in *Archaeometry '94,* ed. S. Demirci, A.M. Özer and G.D. Summers, 461–82. (Ankara).

Meeks, N.D. and Tite, M.S. 1980. The analysis of platinum-group element inclusions in gold antiquities, *J. Archaeol. Sci.* 7, 267–75

Mellor, J.W. 1923. *A Comprehensive Treatise on Inorganic and Theoretical Chemistry,* Vol. III. (London)

Mellor, J.W. 1935. *A Comprehensive Treatise on Inorganic and Theoretical Chemistry,* Vol. XIV. (London)

Mellor, J.W. 1961. *A Comprehensive Treatise on Inorganic and Theoretical Chemistry,* Vol II, Supplement II. (London)

Meyers, P. 1983. Elemental composition of silver objects found by the Princeton Expedition, in Waldbaum 1983, 187–90

Meyers, P., van Zelst, L. and Sayre, E.V. 1973. Determination of major components and trace elements in ancient silver by thermal neutron activation and analysis, *J. Radioanalyt. Chem.* 16, 67–78

Middleton, A.P. 1991. Ceramics: materials for all reasons, in *Science and the Past,* ed. S. Bowman, 16–36. (London)

Moesta, H. 1986. *Erze und Metalle: ihre Kulturgeschichte im Experiment,* 2nd edn. (Berlin)

Moesta, H. and Franke, P.R. 1995. *Antike Metallurgie und Münzprägung: ein Beitrag zur Technikgeschichte.* (Basel)

Moore, D.T. and Oddy, W.A. 1985. Touchstones: some aspects of their nomenclature, petrography and provenance, *J. Archaeol. Sci.* 12, 59–80

Moorey, P.R.S. 1994. *Ancient Mesopotamian Materials and Industries: The Archaeological Evidence.* (Oxford)

Morrisson, C., Barrandon, J-N. and Brenot, C. 1987. Composition and technology of ancient and medieval coinages, *Am. Numis. Soc. Notes* 32, 181–209

Morrisson, C., Brenot, C., Barrandon, J-N., Callu, J-P., Poirier, J. and Halleux, R. 1985. *L'or monnayé 1: Purification et altérations de Rome à Byzance.* (Paris)

Müller, K. 1855. De Mari Erythraeo, in *Geographi Graeci Minores,* Vol. 1, 111–95. (Paris)

Muspratt, J.S. 1860. *Chemistry, Theoretical, Practical and Analytical as Applied and Relating to the Arts and Manufactures,* 2 vols. (Glasgow)

Mu'nis, Ḥusain. (ed.). 1960. *Al-dawḥa al-mosht'abika fi ḍawābit dār al-sikka,* ʿAlī ibn-Yúsuf. (Madrid)

Nassau, K. 1984. *Gemstone Enhancement.* (Oxford)

Needham, J. 1974. *Science and Civilisation in China.* Vol. 5: *Chemistry and Chemical Technology.* Pt II: *Spagyrical Discovery and Invention. Magisteries of Gold and Immortality.* (Cambridge)

Needham, J. 1980. *Science and Civilisation in China.* Vol. 5: *Chemistry and Chemical Technology.* Pt IV: *Spagyrical Discovery and Invention. Apparatus, Theories and Gifts.* (Cambridge)

Needham, J. 1983. *Science and Civilisation in China,* Vol. 5: *Chemistry and Chemical Technology.* Pt V: *Spagyrical Discovery and Invention. Physiological Alchemys.* (Cambridge)

Newman, W.R. (trans. and ed.). 1991. *The Summa Perfectionis of Pseudo-Gerber.* (Leiden)

Nicolini, G. 1990. *Techniques des ors antiques: la bijouterie ibérique du VII au IVe siècle,* Vol. 1. (Paris)

Nihon Gakushiin. 1958. *Meiji-zen Nihon kōgyō gitutsu hattatsushi.* (Tokyo)

Notton, J.F.H. 1971. The use of common salt for removing silver from native gold, *Tech. Memo. (Metallurgy Grp) Johnson Matthey* 87, 1–7

Notton, J.F.H. 1974. Ancient Egyptian gold refining, *Gold Bull.* 7(2), 50–6

Nriagu, J.P. 1983. *Lead and Lead Poisoning in Antiquity.* (New York)

Oddy, W.A. 1981. Gilding through the ages, *Gold Bull.* 14(2), 75–9

Oddy, W.A. 1983. Assaying in antiquity, *Gold Bull.* 16(2), 52–9

Oddy, W.A. 1986. The touchstone: the oldest colorimetric method of analysis, *Endeavour,* new ser. 10(4), 164–6

Oddy, W.A. 1993a. Gilding of metals in the Old World, in La Niece and Craddock 1993, 171–81

Oddy, W.A. 1993b. The assaying of gold by touchstone in antiquity and the medieval world, in *Outils et ateliers d'orfèvres des temps anciens,* ed. Ch. Éluère, 93–100. (St Germain-en-Laye)

Oddy, W.A., Cowell, M.R., Craddock, P.T. and Hook D.R. 1990. The gilding of bronze sculpture in the Classical World, in *Small Bronze Sculpture from the Ancient World,* ed. M. True and J. Podany, 103–34. (Malibu)

Oddy, W.A. and Schweizer, F. 1972. A comparative analysis of some gold coins, in *Methods of Chemical and Metallurgical Investigation of Ancient Coinage,* ed. E.T. Hall and D.M. Metcalf, 171–82. (London)

Oddy, W.A. and Swaddling, J. 1985. Illustrations of metalworking furnaces on Greek vases, in *Furnaces and Smelting Technology in Antiquity,* ed. P.T. Craddock and M.J. Hughes, 43–58. (London)

Ogden, J.M. 1976. The so-called 'platinum' inclusions in Egyptian goldwork, *J. Egyptian Archaeol.* 62, 138–44

Ogden, J.M. 1977. Platinum group metal inclusions in ancient gold artifacts, *J. Hist. Metall. Soc.* 11(2), 53–72

Ogden, J.M. 1992. *Ancient Jewellery.* (London)

Ogden, J.M. 1993. Aesthetic and technical considerations regarding the colour and texture of ancient goldwork, in La Niece and Craddock 1993, 39–49

Oldfather, C.H. (trans.). 1935. *Diodorus of Sicily,* Vol. 2. (London: Loeb Classical Library)

Ottaway, P. 1992. *Anglo-Scandinavian Ironwork from 16–22 Coppergate.* Archaeology of York Fascicule 17/6: The Small Finds. (London)

Oxland. 1845. Patent Application No. 10528: *Improvements in the Manufacture of Chlorine.* London: 20 Feb.

Özgen, I. and Öztürk, J. 1996. *The Lydian Treasure.* Ministry of Culture General Directorate of Monuments and Museums. (Ankara)

Park, M. 1907. *The Travels of Mungo Park.* (London)

Partington, J.R. 1935. *The Origins and Development of Applied Chemistry.* (London)

Partington, J.R. 1960. *A History of Greek Fire and Gunpowder.* (Cambridge)

Partington, J.R. 1961. *A History of Chemistry,* Vol. 2: *1500–1700.* (London)

Partington, J.R. 1970. *A History of Chemistry,* Vol.1, Pt 1: *Theoretical Background.* (London)

Pászthory, E. 1980. Investigations of the early electrum coins of the Alyattes type, in *Metallurgy in Numismatics,* Vol. 1, 151–6. (Oxford)

Pászthory, E. 1982. Die Legierung des Frankfurter Phanes-Stater, in *Studia Paulo Naster oblata,* ed. S. Scheers, Vol. 1, 7–11. (Leuven)

Pedley, F.J. 1972. *Ancient Literary Sources on Sardis.* Archaeological Exploration of Sardis, Monograph 2. (Cambridge, Mass)

Percy, J. 1870. *The Metallurgy of Lead, including Desilverisation and Cupellation.* (London)

Percy, J. 1880. *Metallurgy: Silver and Gold,* Part I. (London)

Petrie, W.M.F. 1915. *Ancient Egypt.* (London)

Petrie, W.M.F. and Quibell, J.E. 1896. *Naqada and Ballas.* (London)

Phillott, D.C. (ed.). 1927. *The Ā-īn-i Akbarī,* Abū L-Faẓl Allamī, Vol. 1, trans. H. Blochmann (1873), Vol. 1. (New Delhi)

Potts, P.J. 1987. *A Handbook of Silicate Rock Analysis.* (Glasgow and London)

Price, M. 1984. Croesus or pseudo-Croesus? Hoard or hoax? Problems concerning the single and double-sigloi of the Croeseid type, in *Festschrift für Leo Mildenberg: Numismatik, Kunstgeschichte, Archäologie,* ed. A. Houghton et al., 211–21. (Wetteren)

Probert, A. 1969. Bartolome de Medina, *J. West* 8(1), 90–124

Projektgruppe Plinius. 1993. *Gold und Vergoldung bei Plinius dem Älteren*. (Tübingen)

Rackham, H. (trans.). 1952. *Pliny: Natural History*, Vol. 9. (London: Loeb Classical Library)

Ramage, A. 1970. Pactolus North, *Bull. Am. Sch. Orient. Res.* 199, 16–28

Ramage, A. 1978a. Gold refining in the time of the Lydian kings at Sardis, in *Proc. Xth Congr. Classical Archaeology 1973*, ed. E. Akurgal, 729–35. (Ankara)

Ramage, A. 1978b. *Lydian Houses and Architectural Terracottas*. Archaeological Exploration of Sardis, Monograph 5. (Cambridge, Mass)

Ramage, A. 1987. Lydian Sardis, in *Sardis: Twenty-Seven Years of Discovery*, ed. E. Guralnick, 6–15. (Chicago)

Ramage, A., Goldstein, S.M. and Mierse, W.E. 1983. Lydian excavation sectors, in Hanfmann 1983, 26–52

Ramage, A. and Ramage, N.H. 1971. The siting of Lydian burial mounds, in *Studies Presented to George M.A. Hanfmann*, ed. D.G. Mitten, J.G. Pedley and J.A. Scott. (Cambridge, Mass)

Raub, Ch. J. 1995. The metallurgy of gold and silver in prehistoric times, in *Prehistoric Gold in Europe: Mines, Metallurgy, and Manufacture*, ed. G. Morteani and J.P. Northover, 243–59. (Dordrecht)

Rây, P. 1956. *History of Chemistry in Ancient and Medieval India*. (Calcutta)

Reed-Hill, R.E. 1973. *Physical Metallurgy Principles*, 2nd edn. (Princeton, NJ)

Rees, A. 1819–20. *The Cyclopaedia or Universal Dictionary of Arts, Sciences and Literature*. (London)

Rehren, T. (Forthcoming). Cupel and crucible: the Xanten silver recovery process, in *Ancient Mining and Metallurgy*, PACT 51

Rehren, T. and Hauptmann, A. 1995. Silberraffinations-Schlacken aus der CUT (Xanten), Insula 39: Mineralogische Untersuchung und archäometallurgische Interpretation, in *Xantener Berichte: Grabung–Forschung–Präsentation*, Vol. 6, 119–37. (Köln)

Ridgeway, W. 1895. How far could the Greeks determine the fineness of gold and silver coins? *Numismatic Chron.*, 3rd ser., XV, 104–9

Robert, L. 1975. Une nouvelle inscription grecque de Sardes: règlement de l'autorité perse relatif à culte de Zeus, *Comptes rendus Académie des inscriptions et belles-lettres* (Apr–Jun), 306–30

Root, M. Cool. 1988. Evidence from Persepolis for the dating of Persian and Archaic Greek coinage, *Numismatic Chron.* 148, 1–22

Rose, T.K. 1915. *The Metallurgy of Gold*, 6th edn. (London)

Rose, T.K. 1915–16. Electrolytic refining of gold, *Bull. Inst. Min. Metall.* (127), 1–21

Rose, V. 1875. Aristoteles de Lapidibus und Arnoldus Saxo, *Z. Deutsches Alterthum* XVIII (6), 321–455

Ruska, J. 1912. *Das Steinbuch des Aristoteles*. (Heidelberg)

Salmang, H. 1961. *Ceramics: Physical and Chemical Fundamentals*, trans. M. Francis. (London)

Schaeffer, J.S., Ramage, N.H. and Greenewalt Jr, C.H. 1997. *The Corinthian, Attic, and Lakonian Pottery from Sardis*. Archaeological Exploration of Sardis, Monograph 10. (Cambridge, Mass)

Scheel, B. 1989. *Egyptian Metalworking and Tools*. (Aylesbury)

Schnabel, C. and Louis, H. 1905. *Handbook of Metallurgy*, 2nd edn, Vol. 1. (London)

Schülter, C.A. 1738. *Gründlicher Unterricht von Hütte-Werken* .... (Braunschweig)

Scott, D.A. and Bray, W. 1980. Ancient platinum technology in South America, *Platin. Metals Rev.* 24(4), 147–57

Scott, D.A. and Bray, W. 1994. Pre-Hispanic platinum alloys: their composition and use in Ecuador and Columbia, in *Archaeometry of Pre-Columbian Sites and Artifacts*, ed. D.A. Scott and P. Meyers, 285–322. (Los Angeles)

Scott, J.A. and Kamilli, D.C. 1981. Late Byzantine glazed pottery from Sardis, in *Actes du XVe Congrès International d'Études Byzantines. II: Art et Archéologie*, 679–96. (Athens)

Searle, A.B. 1929. *The Clayworker's Hand-book: A Manual for All Engaged in the Manufacture of Articles from Clay*. (London)

Searle, A.B. 1940. *Refractory Materials: Their Manufacture and Uses*, 3rd edn. (London)

Seltman, C. 1955. *Greek Coins*, 2nd edn. (London)

Shalev, S. 1993. The earliest gold artifacts in the southern Levant: reconstruction of the manufacturing process, in *Outils et ateliers d'orfèvres des temps anciens*, ed. Ch. Éluère, 9–12. (Paris)

Shear, Th. L. 1924. The gold sands of the Pactolus, *Classical Rev.* 18, 186–8

Singer, F. and Singer, S.S. 1963. *Industrial Ceramics*. (London)

Sisco, A.G. and Smith, C.S. (trans. and ed.). 1949. *Bergwerk- und Probierbüchlein*. (New York)

Sisco, A.G. and Smith, C.S. (trans. and ed.). 1951. *Lazarus

Ercker's Treatise on Ores and Assaying. (Chicago)

Smith, C.S. and Gnudi, M.T. (trans. and ed.). 1942. *The Pirotechnia of Vannoccio Biringuccio.* (New York)

Smith, C.S. and Hawthorne, J.G. (trans. and ed.). 1974. *Mappae Clavicula: a little key to the world of medieval techniques, Trans. Am. Phil. Soc.* 64(4)

Smith, E.A. 1923–24. Early methods of assaying, *Trans. Inst. Min. Metall.* 33, 272–327

Smith, E.A. 1947. *The Sampling and Assay of the Precious Metals, comprising Gold, Silver, Platinum, and the Platinum Group Metals in Ores, Bullion, and Products*, 2nd edn. (London)

Sokolowski, F. 1979. 'TA ENPYRA: On the mysteries in the Lydian and Phrygian cults, *Z. Papyrologie und Epigraphik* 34, 65–9

Starkey, P. 1977. *Saltglaze.* (London)

Stead, I.M. and Meeks, N.D. 1996. The Celtic warrior fibula, *Antiquaries J.* 76, 1–16

Stillman, J.M. 1924. *The Story of Early Chemistry.* Reissued 1960 as *The Story of Alchemy and Early Chemistry.* (New York)

Stock, J.T. 1969. *Development of the Chemical Balance.* (London)

Stronge, S. 1993. *Bidri* ware of India (Technical appendix: P.T. Craddock), in La Niece and Craddock 1993, 135–47

Szabó, Z. 1975. Az aranyfinomításról [gold refining], *Múzeumi Mütárgyvédelem* II (1), 105–19

Taylor, F. Sherwood, 1957. *A History of Industrial Chemistry.* (London)

Themelis, P.G. 1983. An eighth century goldsmith's workshop at Eretria, in *The Greek Renaissance of the Eighth Century BC: Tradition and Innovation,* ed. R. Hägg, 157–65. (Stockholm)

Thölde, J. 1604. *Triumph Wagen Antimonii Fratis Basilii Valentina, ... durch Johann Thölden Hessum.* (Leipzig)

Thorndike, L. 1960. De Lapidibus, *Ambix* 8(1), 6–23

Thorpe, J.F. and Whiteley, M.A. 1939. *Thorpe's Dictionary of Applied Chemistry,* 4th edn, Vol. III. (London)

Thorpe, J.F. and Whiteley, M.A. 1949. *Thorpe's Dictionary of Applied Chemistry,* Vol. IX. (London)

Tite, M.S. 1991. Technology of Rhenish stoneware, in *Archaeometry '90,* ed. E. Pernicka and G.A. Wagner, 337–44. (Basel)

Tite, M.S., Maniatis, Y., Meeks, N.D., Bimson, M., Hughes, M.J. and Leppard, S.C. 1982. Technological studies of ancient ceramics from the Near East, Aegean and Southeast Europe, in *Early Pyrotechnology: The Evolution of the First Fire-using Industries,* ed. T.A. Wertime, and S.F. Wertime, 61–71. (Washington)

Todd, H. 1998. The British Library's Sado mining scrolls, *J. Br. Library* 24(1), 130–43

Toll, C. (trans. and ed.). 1968. *Al-Hamdānī: Die beiden Edelmetalle Gold und Silber.* (Uppsala)

Topkaya, Y.A. 1984. Recent evaluation of Sart placer gold deposit, in *Precious Metals: Mining, Extraction and Processing,* ed. V. Kudryk, D.A. Corrigan and W.W. Laing, 111–21. (Warrendale, Penn)

Tudball, R. 1992. Sodium: a green alternative to salt glazing, *Artists Newsletter* (Jul), 31–2

Tylecote, R.F. 1981. From pot bellows to tuyeres, *Levant* 13, 107–18

Untracht, O. 1982. *Jewelry: Concepts and Technology.* (London)

Uzkut, I. and Semerkant, O. 1980. Salihli–Sart Plaserinde Ağır Mineral Dağılımı ve Değerlendirilmesi, *Maden Mühendisleri Odası Dergisi* 19(4), 5–25

Vallet, G. (ed). 1978. *Les céramiques de la Grèce de l'Est et leur diffusion en occident.* Centre Jean Bérard, Institut Français de Naples

Viccajee, F.K. 1908. *Notes on the Hand Minting of Coins of India.* (Hyderabad)

Vickers, M. 1985. Early Greek coinage: a reassessment, *Numismatic Chron.* 145, 1–44

Villard, F. and Vallet, G. 1955. Megaera Hyblaea V: lampes du VIIe siècle et chronologie des coupes ioniennes, *Mélanges d'archéologie et d'histoire de l'école française de Rome* 67, 7–34

Vismara, N. 1993a. Proposte per un nuovo ordinamento della monetazione arcaica della Lycia, in *Akten des II. International Lykien-Symposios,* ed. J. Borchardt and G. Dobesch, 191–201. (Vienna)

Vismara, N. 1993b. *Monetazione arcaica in elettro dell'Asia Minore nelle civiche raccolte numismatiche: Donazione Winsemann Falghera.* (Milan)

Wälchi, W. 1981. Touching precious metals, *Gold Bull.* 14(4), 154–8

Wälchi, W. and Vuilleumier, P. 1985. Touchstone testing of precious metals, *Aurum* 24, 36–45

Wälchi, W. and Vuilleumier, P. 1991a. Touchstone testing of precious metals, *Gold Technol.* 3, 9–18

Wälchi, W. and Vuilleumier, P. 1991b. Assaying gold in Switzerland, *Gold Technol.* 3, 19–27

Waldbaum, J.C. 1983. *Metalwork from Sardis: The Finds through 1974*. Archaeological Exploration of Sardis, Monograph 8. (Cambridge, Mass)

Waldbaum, J.C. and Magness, J. 1997. The chronology of early Greek pottery: new evidence from seventh-century BC destruction levels in Israel, *Am. J. Archaeol.* 101, 23–40

Wallace, R.W. 1987. The origin of electrum coinage, *Am. J. Archaeol.* 91, 385–96

Walsh, J. (trans.) 1929. Galen visits the Dead Sea and copper mines of Cyprus (AD 166), *Bull. Geog. Soc. Philad.* 25, 92–110.

Wayman, M. and Craddock, P.T. 1993. *Wu tong*, a neglected Chinese decorative technology, in La Niece and Craddock 1993, 128–34

Weidauer, L. 1975. *Probleme der frühen Elektronprägung*. (Fribourg: Office de Livre)

Weisgerber, G. and Roden, C. 1985. Römischer Schmiedeszenen und ihre Gebläse, *Anschnitt* 37(1) 2–21

Weisgerber, G. and Roden, C. 1986. Griechische Metallhandwerker und ihre Gebläse, *Anschnitt* 38(1), 2–26

Whitmore, F.E. and Young, W.J. 1973. Application of the laser microprobe and electron microprobe in the analysis of platiniridium inclusions in gold, in *Application of Science in Examination of Works of Art*, ed. W.J. Young, 88–95. (Boston)

Williams, D. and Ogden, J. 1994. *Greek Gold: Jewellery of the Classical World*. (London)

Wilson, H.H. 1836. *Notes on the Indica of Ctesias*. Ashmolean Society Papers. (Oxford)

Wise, E.M. 1964. Gold alloy systems, in *Gold: Recovery, Properties, and Applications,* ed. E.M. Wise, 97–153. (Princeton, NJ)

Wohlwill, E. 1897–8. Über Goldscheidung auf elecktrolutischem Wege, *Z. Elektrochem.* 4, 379–85, 402–9, 421–3

Woolley, L.C. 1934. *Ur Excavations*. Joint Expedition of the British Museum and the Museum of the University of Pennsylvania to Mesopotamia. (London)

Woolley, L.C. 1937–8. Excavations at Al Mina, Sueidia, *J. Hellenic Studies* 58, 1–30

Wyckoff, D. (trans.). 1967. *Albertus Magnus: Book of Minerals*. (Oxford)

Xénaki-Sakellariou, A. and Chatziliou, C. 1989. *'Peinture en métal' à l'époque mycénienne: incrustation damasquinage niellure*. (Athens)

Young, R.S. 1981. *Three Great Early Tumuli*. With contributions by K. De Vries et al. Ed. E.L. Kohler. The Gordion Excavations, 1950–1973: Final Reports, Vol. I. (Philadelphia)

Young, W.J. 1972. The fabulous gold of the Pactolus Valley, *Bull. Boston Mus. Fine Arts* 70, 5–13

Zimmer, G. 1990. *Griechische Bronzegusswerkstätten: zur Technologieentwicklung eines antiken Kunsthandwerkes*. (Mainz)

Zwicker, U. 1998. An investigation of inclusions of platinum-group metals in ancient coinage, in *Metallurgy in Numismatics*, ed. W.A. Oddy and M.R. Cowell, Vol. 4, 171–201. (London)

Index